**Serono Symposia, USA**
**Norwell, Massachusetts**

# PROCEEDINGS IN THE SERONO SYMPOSIA, USA SERIES

*GLYCOPROTEIN HORMONES: Structure, Function, and Clinical Implications*
    Edited by Joyce W. Lustbader, David Puett, and Raymond W. Ruddon

*GROWTH HORMONE II: Basic and Clinical Aspects*
    Edited by Barry B. Bercu and Richard F. Walker

*TROPHOBLAST CELLS: Pathways for Maternal-Embryonic Communication*
    Edited by Michael J. Soares, Stuart Handwerger, and Frank Talamantes

*IN VITRO FERTILIZATION AND EMBRYO TRANSFER IN PRIMATES*
    Edited by Don P. Wolf, Richard L. Stouffer, and Robert M. Brenner

*OVARIAN CELL INTERACTIONS: Genes to Physiology*
    Edited by Aaron J.W. Hsueh and David W. Schomberg

*CELL BIOLOGY AND BIOTECHNOLOGY: Novel Approaches to Increased
Cellular Productivity*
    Edited by Melvin S. Oka and Randall G. Rupp

*PREIMPLANTATION EMBRYO DEVELOPMENT*
    Edited by Barry D. Bavister

*MOLECULAR BASIS OF REPRODUCTIVE ENDOCRINOLOGY*
    Edited by Peter C.K. Leung, Aaron J.W. Hsueh, and Henry G. Friesen

*MODES OF ACTION OF GnRH AND GnRH ANALOGS*
    Edited by William F. Crowley, Jr., and P. Michael Conn

*FOLLICLE STIMULATING HORMONE: Regulation of Secretion and Molecular
Mechanisms of Action*
    Edited by Mary Hunzicker-Dunn and Neena B. Schwartz

*SIGNALING MECHANISMS AND GENE EXPRESSION IN THE OVARY*
    Edited by Geula Gibori

*GROWTH FACTORS IN REPRODUCTION*
    Edited by David W. Schomberg

*UTERINE CONTRACTILITY: Mechanisms of Control*
    Edited by Robert E. Garfield

*NEUROENDOCRINE REGULATION OF REPRODUCTION*
    Edited by Samuel S.C. Yen and Wylie W. Vale

*FERTILIZATION IN MAMMALS*
    Edited by Barry D. Bavister, Jim Cummins, and Eduardo R.S. Roldan

*GAMETE PHYSIOLOGY*
    Edited by Ricardo H. Asch, Jose P. Balmaceda, and Ian Johnston

*GLYCOPROTEIN HORMONES: Structure, Synthesis, and Biologic Function*
    Edited by William W. Chin and Irving Boime

*THE MENOPAUSE: Biological and Clinical Consequences of Ovarian Failure:
Evaluation and Management*
    Edited by Stanley G. Korenman

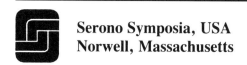

**Serono Symposia, USA**
**Norwell, Massachusetts**

Joyce W. Lustbader      David Puett
Raymond W. Ruddon      Editors

# Glycoprotein Hormones

## Structure, Function, and Clinical Implications

With 110 Figures

Springer-Verlag
New York Berlin Heidelberg London Paris
Tokyo Hong Kong Barcelona Budapest

Joyce W. Lustbader, Ph.D.
Irving Center for Clinical Research
Columbia University College of
  Physicians and Surgeons
New York, NY 10032
USA

David Puett, Ph.D.
Department of Biochemistry
University of Georgia
Athens, GA 30602
USA

Raymond W. Ruddon, M.D., Ph.D.
Eppley Institute for Research
University of Nebraska Medical Center
Omaha, NE 68198
USA

Proceedings of the Symposium on Glycoprotein Hormones: Structure, Function, and Clinical Implications, sponsored by Serono Symposia, USA, held March 11 to 14, 1993, in Santa Barbara, California.

For information on previous volumes, please contact Serono Symposia, USA.

Library of Congress Cataloging-in-Publication Data
Symposium on Glycoprotein Hormones: Structure, Function, and
  Clinical Implications (1993: Santa Barbara, Calif.)
    Glycoprotein hormones/edited by Joyce W. Lustbader, David Puett,
  and Raymond W. Ruddon.
      p.  cm.
    "Proceedings of the Symposium on Glycoprotein Hormones: Structure,
  Function, and Clinical Implications, sponsored by Serono Symposia,
  USA, held March 11 to 14, 1993, in Santa Barbara, California"—T.p.
  verso.
    Includes bibliographical references and index.
    ISBN 0-387-94165-7
    1. Glycoprotein hormones—Congresses.  I. Lustbader, Joyce
  Werner, 1955– .  II. Puett, David.
  III. Ruddon, Raymond W., 1936– .
  IV. Serono Symposia, USA.  V. Title.
  QP572.G58S96  1993
  612.6—dc20                        93-33014

Printed on acid-free paper.

Production coordinated by Marilyn Morrison and managed by Francine McNeill; manufacturing supervised by Vincent Scelta.
Typeset by Best-set Typesetter Ltd., Hong Kong.
Printed and bound by Braun-Brumfield, Inc., Ann Arbor, MI.

Printed in the United States of America.

9 8 7 6 5 4 3 2 1

ISBN 0-387-94165-7 Springer-Verlag New York Berlin Heidelberg
ISBN 3-540-94165-7 Springer-Verlag Berlin Heidelberg New York

## SYMPOSIUM ON GLYCOPROTEIN HORMONES: STRUCTURE, FUNCTION, AND CLINICAL IMPLICATIONS

**Scientific Committee**

Joyce W. Lustbader, Ph.D.
Columbia University
New York, New York

David Puett, Ph.D.
University of Georgia
Athens, Georgia

Raymond W. Ruddon, M.D., Ph.D.
University of Nebraska
Omaha, Nebraska

**Organizing Secretary**

Bruce K. Burnett, Ph.D.
Serono Symposia, USA
100 Longwater Circle
Norwell, Massachusetts

# Preface

Advances in the field of glycoprotein hormones necessitated a second international symposium on this topic, held March 11 to 14, 1993, in Santa Barbara, California, and again sponsored by Serono Symposia, USA. The meeting was twofold in its concept: (1) the dissemination of the current research in the field and (2) honoring three scientists who have greatly contributed to this field, Drs. Harold Papkoff, Robert Ryan, and Darrell Ward, upon their retirements. We were honored to have Dr. John Pierre present as a participant at the meeting and also serving as master of ceremonies at the banquet.

Certainly one of the highlights at the first meeting, held in 1989, was the cloning of the CG/LH receptor. This second meeting was also filled with insightful and innovative scientific presentations. Significant advances in the regulation of gene transcription of the gonadotropins were presented by Pamela Mellon, James Hoeffler, John Nilson, Bruce Weintraub, and Joel Habener. These presentations reflected major advances from the concepts offered at the last meeting. Other advances in the understanding of hormone structure/function included the continuing work by Joyce Lustbader on the determination of the three-dimensional structure of hCG, including a new approach using nonisotopic labels for 3-D and 4-D NMR imaging. New insights into biosynthesis of hCG were proposed by Irving Boime. Studies directed at identifying receptor contact regions were presented by Henry Keutmann and David Puett. Other significant advances in understanding the biosynthesis and actions of this molecule were presented by Raymond Ruddon, Diana Blithe, and Jacques Baenziger. Raymond Ruddon and his group have now definitively established the disulfide pairings for the beta subunit of CG. New concepts were presented related to receptor binding and activation by Leonard Kohn, William Moyle, and Tae Ji. The meeting was concluded with an examination of the clinical relevance of these new findings from the laboratories of William Crowley, Dominique Bellet, Robert Canfield, Glenn Braunstein, William Odell, and Bruce Nisula. This text has been arranged to reflect these six areas of study.

The importance of this symposium was underscored by the attendance of representatives from over twenty countries. We were also honored to have in attendance Dr. Piero Donini of Serono, who was responsible for the first commercial preparation of human menopausal gonadotropins. It was fascinating to hear him reflect on his scientific career. We gratefully acknowledge the March of Dimes and Serono Symposia, USA for awarding travel grants that made it possible for a number of young investigators to attend the meeting. Last, we would like to thank Dr. Bruce Burnett and his staff for their thoroughness and their penchant for details, which culminated in a wonderful meeting. The atmosphere was truly convivial and engaging, and we hope that Serono Symposia, USA will continue to support future meetings on the glycoprotein hormones as part of their interest in the regulation of human fertility.

Joyce W. Lustbader
David Puett
Raymond W. Ruddon

# Contents

**Part VI.   Clinical Implications**

# Contributors

TAKASHI AKAMIZU, Section on Cell Regulation, Laboratory of Biochemistry and Metabolism, National Institute of Diabetes and Digestive and Kidney Diseases, National Institutes of Health, Bethesda, Maryland, USA.

ELAINE T. ALARID, Departments of Reproductive Medicine and Neurosciences, School of Medicine, University of California, La Jolla, California, USA.

JACQUES U. BAENZIGER, Department of Pathology, Washington University School of Medicine, St. Louis, Missouri, USA.

TOSHIAKI BAN, Section on Cell Regulation, Laboratory of Biochemistry and Metabolism, National Institute of Diabetes and Digestive and Kidney Diseases, National Institutes of Health, Bethesda, Maryland, USA.

KERRY M. BARNHART, Departments of Reproductive Medicine and Neurosciences, School of Medicine, University of California, La Jolla, California, USA.

ELLIOTT BEDOWS, Eppley Institute for Research in Cancer and Allied Diseases, University of Nebraska Medical Center, Omaha, Nebraska, USA.

DOMINIQUE BELLET, Department of Immunology, CNRS, School of Pharmacy, Paris, and Department of Molecular Immunology, Gustave Roussy Institute, Villejuif Cedex, France.

JEAN-MICHEL BIDART, Department of Immunology, CNRS, School of Pharmacy, Paris, and Department of Molecular Immunology, Gustave Roussy Institute, Villejuif Cedex, France.

STEVEN BIRKEN, Irving Center for Clinical Research, Department of Medicine, Columbia University College of Physicians and Surgeons, New York, New York, USA.

DIANA L. BLITHE, Developmental Endocrinology Branch, National Institute of Child Health and Human Development, National Institutes of Health, Bethesda, Maryland, USA.

MASAKI BO, Department of Obstetrics and Gynecology, Kobe University School of Medicine, Kobe, Japan.

DONALD L. BODENNER, Molecular and Cellular Endocrinology Branch, NIDDK, National Institutes of Health, Bethesda, Maryland, USA.

IRVING BOIME, Department of Molecular Biology and Pharmacology, Department of Obstetrics and Gynecology, Washington University School of Medicine, St. Louis, Missouri, USA.

GLENN D. BRAUNSTEIN, Department of Medicine, Cedars-Sinai Medical Center–UCLA School of Medicine, Los Angeles, California, USA.

JONATHAN MILES BROWN, Martek Biosciences Corporation, Columbia, Maryland, USA.

MARITA BÜSCHER, IMCB, National University of Singapore, Singapore, Republic of Singapore.

ROBERT E. CANFIELD, Irving Center for Clinical Research, Columbia University College of Physicians and Surgeons, New York, New York, USA.

YI CHEN, Irving Center for Clinical Research, Columbia University College of Physicians and Surgeons, New York, New York, USA.

C.M. CLAY, Department of Physiology, Colorado State University, Ft. Collins, Colorado, USA.

PAUL COTTU, Department of Medicine, Gustave Roussy Institute, Villejuif Cedex, France.

WILLIAM F. CROWLEY, Jr., Reproductive Endocrine Unit, Harvard Medical School, Massachusetts General Hospital, Boston, Massachusetts, USA.

JOEL S. FINKELSTEIN, Endocrine Unit, Harvard Medical School, Massachusetts General Hospital, Boston, Massachusetts, USA.

KRISTEN FRIEDMAN, School of Public Health, Columbia University College of Physicians and Surgeons, New York, New York, USA.

MARIE C. GELATO, Department of Endocrinology, SUNY at Stony Brook, Stony Brook, New York, USA.

CESIDIO GIULIANI, Section on Cell Regulation, Laboratory of Biochemistry and Metabolism, National Institute of Diabetes and Digestive and Kidney Diseases, National Institutes of Health, Bethesda, Maryland, USA.

JEANINE GRIFFIN, Department of Medicine, University of Utah School of Medicine, Salt Lake City, Utah, USA.

ARTHUR GUTIERREZ-HARTMANN, Department of Medicine, Division of Endocrinology, and Department of Biochemistry, Biophysics, and Genetics, UCHSC, Denver, Colorado, USA.

JOEL F. HABENER, Laboratory of Molecular Endocrinology, Massachusetts General Hospital, Howard Hughes Medical Institute, Harvard Medical School, Boston, Massachusetts, USA.

JANET E. HALL, Reproductive Endocrine Unit, Harvard Medical School, Massachusetts General Hospital, Boston, Massachusetts, USA.

D.L. HAMERNIK, Department of Veterinary Science, University of Nebraska, Lincoln, Nebraska, USA.

MAUREEN C. HATCH, School of Public Health, Columbia University College of Physicians and Surgeons, New York, New York, USA.

JONATHAN H. HECHT, Departments of Reproductive Medicine and Neurosciences, School of Medicine, University of California, La Jolla, California, USA.

L.L. HECKERT, Department of Pharmacology, Case Western Reserve University, Cleveland, Ohio, USA.

AKINARI HIDAKA, Section on Cell Regulation, Laboratory of Biochemistry and Metabolism, National Institute of Diabetes and Digestive and Kidney Diseases, National Institutes of Health, Bethesda, Maryland, USA.

JAMES P. HOEFFLER, Department of Medicine, Division of Medical Oncology, and Department of Biochemistry, Biophysics, and Genetics, UCHSC, Denver, Colorado, USA.

FRIEDEMANN HORN, Biochemistry Institute at RWTH, Aachen, Neuklinikum Pauwelsstrasse, Aachen, Germany.

FRANCK HOUSSEAU, Department of Immunology, CNRS, School of Pharmacy, Paris, France.

JIANING HUANG, Department of Biochemistry, University of Georgia, Athens, Georgia, USA.

JEFFREY R. HUTH, Eppley Institute for Research in Cancer and Allied Diseases, University of Nebraska Medical Center, Omaha, Nebraska, USA.

SHOICHIRO IKUYAMA, Section on Cell Regulation, Laboratory of Biochemistry and Metabolism, National Institute of Diabetes and Digestive and Kidney Diseases, National Institutes of Health, Bethesda, Maryland, USA.

STEPHEN M. JACKSON, Department of Biochemistry, Biophysics, and Genetics, UCHSC, Denver, Colorado, USA.

INHAE JI, Department of Molecular Biology, University of Wyoming, Laramie, Wyoming, USA.

TAE H. JI, Department of Molecular Biology, University of Wyoming, Laramie, Wyoming, USA.

LATA JOSHI, Molecular and Cellular Endocrinology Branch, NIDDK, National Institutes of Health, Bethesda, Maryland, USA.

AMALIA CHRISTINA KELLY, Columbia University College of Physicians and Surgeons, New York, New York, USA.

R.A. KERI, Department of Pharmacology, Case Western Reserve University, Cleveland, Ohio, USA.

HENRY T. KEUTMANN, Department of Medicine, Endocrine Unit, Massachusetts General Hospital, and Harvard Medical School, Boston, Massachusetts, USA.

MYUNG K. KIM, Molecular and Cellular Endocrinology Branch, NIDDK, National Institutes of Health, Bethesda, Maryland, USA.

LEONARD D. KOHN, Section on Cell Regulation, Laboratory of Biochemistry and Metabolism, National Institute of Diabetes and Digestive and Kidney Diseases, National Institutes of Health, Bethesda, Maryland, USA.

SHINJI KOSUGI, Section on Cell Regulation, Laboratory of Biochemistry and Metabolism, National Institute of Diabetes and Digestive and Kidney Diseases, National Institutes of Health, Bethesda, Maryland, USA.

JOYCE W. LUSTBADER, Irving Center for Clinical Research, Columbia University College of Physicians and Surgeons, New York, New York, USA.

JAMES A. MAGNER, Department of Medicine, East Carolina University, Greenville, North Carolina, USA.

PENELOPE K. MANASCO, Laboratory of Molecular and Integrative Neurosciences, National Institute of Environmental Health Sciences, National Institutes of Health, Research Triangle Park, North Carolina, USA.

ISABELLE MARCILLAC, Department of Molecular Immunology, Gustave Roussy Institute, Villejuif Cedex, France.

KATHRYN A. MARTIN, Reproductive Endocrine Unit, Harvard Medical School, Massachusetts General Hospital, Boston, Massachusetts, USA.

CRISTINA MATERA, Columbia University College of Physicians and Surgeons, New York, New York, USA.

PAMELA L. MELLON, Departments of Reproductive Medicine and Neurosciences, School of Medicine, University of California, La Jolla, California, USA.

TERRY E. MEYER, Pioneer Hi-Bred International, Johnston, Iowa, USA.

KIMBERLY MOUNTJOY, Eppley Institute for Research in Cancer and Allied Diseases, University of Nebraska Medical Center, Omaha, Nebraska, USA.

WILLIAM R. MOYLE, Department of Obstetrics and Gynecology, Robert Wood Johnson Medical School, Piscataway, New Jersey, USA.

YOKO MURATA, Molecular and Cellular Endocrinology Branch, NIDDK, National Institutes of Health, Bethesda, Maryland, USA.

GIORGIO NAPOLITANO, Section on Cell Regulation, Laboratory of Biochemistry and Metabolism, National Institute of Diabetes and Digestive and Kidney Diseases, National Institutes of Health, Bethesda, Maryland, USA.

J.H. NILSON, Department of Pharmacology, Case Western Reserve University, Cleveland, Ohio, USA.

BRUCE C. NISULA, Developmental Endocrinology Branch, National Institute of Child Health and Human Development, National Institutes of Health, Bethesda, Maryland, USA.

JOHN F. O'CONNOR, Irving Center Core Laboratory, Irving Center for Clinical Research, Columbia University College of Physicians and Surgeons, New York, New York, USA.

WILLIAM D. ODELL, Department of Medicine, University of Utah School of Medicine, Salt Lake City, Utah, USA.

FUMIKAZU OKAJIMA, Section on Cell Regulation, Laboratory of Biochemistry and Metabolism, National Institute of Diabetes and Digestive and Kidney Diseases, National Institutes of Health, Bethesda, Maryland, USA.

HAROLD PAPKOFF, Department of Animal Science, University of California, Davis, California, USA.

FULVIO PERINI, Eppley Institute for Research in Cancer and Allied Diseases, University of Nebraska Medical Center, Omaha, Nebraska, USA.

SUSAN POLLAK, Irving Center for Clinical Research, Columbia University College of Physicians and Surgeons, New York, New York, USA.

David Puett, Department of Biochemistry, University of Georgia, Athens, Georgia, USA.

S.N. VENKATESWARA RAO, Department of Obstetrics and Gynecology, Robert Wood Johnson Medical School, Piscataway, New Jersey, USA.

SUSAN R. ROSE, Division of Pediatric Endocrinology, University of Tennessee, Memphis, Tennessee, USA.

NATHALIE ROUAS, Department of Immunology, CNRS, School of Pharmacy, Paris, France.

RAYMOND W. RUDDON, Eppley Institute for Research in Cancer and Allied Diseases, University of Nebraska Medical Center, Omaha, Nebraska, USA.

ROBERT J. RYAN, Department of Biochemistry and Molecular Biology, Mayo Clinic, Rochester, Minnesota, USA.

MOTOYASU SAJI, Section on Cell Regulation, Laboratory of Biochemistry and Metabolism, National Institute of Diabetes and Digestive and Kidney Diseases, National Institutes of Health, Bethesda, Maryland, USA.

ARLEEN L. SAWITZKE, Department of Medicine, University of Utah School of Medicine, Salt Lake City, Utah, USA.

HIROKI SHIMURA, Section on Cell Regulation, Laboratory of Biochemistry and Metabolism, National Institute of Diabetes and Digestive and Kidney Diseases, National Institutes of Health, Bethesda, Maryland, USA.

YOSHIE SHIMURA, Section on Cell Regulation, Laboratory of Biochemistry and Metabolism, National Institute of Diabetes and Digestive and Kidney Diseases, National Institutes of Health, Bethesda, Maryland, USA.

CLAUDIA STAUBER, Center for Molecular Biology, Autonomous University, Madrid, Spain.

DAVID J. STEGER, Departments of Reproductive Medicine and Neurosciences, School of Medicine, University of California, La Jolla, California, USA.

BRIAN L. STRAUSS, Department of Molecular Biology and Pharmacology, Washington University School of Medicine, St. Louis, Missouri, USA.

MARIUSZ W. SZKUDLINSKI, Molecular and Cellular Endocrinology Branch, NIDDK, National Institutes of Health, Bethesda, Maryland, USA.

KAZUO TAHARA, Section on Cell Regulation, Laboratory of Biochemistry and Metabolism, National Institute of Diabetes and Digestive and Kidney Diseases, National Institutes of Health, Bethesda, Maryland, USA.

ANN E. TAYLOR, Reproductive Endocrine Unit, Harvard Medical School, Massachusetts General Hospital, Boston, Massachusetts, USA.

N. RAO THOTAKURA, Molecular and Cellular Endocrinology Branch, NIDDK, National Institutes of Health, Bethesda, Maryland, USA.

FRÉDÉRIC TROALEN, Department of Molecular Immunology, Gustave Roussy Institute, Villejuif Cedex, France.

CHUAN WANG, Department of Chemistry, Rutgers University, Piscataway, New Jersey, USA.

DARRELL N. WARD, Department of Biochemistry and Molecular Biology, The University of Texas M.D. Anderson Cancer Center, Houston, Texas, USA.

BRUCE D. WEINTRAUB, Molecular and Cellular Endocrinology Branch, NIDDK, National Institutes of Health, Bethesda, Maryland, USA.

RANDALL C. WHITCOMB, Research Division, Park-Davis Pharmaceuticals, Ann Arbor, Michigan, USA.

HAO WU, Department of Biochemistry, Columbia University College of Physicians and Surgeons, New York, New York, USA.

HAIYING XIA, Department of Biochemistry, University of Georgia, Athens, Georgia, USA.

DAVID YARMUSH, Department of Biology, Columbia University College of Physicians and Surgeons, New York, New York, USA.

HUAWEI ZENG, Department of Molecular Biology, University of Wyoming, Laramie, Wyoming, USA.

XIAOLU ZHANG, Ciba-Geigy Corporation Pharmaceuticals Division, Summit, New Jersey, USA.

# Part I

## Reflections Past and Present

# 1

# Reflections on Purifying Gonadotropins in the 1960s

Harold Papkoff

In this short retrospective I have elected to confine myself to a few recollections on some of the problems faced by myself and the many others who were working on the purification of the glycoprotein hormones in the 1960s. There is little question that the isolation of highly purified gonadotropins (and thyrotropin) in the late 1950s and the 1960s, and the preparation of their subunits, provided a powerful stimulus to physiological and clinical studies of these hormones. Thus, it became possible to establish specific, sensitive, and reliable radioimmunoassays and, later, radioreceptor assays, the results of which provided the groundwork for the spectacular gene-related studies of recent years.

Initially, the availability of purified glycoprotein hormones was made possible by the generosity of the individual investigators who purified the materials and later by programs sponsored by the National Institutes of Health, as well as the United States Department of Agriculture. This short narrative addresses some of my early experiences and observations in attempting to isolate in highly purified form the gonadotropins *luteinizing hormone* (LH) and *follicle stimulating hormone* (FSH), mainly from sheep pituitary glands. My reflections may be unique to me and my laboratory at the time, but I suspect that the many other scientists concurrently trying to do the same things had similar experiences. Rather than describe who did what when, I would hope to give the readers, especially those who are younger, a feel for what went on at the bench level. I am relying solely on memory since I never kept a diary, possibly because I did not think that what I was doing was of such import and possibly because there was much that I did not want recorded for posterity, such as many ill-advised experiments that resulted in failure.

By the 1960s, ion exchangers (like CM- and DEAE-cellulose) and gel-filtration media (initially, Sephadex gels ) had become available, which greatly facilitated protein purification. Prior to this time, the protein purifier had to rely mainly on solubility techniques involving precipitations at various conditions of pH or precipitations at different alcohol or salt (sodium chloride, ammonium sulfate, etc.) concentrations. Despite these limitations, reasonably purified preparations of prolactin and growth hormone had been obtained by 1950, in part because of the high pituitary content of these hormones (at least in bovine and ovine glands) and because of their solubility characteristics, which facilitated separations. The glycoprotein hormones, however, were quite soluble throughout the pH range, as well as in solutions of low alcohol and salt concentration, and purification ultimately required the use of ion exchange and gel-filtration columns, as well as classical precipitation techniques. Pituitaries were easy to obtain commercially in large quantity if one were content to work on the sheep, cow, or pig. It was usual for me to use at least 100 g of tissue for a pilot experiment and 0.5–1.0 kg for a more serious effort.

Given the above, it sounds as if we had all we needed to purify readily whatever glycoprotein hormone we were after. We had but to apply the above techniques and assay our samples. It was not all that easy. I will deal with problems of bioassay later, but first some memories related to purifications.

It was believed at the time that everything had to be done in a cold room to avoid hormonal inactivation. Cold rooms then were small and cramped, damp and musty, cluttered, and not very comfortable places in which to work. Nobody minded, however, if I smoked my pipe. Tissue homogenizers were primitive: a glass homogenizer or Waring blender for small amounts of tissue and a butcher's meat grinder for kilogram quantities. Centrifuges were primitive. It was not until the mid-1960s that we had anything resembling a modern Sorval. We relied primarily on a refrigerated International that could barely reach 3000 rpm. We needed long times of centrifugation; even then, our precipitates were poorly packed, and if not careful, we often poured off both supernatant fluid and precipitate, which required centrifuging it again. Also, our centrifuge tubes were made of glass (plastic came later) and, of course, they would frequently shatter. Performing large-scale extractions and initial salt precipitations hardly made one feel like a scientist because so much time was spent in doing brute physical labor.

Another problem was the time consumed in dialyzing samples to rid them of salt and in lyophilizing samples. Dialysis membranes were of poor quality and would often spring a leak with resultant loss of sample (invariably, the important one), and until the mid-1960s our lyophilizers were homemade of glass attached to a vacuum pump, capable of handling only four samples. Since they did not have a manometer in the system, we all became very adept at determining vacuum pressure by the sound of the pump, but even so, with some degree of frequency, a leak would develop and result in the thawing of all the samples. This is a problem that has survived into the 1990s, even with modern lyophilizers. If all went well with no mishaps, it would take me one to two weeks to fractionate a kilo of sheep pituitaries and prepare crude LH and FSH fractions.

Having succeeded by sheer stubborness in preparing some crude fractions, we were now ready to exploit some of the new ion exchangers and gel-filtration columns. Countless hours were spent in sieving ion exchangers and gels to a uniform range and packing columns with slow defined flow rates and with nary an air bubble. I think we really overdid it, as years later I found we could be much more sloppy and still get suitable results.

The biggest bane, however, was the fraction collector. The first collectors on the market were expensive, huge, and malfunctioned with a high rate of frequency. Many an experiment poured into a single tube. Since we had only a couple of collectors for a laboratory of about 20 people, we frequently collected tubes by hand, even if it meant spending all night in the lab. Considering all of these and other petty problems, it is somewhat remarkable to me that we were able to accomplish as much as we did.

The greatest difficulty for the protein chemist, however, was coping with and coming to terms with the bioassays that ultimately determined our success or failure. My laboratory relied on two in vivo rat assays to test our fractions for LH and FSH. For LH we used what was called the ventral prostate assay, which I believe was originally devised by Roy O. Greep. In this assay, hypophysectomized young (21-day) male rats were injected for a couple of days with hormone solution, and on autopsy the ventral prostates were weighed. Increases in weight were dose related, and it later turned out that this assay, which took about a week to perform, was able to detect about 1.0 µg of purified ovine LH (and somewhat lower levels of *human chorionic gonadotropin* [hCG] and human LH).

In addition to the relative insensitivity of the assay, the requirement for hypophysectomized rats was also troublesome. We were fortunate in having people who could perform this operation efficiently, but we were limited to having only 30 rats a week. One can see that if we divided the rats into groups of four, we could test seven solutions, and if we used one group for a control, two groups for two doses of a standard, we were left with four groups that could be used to test two experimental fractions at two doses each. Not many! As a result, we often resorted to the hazardous practice of using one-point assays.

Our assay for FSH was not much better. It was the hCG augmentation assay described by Steelman and Pohley in 1953. Immature (21-day) female rats were injected with a mixture of hCG and test solution for several days, and on autopsy the ovaries were weighed. The use of the hCG was to swamp out any LH contamination in the preparation being tested. Since the animals did not have to be hypophysectomized, we had somewhat more available to us each week than for the LH assay, about 40, but we were still constrained in the number of samples we could test in a given assay. This assay had a sensitivity similar to the LH assay and could detect about 0.5 µg of highly purified ovine FSH. Given other problems, such as entire assays in which the animals were all unresponsive and animals dying before completion of the assay, we frequently had to assay a sample two or three times before we could rely on the results.

For me, as a protein biochemist, coping with the bioassays was a constant source of frustration, as I could generate samples from fractionation experiments far faster than they could be reliably assayed. My response was to devise methods that generated only a few fractions at each step. This was easily done where precipitation steps were employed. With ion exchange columns, however, it meant sacrificing the higher resolution that one could obtain with gradient systems in favor of stepwise elutions yielding three to four fractions.

Having obtained what one thought was a highly purified preparation, usually because of high biopotency, there was the problem of demonstrating that it was indeed pure or highly purified. It was easy enough to show biological purity by virtue of bioassays for expected contaminant hormones, although this often required use of large amounts of material because of the insensitivity of most of the assays. Chemical purity, however, was to become a Holy Grail, as most of the procedures available to us, such as N-terminal analysis, showed the gonadotropins to be impure. In other words, we could never demonstrate purity; we could always demonstrate impurity. For this reason, I never referred to any of the preparations I had prepared as "pure," but rather as "highly purified." Countless experiments were performed trying to rid our preparations of evidence of impurity. We know in retrospect, of course, that much of this problem resulted from the now-expected microheterogeneity exhibited by the glycoprotein hormones with respect to the carbohydrate and "nibbling" at the termini of the polypeptide chains. I believe it is John G. Pierce who often said, in effect, that if we had waited for preparations that satisfied conventional criteria of purity, we would never have gotten to the point of actually sequencing the polypeptide chains. Physiologists, however, were so grateful for the use of good gonadotropin preparations that they rarely if ever questioned purity, and only if the material did not behave as they thought it should would they berate us. So, for many of them, the criterion of purity was the fact that I had said it was highly purified.

Despite the difficulties I and others encountered in purifying gonadotropins, the end result has been a great deal of gratification in knowing that this work played a vital role in the development of our present-day knowledge of the gonadotropins. It has also been of great gratification to see materials prepared by me used by other investigators, especially physiologists, in many exciting experiments. I am sure that the younger investigator of today will someday look back and have a similar tale to tell.

# 2

# Where to from Here?

ROBERT J. RYAN

Where to from here? is a necessary question for scientists. It is most often asked at the time of preparation of a grant application. At these times, however, the answers are apt to be biased by what you judge to be salable to a review committee. I think better answers come after the application has been submitted. This allows time to obtain data and make the goal salable at the next application three to five years later.

Answering this question has been crucial to my career. One of the first times was during my fellowship in E.B. (Ted) Astwood's lab in 1957. I started in Ted's lab in July of that year with the idea that I was going to culture pituitary cells and clone them in the manner of Theodore Puck. I wanted separate flasks of cloned cells, each making a separate hormone. Ted let me pursue this goal with his encouragement, even though I am sure he thought I would fail. He believed that everyone should have the chance to fail. It took me six months to realize that my ideas were naive for the time and place and to recognize that I would fail. I needed to address the question, Where to from here?

The answer, to purify and develop assays for the human gonadotropins, occupied me for the next dozen years. I came to this answer for several reasons. First, it was an important element in understanding pituitary-gonadal function. Second, no one was working on the human hormones as far as I could tell. Work was progressing on the purification of domestic animal LH, and techniques for protein purification were developing rapidly. Third, there was considerable expertise in protein purification in the Astwood lab. Fourth, I felt I was able to learn and use these techniques. Fifth, being an M.D. by formal training, the human hormones were of more interest to me than domestic animal hormones.

Over the next few years, several things happened that convinced me that I had set out on the right road. Parlow's ascorbic acid depletion assay for LH became available: Berson and Yalow published the radioimmunoassay technique, and it was obviously applicable to the gonadotropins once they were purified. The National Pituitary Agency was developed and made the starting material for purifying human pituitary hormones available. Great progress was made in the purification of the ovine and bovine hormones, and the techniques were applicable to the human hormones after suitable modifications.

By the late 1960s these goals were achieved, and serum concentrations of LH and FSH in humans were being measured under various physiologic and pathologic circumstances. It was becoming increasingly boring to me and, therefore, time for a new Where to from here?

Two new long-term goals rather than one were chosen because they were complementary and because the move to Mayo in 1967 increased the funding for the laboratory. The first goal was to study the structure of the gonadotropins, and the second was to look at the mechanism of action of the gonadotropins. With respect to the structure of gonadotropins, I deliberately decided not to get directly involved in amino acid sequencing. I did so for several reasons. First, it would require several years to tool up and learn the techniques. Second, others in the field were making good progress in this area. And last, a collaborative effort with Henry Keutmann was begun, and he was expert in amino acid sequencing. This was a most fortunate event, as it led to a twenty-year collaboration that became increasingly important with the passage of time. Instead of sequencing, I decided to focus on the physical properties of the molecules, the role of the carbohydrates, and modification of specific residues. These goals were pursued for a decade.

With respect to the mechanism of action of the gonadotropins, it was obvious, even in the late 1960s, that there had to be a receptor for these molecules on target cells. It was also obvious that a good tissue source would be the same ovaries used for the ascorbic acid depletion assay for LH and hCG. Within a matter of a few weeks we were able to demonstrate significant and specific binding of radiolabeled human LH to pseudopregnant rat ovaries. Over the next 10–15 years the characteristics, distribution, and localization of these receptors and their relationship to the adenylate cyclase system were studied. Although the purification of the LH-CG receptor was started about 1971, this goal was not reached until 1989.

In the early 1980s progress toward these goals was slowing, and again we had to ask Where to from here? A new technical approach was needed to define the receptor binding sites on the hormone and, down the road, the hormone binding sites on the receptor. The molecular biologic approach of site-specific mutagenesis was considered, but discarded for several reasons. First, it would require a several-year period to learn and apply the techniques. Second, this approach had a number of limitations. The amounts of mutated protein obtained would be quite small and, therefore, studies would be difficult, particularly with respect to low-affinity binding and whether a change was due to direct effect on the binding site or an indirect effect through change in conformation.

Because of our collaboration with Henry Keutmann and the increasing use of synthetic peptides to define important sites on other proteins, a synthetic peptide strategy was chosen. There was a several-year lag to obtain funds (from the Mellon Foundation, NIH, and Mayo) and establish a peptide synthesis and sequencing facility at Mayo. Progress was then very rapid. Henry concentrated on the beta subunits of hCG and LH, and we concentrated on the common alpha subunit and TSH beta. More recently, the same approach has been used to look at potential hormone binding sites on the LH-CG and TSH receptors. It has been great fun.

This brief narrative sounds like a straight-line approach to each successive goal. It was not. Over a span of thirty years, there were approximately fifty postdoctoral fellows or research associates who worked in the lab. The success was due to their efforts. However, each had different talents and interests. Because I shared some of Ted Astwood's philosophy about training, I tried to allow as much freedom as possible. This, of course, resulted in fits and starts in progress and a number of tangents. Several of the tangents were of particular interest and were exploited for several years.

As I approached retirement in 1990, I thought it was again time for two Where to from here? questions. For me, the answer was away from grant applications, manuscripts, reviewing, and committees and toward the golf course, garden, stamp collecting, novels, and pleasant times with my wife and family. For the field of gonadotropin research, I thought the answer should be the goals of determining the three-dimensional structures of the hormone and receptors and detailing knowledge of their sites at interaction. These again are long-term goals. It will require the talents of people trained in crystallography, NMR, molecular biology, protein-peptide chemistry, and molecular modeling. Because no one individual has all of these talents, it will mean a team effort or strong interactive collaborations such as Henry Keutmann and I enjoyed. Perhaps the NIH should consider a multisite program project. At the very least, I hope that those working on gonadotropins in the future will be as open and congenial with each other as I found my peer group to be.

# 3

# Remembrance of Things Foregone

Darrell N. Ward

The organizers of this symposium requested the three honorees at the symposium to consider a remembrance somewhat along the lines of the highly successful series published in *Endocrinology* for the 75th anniversary of The Endocrine Society. I know one of my fellow honorees, Harold Papkoff, contributed a remembrance to that series (1) that I much admired, and I am sure he can do it again. The other fellow honoree, Robert J. Ryan, can always be counted on for that last word, and I will expect as much in this series. But that leaves me, as the third in this series of honorees (and also fellow retirees), with the dilemma of trying to produce some sort of remembrance that will be of interest to the readers of this symposium volume. After some considerable thought I decided that rather than simple remembrances, I might best interest the readers in some of the things I have not forgotten, but for one reason or another I have foregone. Perhaps these things may strike an interested listener who still has laboratory and grant support at his disposal.

As an example of something foregone by me, the investigation of the rabbit LH will serve. We first got interested in that from a conversation with Claude Desjardins, who was utilizing many rabbits in his research and figured it would be little extra effort to collect the pituitaries at the appropriate times in their research. He was also interested in the LH for their radioimmunoassays. From that conversation an extensive collection ensued, and eventually the LH was isolated and characterized and the study was published (2). We then followed it up with determination of the amino acid sequence of the alpha and beta subunits (3, 4).

The finding from that study that still piques my interest concerns the N-terminus of the beta subunit. (Others found it less interesting, as we only got a handful of reprint requests.) The N-terminus was blocked on most of the molecules, but there were some that contained an Ala on the N-terminus. Thus, the N-terminal heterogeneity defined two forms of rabbit LH. The Ala proved to be in the same position as the Ser on all the other mammalian beta subunits (+1) except whale LH, which has a Pro.

(Position 1 is defined with respect to the half-cystine at position 9, to put everything in common register.) However, the striking thing was that the majority of the rabbit LH beta subunits had two additional residues (pyroGlu-Pro) preceding the Ala-Arg- of positions 1 and 2. We observed this with several rabbit LH preparations. Moreover, B. Shome and A.F. Parlow also found the blocked N-terminus of their rabbit LH preparations (personal communication, 4).

We do not know the signal sequence for rabbit lutropin, but for those we do know in the mammalian series the -2, -1 residues are -W-A, so this does not seem to be simply an aberrant signal peptidase cleavage in the rabbit. I have often wondered whether this unusual structure feature in the rabbit was somehow related to the mechanism of reflex ovulation in the rabbit. At the very least, I have wondered what the complete signal peptide looks like in the rabbit. I have wondered this to a couple of study sections, in fact. They approved the idea, but did not fund that part of the budget to do the required recombinant DNA work. There seems to be an a priori understanding that peptide chemists cannot learn nucleic acid work. Thus, I had to forego this inquiry.

The question of the disulfide closures of the glycoprotein hormone subunits has never been satisfactorily resolved, since there are several disconsonant reports. We addressed this problem at an earlier Serono Symposium (5). The part that we have foregone on this problem has been a desire to develop new, more facile methods of analysis. When the time to retire arrived, we were working on methods to derivatize the newly formed sulfhydryl groups with azodye reactants. The idea was to get a sufficiently rapid reaction with the sulfhydryl group shortly after formation by a partial-reduction reaction. This would then trap the newly opened disulfide with color-labeled derivative that could be rapidly analyzed or followed during fractionation of cleavage products from the subunit protein. This is not too different from the selective radioactive labeling that John Pierce used to open this field (6), but we anticipated advantages in the handling and analysis of the peptides concerned. The studies were promising as far as we had gone, but time ran out and retirement closed the door.

John Pierce introduced us to another problem related to the oxidative closure of the subunits. The alpha subunit may be reduced and reoxidized with recovery of the ability to form a functional alpha-beta subunit (7). However, neither John Pierce nor all those who have tried to reduce and reoxidize a beta subunit in a test tube have been able to do it in a manner that allows recovery of the activity in the appropriate recombinant (8). One of the present symposium organizers, Raymond Ruddon, has worked out superb systems to follow this process in vitro with cell-derived systems. His colleagues have provided the rather complete details of the sequence of events in nature (for example, 9). In this connection, I have so far had to forego the hope that one could establish conditions in the test tube for the utilization of enzymatic procedures (e.g., protein disulfide isomerase, or PDI) to produce a native form of the reoxidized beta subunit. This tool would provide a marvelous approach to study the folding of the subunits in "pristine" environments that would allow the facile analysis of the folded proteins obtained. I hope some day some one takes up this challenge. (See "Addendum.")

Finally, as my remembrance of things I have had to forego concludes, I should like to mention the unusual structural conditions to be attributed to the equine subunits. We showed in a paper with George Bousfield (10) that when the subunits of the equine were used to produce hybrids with other species—for example, equine alpha with porcine beta LH or the converse—the alpha-beta subunits readily combined, but the activity of the recombinants was not what we had been led to expect. What happened, we observed, was that the hybrid containing the equine alpha always showed an equal or often greater potency than the parent hormone from which the beta subunit was obtained. But, surprisingly, the beta subunit from equine showed an absolute requirement for the equine alpha subunit in order to obtain an active alpha-beta dimer. The implication for some very important structural details is apparent, and it is made more intriguing when one compares the high degree of homology in the molecules concerned. With the closing of the laboratory, this is yet another of the things I have had to forego. Perhaps this will be one for which I will not have long to lament, for my good colleague George Bousfield is continuing the equine hormone studies, and he informs me of some very interesting chimeric molecules he has been able to generate that are beginning to shed light on all of this. Perhaps some of the things I remember as foregone today will become inspiration for readers of this symposium proceedings, and we will hear about them in more concrete terms at future symposia.

## Addendum

When these remarks were presented at the Symposium on Glycoprotein Hormones from which this text was derived, it was necessary to note that some of the items on the foregoing "wish list" had already been answered by presentations at the symposium. I now believe the placement of the disulfide bonds in the beta subunits has been substantially established by two different methodologies. The studies of Tsunsawa et al. (11) for the oLHβ and Mise and Bahl (12) for the hCGβ have shown agreement for placements obtained by selective degradation-labeling procedures, while the studies from Ruddon's laboratory have substantiated these placements by integrative procedures developed over several years; see, for example, the overview of these studies as presented by Ruddon et al. (13) at this symposium. The work from Ruddon's laboratory regarding these disulfide placements was conclusively established by site-directed mutations and their effect on the folding patterns of the β-subunit of hCG taken in consideration of the foregoing studies from their laboratory, as summarized in the poster presentation by Bedows et al. (14). These studies specifically concern the LH and CG molecules. The only remaining caveat concerns the assumption that because the half-cystines are identically placed throughout the glycoprotein hormone sequences, the disulfide bonds will be identical. At the present time, this is a reasonable assumption, but it would be well if new methodology were to be developed that would allow direct determinations to be done on FSH and TSH, for example. Thus, that part of my foregone wish list remains a viable item to be achieved.

The problem of the conditions required for reoxidation of the reduced beta subunit to recover native conformation and the ability to recombine with the α-subunit to produce an active α-β dimer was also solved in another presentation from Ruddon's laboratory at this meeting, namely the poster presentation by Huth et al. (15). In their study, the reduced β-subunit could be induced to fold properly during reoxidation in the presence of the enzyme protein disulfide isomerase and proper control of the redox potential of the system. It is gratifying indeed that so many of the problems I had to forego are already addressed. I appreciate the effort, support, and organization provided by Serono for the forum for these marvelous revelations about the glycoprotein hormones, for which I have an undying interest. Thank you.

## *References*

1. Papkoff H. Glycoprotein hormones were always composed of subunits—we just had to find out the hard way. Endocrinology 1991;129:579–81.
2. Ward DN, Desjardins C, Moore WT Jr, Nahm HS. Rabbit lutropin: preparation, characterization of the hormone, its subunits and radioimmunoassay. Int J Pept Protein Res 1979;13:62–70.
3. Glenn SD, Nahm HS, Ward DN. The amino acid sequence of the rabbit glycoprotein hormone alpha subunit. J Prot Chem 1984;3:143–56.
4. Glenn SD, Nahm HS, Ward DN. The amino acid sequence of the rabbit lutropin beta subunit. J Prot Chem 1984;3:259–73.
5. Ward DN, Bousfield GR, Mar AO. Chemical reduction-reoxidation of the glycoprotein hormone disulfide bonds. In: Bellet D, Bidart JM, eds. Structure-function relationship of gonadotropins. New York: Raven Press, 1989:1–19.
6. Cornell JS, Pierce JG. Studies on the disulfide bonds of glycoprotein hormones: locations in the α chain based on partial reductions and formation of $^{14}$C-labeled S-carboxymethyl derivatives. J Biol Chem 1974;249:4166–74.
7. Guidice LC, Pierce JG. Studies on the disulfide bonds of glycoprotein hormones: complete reduction and reoxidation of the disulfide bonds of the α subunit of bovine luteinizing hormone. J Biol Chem 1976;251:6392–9.

8. Reeve JR Jr, Pierce JG. Disulfide bonds of glycoprotein hormones: their selective reduction in the β subunits of bovine lutropin and thyrotropin. Int J Pept Protein Res 1981;18:79–87.

9. Beebe JS, Huth JR, Ruddon RW. Combination of the chorionic gonadotropin free β-subunit with α. Endocrinology 1990;126:384–91.

10. Bousfield GR, Liu WK, Ward DN. Hybrids from equine LH: alpha enhances, beta diminishes activity. Mol Cell Endocrinol 1985;40:69–77.

11. Tsunasawa S, Liu W-K, Burleigh BD, Ward DN. Studies of disulfide bond location in ovine lutropin β-subunit. Biochim Biophys Acta 1977;492:340–56.

12. Mise T, Bahl OP. Assignment of the disulfide bonds in the β-subunit of human chorionic gonadotropin. J Biol Chem 1981;256:6587–92.

13. Ruddon RW, Huth JR, Bedows E, Mountjoy K, Perini F. Folding of the β subunit of hCG and its role in the assembly of the αβ heterodimer. In: Lustbader JW, Puett JD, Ruddon RW, eds. Glycoprotein hormones: structure, function, and clinical implications. New York: Springer-Verlag, 1994:137–55. (*See* Chapter 13, this volume).

14. Bedows E, Huth JR, Suganuma N, Boime I, Ruddon RW. Disulfide bond mutations affect the folding of hCG-β in transfected CHO cells. Glycoprotein hormones: structure, function, and clinical implications. New York: Springer-Verlag, 1994:351. (*See* Appendix, this volume.)

15. Huth JR, Perini F, Bedows E, Ruddon RW. The in vitro, protein disulfide isomerase (PDI)-catalyzed, and intracellular hCG-β folding/hCG subunit assembly pathways are indistinguishable. Glycoprotein hormones: structure, function, and clinical implications. New York: Springer-Verlag, 1994:357. (*See* Appendix, this volume.)

# Part II

## Regulation of Gene Expression

# 4

# Endocrine Regulation of Glycoprotein Hormone Alpha Subunit Gene Expression in Transgenic Mice

C.M. CLAY, R.A. KERI, D.L. HAMERNIK, L.L. HECKERT, AND J.H. NILSON

*Luteinizing hormone* (LH), *follicle stimulating hormone* (FSH), and *thyroid stimulating hormone* (TSH) are members of the glycoprotein hormone family and, as such, are heterodimers composed of an identical alpha subunit and distinct beta subunits that confer biological specificity (1). The common alpha subunit and each of the beta subunits are encoded by unique single copy genes that are expressed in specific cell types (gonadotropes: LH and FSH; thyrotropes: TSH) of the anterior pituitary gland (1, 2). LH and FSH are referred to as gonadotropins because they stimulate normal gonadal development and function, whereas TSH is necessary for normal development and function of the thyroid gland.

The synthesis and secretion of LH and FSH are exquisitely regulated via negative feedback by gonadal estrogen and androgen acting either directly at the pituitary gland or at the hypothalamus to affect secretion of the hypothalamic releasing peptide, *gonadotropin releasing hormone* (GnRH) (2–4). Accordingly, castration typically results in an elevation of both pituitary and circulating concentrations of LH and FSH that can be attenuated by replacement with either androgen or estrogen. Negative feedback effects of gonadal steroids include transcription of both the common alpha subunit and unique beta subunit genes (2). Thus, changes in the synthesis and secretion of intact heterodimer result from coordinated and regulated expression of both subunits, although the molecular mechanisms that control expression may be distinct for the genes encoding the alpha and beta subunits. As the dynamics of such a system are difficult to recapitulate in vitro, we have utilized transgenic mice harboring chimeric genes as in vivo models for studying molecular mechanisms underlying tissue-specific, and hormonally regulated, expression of the gene encoding the common alpha subunit of the glycoprotein hormones. Herein we describe two surprising results that emerged from these studies. First, although estrogen and androgen both suppress expression of α-CAT chimeric genes in transgenic mice, the molecular mechanisms underlying the attenuation of transcription appears quite distinct. Second, while 1500 bp of proximal 5' flanking region derived from the human alpha subunit gene is sufficient to confer gonadotrope-specific expression to a heterologous reporter gene, this same region lacks necessary *cis*-acting element(s) for thyrotrope-specific expression. Thus, distinct *cis*-acting elements are required for gonadotrope- and thyrotrope-specific expression of the glycoprotein hormone alpha subunit gene.

# Estrogen and Androgen Regulate Expression of the Glycoprotein Hormone Alpha Subunit Gene by Different Mechanisms

Our initial studies with transgenic mice had established that either 1500 bp or 315 bp of proximal 5' flanking region derived from the human or bovine alpha subunit genes are sufficient to confer pituitary-specific expression to the bacterial gene encoding *chloramphenicol acetyl transferase* (CAT) (5). Besides length, the chief difference between the human and bovine alpha subunit promoter regions is the presence of tandemly repeated *cAMP response elements* (CRE) in the proximal promoter of the human gene (Fig. 4.1). A single base transition renders the CRE homologue in the bovine gene nonfunctional and prevents expression from occurring in placenta (5, 6).

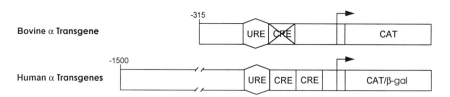

FIGURE 4.1. Chimeric human and bovine alpha subunit genes used to produce transgenic mice. The proximal promoter sequence (−315/+45) of the bovine alpha subunit gene was linked to CAT and the (−1500/+45) region of the human alpha subunit gene was linked to either CAT or lacZ. Besides length, the major difference between the two promoter regions is the presence of two functional cAMP responsive elements within the human gene, while the corresponding bovine sequence contains a single homologue of the CRE that fails to bind CREB.

Demonstration of pituitary-specific expression afforded a unique opportunity to assess the role of these promoter regulatory regions in
conferring hormonal responsiveness. In particular, we sought to determine
if the well-established attenuation of expression of the alpha subunit gene
by gonadal estrogen and androgen could be recapitulated in these two
lines of transgenic mice. This question was addressed using a castration
and steroid replacement paradigm.

Chronic administration of estradiol suppresses expression of the alpha
subunit gene in all species examined to date (2, 7). Thus, adult female
transgenic mice harboring either the human $\alpha(-1500)$CAT or bovine
$\alpha(-315)$CAT transgene were ovariectomized and treated with 17$\beta$-
estradiol for 14 days or treated with vehicle alone (7). On the last day of
treatment, pituitaries, brains, and livers were harvested for analysis of
CAT activity. Trunk blood samples were also collected and assayed for
serum concentrations of LH. Estradiol treatment suppressed pituitary
CAT activity by 82% and 72% in human $\alpha$-CAT and bovine $\alpha$-CAT
transgenic mice, respectively (Fig. 4.2). As expected, serum concentrations of LH were severely attenuated in mice receiving estradiol. Thus,
sufficient DNA regulatory sequence is present in the flanking regions
of both transgenes to confer responsiveness to estradiol. Based on the
similar responses of both transgenes to estradiol and the high degree
of homology between the human and bovine genes, we suggest that
element(s) responsible for estradiol regulation reside within 315 bp of the
transcriptional start sites of the two genes.

Although the human $\alpha$-CAT chimeric transgene could be regulated by
estradiol in transgenic mice, this same effect was not observed in transient
transfection assays into cell lines that contain *estrogen receptor* (ER)
(MCF-7) or by cotransfection of an estrogen receptor expression vector
into non-ER-containing cell lines (BeWo) (7). In addition, using a filter
binding assay we were unable to detect any high-affinity binding sites for
ER located in 17 kbp of the human alpha subunit gene. Thus, while the
human alpha subunit gene is regulated by estradiol, this regulation is
mediated by a mechanism that does not involve binding of activated ER
to a canonical ERE located within the human-alpha subunit gene proper
(7). At least one possibility that could explain this phenomenon is that
estrogen is exerting its effect indirectly by suppressing the secretion of
GnRH at the level of the hypothalamus. Alternatively, estradiol might
affect the expression of some other gene product within gonadotropes
that then acts to interfere with the GnRH signal transduction pathway.

Like estrogen, testicular androgens also suppress synthesis and secretion
of the pituitary gonadotropins (2, 4). This suppressive effect includes
transcription of both the common alpha subunit gene and the unique beta
subunit genes. Thus, we applied a similar castration/replacement paradigm
as above, but steroid replacement included either estradiol, testosterone,
or dihydrotestosterone. Adult male transgenic mice harboring either the

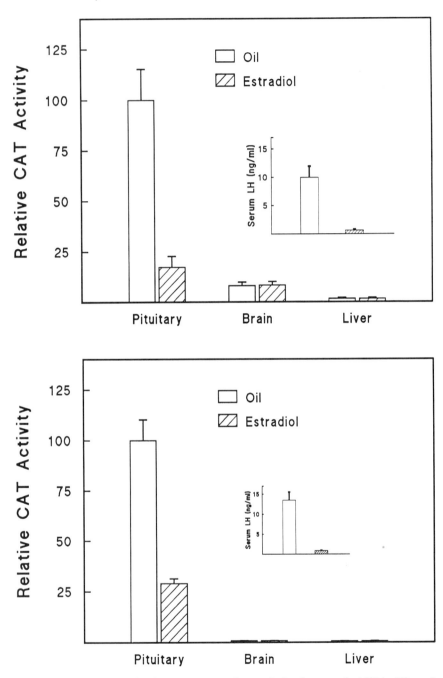

FIGURE 4.2. The proximal promoter regions of the human (−1500/+45) and bovine (−315/+45) alpha subunit genes confer estradiol responsiveness to CAT in pituitaries of female transgenic mice. Adult female transgenic mice were ovariectomized and treated with subcutaneous injections of 300 ng estradiol 17β/

human $\alpha(-1500)$CAT or bovine $\alpha(-315)$CAT were castrated and fitted with subcutaneous, constant-release pellets containing either T, DHT, or E. Control animals were castrated and received pellets containing carrier alone. At 2 weeks postcastration, tissues were analyzed for CAT activity as above.

The CAT activity assayed in pituitaries of either the human $\alpha$ CAT or bovine $\alpha$ CAT transgenic mice receiving T implants was significantly lower than that measured in pituitaries of castrate controls (Fig. 4.3). Physiologically consistent with attenuation in pituitary CAT activity, treatment with T also resulted in severely suppressed serum concentrations of LH.

The suppressive effects of DHT on transgene expression were not as pronounced as those of T. In light of the relative instability of DHT in blood, one explanation is simply that the dose of DHT was only minimally effective. Another possibility is that suppression of alpha transgene expression in T-implanted mice was a consequence of both a direct effect of androgen (DHT) and an indirect effect that occurs after aromatization to estrogen. This would explain the partial suppression of pituitary CAT activity that occurred when animals were treated with either DHT (a nonaromatizable androgen) or E alone. Nevertheless, androgenic regulation of the alpha subunit promoter in transgenic mice is demonstrable in two distinct lines of mice harboring reporter genes under the control of alpha subunit promoter fragments from different species.

Based on these experiments, we concluded that sequences contained within the proximal promoter regions of the H$\alpha$ and B$\alpha$ subunit genes can confer both androgenic and estrogenic regulation on a bacterial reporter gene in transgenic mice. The application of both transient transfection assays and DNA protein binding assays revealed that the effects of estrogen are exerted via a mechanism independent of ER binding to the human alpha subunit gene (7). Similar approaches, however, revealed that the molecular mechanisms underlying androgen suppression of alpha subunit gene expression may be quite different.

Our first indication of this divergent mechanism was obtained with coexpression assays in $\alpha$T3 cells, a pituitary cell line of gonadotrope origin (8). Figure 4.4 shows the results of a coexpression assay in which

---

100 µl safflower oil for 14 days. Results from H$\alpha$CAT and B$\alpha$CAT transgenic mice are shown in the upper and lower panels, respectively. CAT activity was assessed for pituitary, brain, and liver and is expressed relative to activity in pituitaries from oil-treated ovariectomized (control) mice, which is arbitrarily assigned a value of 100. Serum levels of LH were quantitated for oil- and estradiol-treated mice (inset). Reported values are mean ± SEM from 16 (H$\alpha$CAT) or 21 (B$\alpha$CAT) mice/group and represent two separate experiments for each construct. Reprinted with permission from Keri, Andersen, Kennedy, et al. (7).

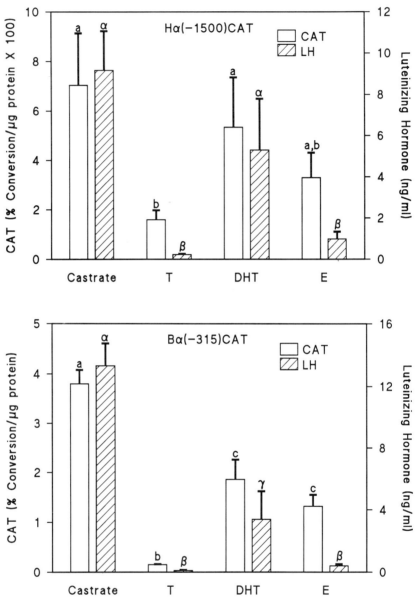

FIGURE 4.3. The proximal promoter regions of the human (−1500/+45) and bovine (−315/+45) alpha subunit genes confer androgen and estrogen responsiveness to CAT in pituitaries of male transgenic mice. Adult male transgenic mice were castrated and fitted with constant-release pellets containing either testosterone (T, 5 mg/pellet), 5α-dihydrotestosterone (DHT, 5 mg/pellet), or estradiol 17β (E, 0.25 mg/pellet). Control animals were castrated and received pellets containing inert carrier. Each treatment group contained a minimum of 7 animals.

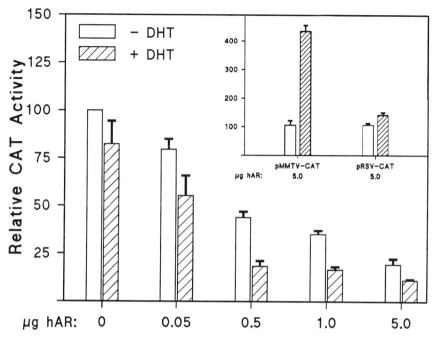

FIGURE 4.4. Coexpression of human AR in αT3 pituitary cells suppress Hα(−1500/+45)CAT expression in a steroid-dependent manner. Ten micrograms of pHα(−1500/+45)CAT were transiently cotransfected with the indicated amounts of human androgen expression vector (AR) into αT3 cells. On the day after transfection, cells were treated with medium containing $10^{-6}$ M DHT in ethanol or ethanol alone. Cells were harvested approximately 48 h later, and CAT assays were performed. Relative CAT activity is the percentage of CAT activity measured in lysates from cells transfected with 10 μg of pHα(−1500/+45)CAT in the absence of DHT and AR. Values represent mean ± SEM of triplicate plates in 4 independent transfections. CAT activities for pMMTV-CAT and pRSV-CAT cotransfected with 5.0 μg of pCMVhAR are expressed as the percentage of CAT activity measured for vector alone (inset). Reprinted with permission from Clay, et al., J Biol Chem 1993; vol 268.

At 2 weeks postcastration, pituitaries were harvested and analyzed for CAT activity. Trunk blood was collected and serum was assayed for LH concentrations. Vertical bars with different lettered superscripts represent statistical differences in CAT activity ($P < 0.05$). Similarly, vertical bars with different Greek symbols (α, β, γ) represent significantly different levels of serum LH ($P < 0.05$). Reprinted with permission from Clay, et al., J Biol Chem 1993; vol 268.

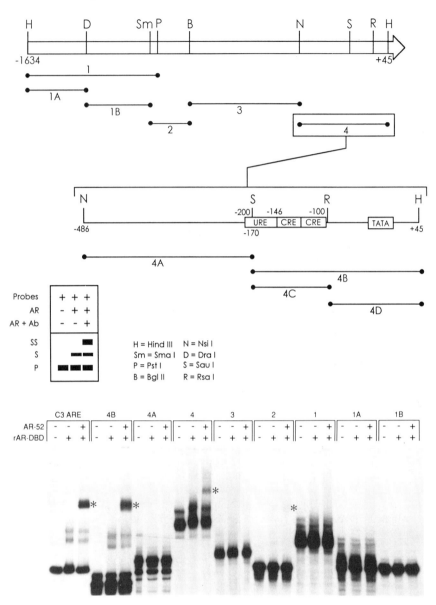

FIGURE 4.5. A high-affinity binding site for AR resides within the proximal 170 bp of the human alpha subunit gene promoter. The upper panel indicates the restriction fragments that were prepared from approximately 1500 bp of proximal 5′ flanking region of the human α-subunit gene. In the inset we show an expected antibody-dependent shift (super shift or SS). Also shown is the unshifted probe (P) and an antibody-independent shift (S). Each fragment was end-labeled with $^{32}P$ and used as a probe in a gel-mobility/super-shift assay (lower panel). Each fragment is represented in the mobility-shift assay by three lanes: $-/-$ = probe

10 μg of pHα(−1500/+45)CAT was transiently cotransfected with the indicated amounts of a human androgen receptor expression vector (pCMVhAR) (9) and then incubated in the absence or presence of $10^{-6}$ M DHT. At 50 ng of cotransfected pCMVhAR, there was no effect of DHT on CAT expression; however, at 0.5, 1.0, and 5.0 μg of pCMVhAR, treatment of cells with DHT resulted in a significant reduction in expression of pHα(−1500)CAT. Also, consistent with the presence of multiple ARE within the LTR of the MMTV promoter (10), a significant steroid-dependent induction of CAT activity was observed when 5.0 μg of pCMVhAR was cotransfected with pMMTV-CAT. Finally, suppression of Hα(−1500)CAT activity was promoter specific, as this effect of AR was not evident when pCMVhAR was cotransfected with pRSVCAT. These data indicated to us that, unlike estrogen, transcriptional repression of the human alpha subunit gene by androgen may involve direct binding of activated AR to its cognate promoter.

To investigate whether a high-affinity binding site for AR exists in the proximal 1500 bp of the human alpha subunit gene, we prepared a series of eight restriction fragments spanning this region. Each fragment was used as a probe in gel-mobility-shift assays using bacterially expressed and purified DNA binding domain of the rat AR (rAR-DBD) (9). Androgen receptor bound to probe was identified by including a specific antibody to the rAR DNA binding domain (AR-52) (11). Inclusion of both rAR-DBD and AR-52 further retards the migration of the labeled probe through the gel if a binding site for AR is present within the probe sequence (Fig. 4.5).

The results of a gel-mobility-shift assay using the eight restriction fragments spanning 1500 bp of human alpha 5′ flanking region are shown in Figure 4.5. Also included in the assay was a fragment containing a previously defined ARE derived from the androgen-responsive C3 gene (11). An antibody-dependent band with altered mobility was readily detected when the C3 ARE fragment was incubated with rAR-DBD and AR-52 (Fig. 4.5, lane 3). Likewise, two fragments from the human alpha subunit promoter appeared to bind AR. Fragment 2, located between −1100 and −850 probably represent a weak interaction, as judged by the faint band (Fig. 4.5, lane 18). In contrast, fragment 4, located between −486 and +45 appeared to possess the highest affinity binding site for AR based on the intensity of the antibody-dependent "super-shift"

---

alone; −/+ = probe + rat AR DNA binding domain; +/+ = probe + rAR-DBD + anti-rAR-DBD (AR-52). C3-ARE is an authentic androgen-responsive element derived from the C3 gene (11). Asterisks indicate specific AR-DNA complexes as identified by their altered migration relative to binding assays carried out in the absence of antibody. Reprinted with permission from Clay, et al., J Biol Chem 1993; vol 268.

(Fig. 4.5, lane 12). In fact, when fragment 4 was cleaved with SauI, a pronounced super-shifted band was evident for fragment 4B ($-170/+45$, lane 6), whereas no binding was detected for fragment 4A ($-486/-170$, lane 9). Thus, the binding of AR to fragment 4 can be fully accounted for by sequences located within the proximal 170 bp of human alpha 5′ flanking region.

Contained within the proximal 170 bp of the human alpha promoter are a number of tightly packed *cis*-acting elements. In light of the data presented above, we felt that at least one mechanism whereby activated AR may attenuate expression of the human alpha subunit gene is by binding at, or near, one or more of these elements, thus effectively blocking access of the appropriate *trans*-acting factor. Therefore, we conducted oligonucleotide competitions with four different oligonucleotides corresponding to sequences representing previously defined *cis*-acting elements: (i) the tandem CRE ($\alpha$36, $-146/-111$) (5); (ii) the *junctional regulatory element* (JRE) (FGH, $-120/-90$) (12); (iii) CCAAT box (GHI, $-110/-80$) (13); and (iv) a nonfunctional element (IJK, $-90/-60$) (13) (Fig. 4.6). The radiolabeled probe was human alpha $-170/+45$. The only oligonucleotide capable of displacing binding to the probe was the JRE. Thus, we propose that activated AR may attenuate transcriptional activity of the human alpha promoter by blocking binding of a regulatory protein (JRF) previously defined as binding to the JRE and important for basal expression of the human alpha subunit gene (12). If correct, such an interaction would occur directly at the level of the pituitary and may represent one of several physiological avenues through which androgens regulate gonadotropin gene expression.

Invoking such a mechanism for androgen regulation clearly does not preclude a hypothalamic site of action (to reduce secretion of GnRH) of androgen as well. In addition, the absence of a direct effect of estrogen on alpha subunit gene expression strongly implicates a hypothalamic site of action of estrogen in attenuating expression of the chimeric $\alpha$-CAT genes in transgenic mice. An underlying assumption, however, is that the chimeric transgenes are in fact responsive to GnRH. This is addressed in the following experiments.

## GnRH Regulation of the Human and Bovine Glycoprotein Hormone Alpha Subunit Gene in Transgenic Mice

To determine whether the human $\alpha$(1500)CAT and bovine $\alpha(-315)$CAT transgenes were regulated by GnRH, an exogenous, pulsatile GnRH treatment was imposed on mice whose endogenous secretion of GnRH was physiologically clamped by chronic administration of estrogen (14).

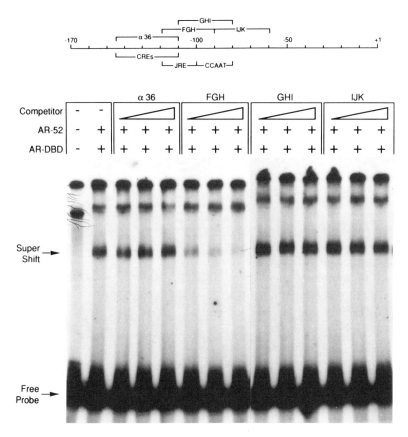

FIGURE 4.6. Oligo competition maps the AR binding element to sequences residing between −120 and −90 in the human alpha promoter. Human α(−170/+45) (fragment 4B) was end-labeled with $^{32}$P and used in a gel-mobility-shift assay with rAR-DBD and AR-52. Four separate unlabeled oligonucleotides were added to the binding reactions in increasing concentrations as indicated by the wedge. The competitor oligonucleotides were added at 12-, 50-, and 100-fold over the concentration of probe. A schematic diagram of the promoter is given at the top of the figure, depicting previously characterized response elements ot the bottom and the oligonucleotides used in the competition above. Reprinted with permission from Clay, et al., J Biol Chem 1993; vol 268.

Thus, OVX mice were administered estradiol for 7 days and then injected with GnRH every other hour for an additional 7 days. Daily injections of estradiol continued throughout the 14-day experiment.

As expected, chronic treatment with estradiol suppressed CAT activity in pituitaries of HαCAT or BαCAT transgenic mice compared to CAT activity in pituitaries of OVX transgenic mice (Fig. 4.7). However, pulsatile administration of GnRH in the continued presence of estradiol

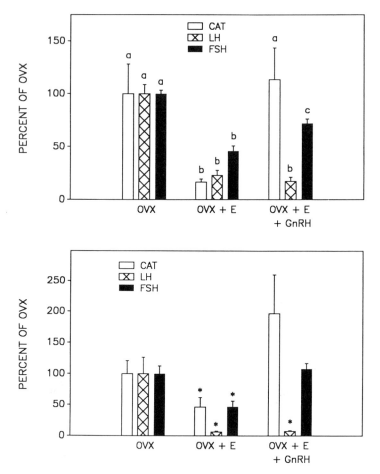

FIGURE 4.7. The proximal promoter regions of the human (−1500/+45) and bovine (−315/+45) alpha subunit genes confer GnRH responsiveness in transgenic mice. After ovariectomy, transgenic mice received daily injections of either oil or estradiol 17β (E) for a total of 14 days. After 1 week of estradiol treatment, saline or GnRH was administered to HαCAT (upper panel; n = 13/group) or BαCAT (lower panel; n = 10/group) transgenic mice every other hour for an additional 7 days. Serum and pituitaries were collected 30 min after the last injection of GnRH. Pituitary CAT activity and serum LH and FSH are expressed as a percentage of the mean of that measured in OVX mice (OVX = 100%) ± SEM. CAT activity measured in pituitaries of nontransgenic animals was less than 1% of that measured in pituitaries of OVX transgenic mice (data not shown). Upper panel: vertical bars with different superscripts are different ($P < 0.05$). Lower panel: * = different from OVX ($P < 0.05$). Reprinted with permission from Hamernik, Keri, Clay, et al. (14).

completely restored CAT activity to values similar to those measured in non-GnRH, nonestradiol, treated OVX mice. The stimulatory effects of GnRH were equally demonstrable in both the HαCAT and BαCAT lines of transgenic mice. Because the human and bovine alpha subunit genes share 85% sequence homology within the proximal 315 bp of proximal 5' flanking region (5), we suggest that elements necessary for GnRH responsiveness may reside within the proximal 315 bp of the 5' flanking sequence of both genes. As the only cell types within the anterior pituitary known to contain GnRH receptors are gonadotropes, we further suggest that the element(s) necessary for gonadotrope-specific expression also reside within 315 bp of proximal 5' flanking region.

As discussed above, the most striking difference between the human and bovine promoters is the absence of a canonical CRE in the bovine promoter, whereas the human promoter contains two tandemly repeated CRE. We (5) and others (15) have shown that the imperfect CRE homologue in the bovine promoter cannot bind homodimers of CREB, nor can it confer cAMP responsiveness. Thus, a functional CRE (i.e., an element capable of binding CREB) is not, apparently, requisite for regulation of alpha subunit gene expression by GnRH.

## Distinct *cis*-Acting Elements Required for Gonadotrope- and Thyrotrope-Specific Expression of the Human and Bovine Alpha Subunit Genes

The GnRH replacement paradigm discussed above suggests that both the human and bovine alpha subunit transgenes were active in gonadotropes and contain a GnRH regulatory element(s) (14). These experiments do not, however, directly address the question as to whether the transgenes might be expressed in other pituitary cell types. Of particular interest was whether the chimeric genes were expressed in thyrotropes, which, besides gonadotropes, are the only other cell type in the pituitary gland known to express the glycoprotein hormone alpha subunit gene (1, 2). This possibility was directly addressed by constructing a new line of transgenic mice that harbored a transgene consisting of 1500 bp of proximal 5' flanking region from the human alpha subunit gene linked to the bacterial lacZ gene (14).

To examine colocalization of bacterial lacZ with the pituitary glycoprotein hormones, pituitary glands from Hαβgal transgenic mice were perfusion fixed and assayed for lacZ expression by staining with X-gal, followed by immunohistochemical staining with antisera against either LHβ or TSHβ. Although there were many LH staining cells that did not express lacZ, all X-gal staining cells observed colocalized with LH-containing cells (14). Thus, full transcriptional activity of the alpha subunit

gene in all gonadotropes must require additional regulatory elements
residing outside of this 1.5-kbp human alpha promoter fragment. In stark
contrast, however, there were no X-gal staining cells that displayed
immunoreactivity with TSH antisera (14). These results suggest that
1.5 kbp of the 5′ flanking region of the human alpha subunit gene contain
the minimal elements necessary for activity in gonadotropes, but lack a
regulatory element(s) required for thyrotrope expression.

The lack of expression of the Hα(−1500)lacZ transgene in thyrotropes
was surprising in light of transient expression results reported by other
investigators. Using transient expression assays in primary cell cultures,
Burrin and Jameson (16) and Ocran et al. (17) have implicated the

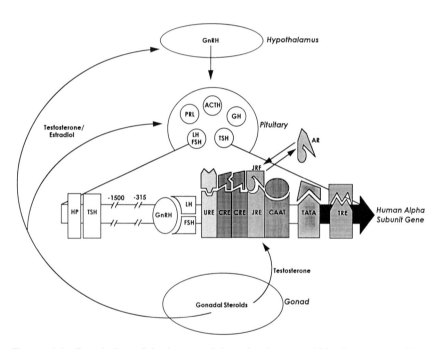

FIGURE 4.8. Regulation of the human alpha subunit gene within the context of the
hypothalamic/pituitary/gonadal axis. GnRH from the hypothalamus stimulates
expression of the alpha subunit gene via an element(s) that resides within the
proximal 315 bp of 5′ flanking sequence. In addition, this sequence contains
elements that confer estrogen and androgen regulation. While estrogen regulates
the alpha subunit promoter through an indirect manner that does not involve
ER binding to the gene, androgen regulation may occur via both indirect (hypothalamic regulation of GnRH) and direct (AR displacement of JRF binding)
mechanisms. Elements necessary for expression in the developing pituitary (HP)
and in thyrotropes must reside outside of the 1500 bp used to construct transgenic
mice.

presence of DNA elements regulating TRH responsiveness and thyrotrope expression residing within 1.5 kbp of the 5' flanking region in both the human and murine alpha subunit genes. The discrepancy between the transfection studies and the pattern of gonadotrope-restricted expression we observed in transgenic mice is intriguing. One explanation is that the latter carries with it more stringent requirements for cell-specific expression. This explanation is supported by several other lines of evidence. First, immortalized cell lines have been obtained from transgenic mice harboring approximately 1.8 kbp of proximal 5' flanking sequence from the human alpha subunit gene linked to the SV-40 T-antigen (8). These cell lines express endogenous alpha subunit and are responsive to GnRH, but fail to respond to TRH, thus indicating an inability of this promoter fragment to direct thyrotrope-specific expression. Second, transgenic mice harboring a transgene consisting of 315 bp of proximal 5' flanking region derived from the bovine alpha subunit gene linked to the diphtheria toxin A-chain structural gene display a marked hypogonadal phenotype due to the almost complete ablation of gonadotropes. However, all other cell types in the pituitary gland, including thyrotropes, were present in the expected numbers and ratio (18). Collectively, these data all indicate that the minimal *cis*-acting elements required for gonadotrope-specific expression reside within 315 bp of proximal 5' flanking region of the alpha subunit gene, but that additional *cis*-acting elements necessary for thyrotrope-specific expression must reside outside of 1500 bp of proximal 5' flanking region. The discrepancy between transfection analysis and the results with transgenic mice underscores the importance of transgenic mice for the functional evaluation of *cis*-acting elements and *trans*-acting factors involved in regulation of gene expression in the endocrine system.

# The Proximal Promoter of the Glycoprotein Hormone Alpha Subunit Gene Contains a Complex Array of Regulatory Elements That Contribute to Cell-Specific Expression and Hormonal Regulation

Presented in Figure 4.8 is our current understanding of the *cis*-acting elements involved in cell-specific, and hormonally regulated, expression of the human glycoprotein hormone alpha subunit gene. Contained within this region are several elements that have been defined as being important for placenta-specific expression of the human alpha subunit gene, including the *upstream regulatory element* (URE) (5, 19) and the tandem CRE (5). The JRE (12) is so named due to its juxtaposition between the CRE and a canonical CCAAT box (13). Others have identified a *thyroid hormone response element* (TRE) located immediately distal to the ubiquitous TATA element (20). While not yet defined, our results with transgenic mice suggest that element(s) sufficient for gonadotrope-specific expression reside within at least 1500 bp of proximal 5' flanking region of the human alpha subunit gene. Assuming conserved mechanisms for gonadotrope-specific expression, then the putative gonadotrope-specific element(s) would, presumably, reside within 315 bp of proximal 5' flanking region. Similarly, the DNA sequences comprising the GnRH response element(s) also would reside within 315 bp. Finally, androgen and estrogen both act to attenuate expression of the chimeric αCAT transgenes; however, distinct molecular mechanisms may be responsible for androgenic and estrogenic regulation. The absence of a high-affinity binding site for ER in the human alpha subunit gene suggests that estrogen is exerting a negative effect by suppressing the secretion of GnRH from the hypo-thalamus or interfering with GnRH signal transduction at the level of the pituitary. In contrast, a high-affinity binding site for AR has been mapped to sequences comprising the JRE. Thus, binding of activated AR could, presumably, block the accessibility of the JRE to its cognate *trans*-acting factor and so attenuate transcriptional activity of the alpha subunit gene promoter.

## References

1. Fiddes JC, Talmadge K. Structure, expression, and evolution of the genes encoding the human glycoprotein hormones. Recent Prog Horm Res 1984; 40:43–78.
2. Gharib SD, Wierman ME, Shupnik MA, Chin WW. Molecular biology of the pituitary gonadotropins. Endocr Rev 1990;11:177–90.
3. Brinkley HJ. Endocrine signalling and female reproduction. Biol Reprod 1981;24:22–43.
4. Desjardins C. Endocrine signalling and male reproduction. Biol Reprod 1981;24:1–21.
5. Bokar JA, Keri RA, Farmerie TA, et al. Expression of the glycoprotein hormone alpha subunit gene in the placenta requires a functional cyclic AMP response element, whereas a different *cis*-acting element mediates pituitary specific expression. Mol Cell Biol 1989;9:5113–22.
6. Bokar JA, Roesler WJ, Vandenbark GR, Kaetzel DM, Hanson RW, Nilson JH. Characterization of the cAMP responsive elements from the genes for the alpha subunit of the glycoprotein hormones and phosphoenolpyruvate carboxykinase (GTP). J Biol Chem 1988;263:19740–7.
7. Keri RA, Andersen B, Kennedy GC, et al. Estradiol inhibits transcription of the human glycoprotein hormone alpha subunit gene despite the absence of a high affinity binding site for estrogen receptor. Mol Endocrinol 1991;5: 725–33.
8. Windle JJ, Weiner RI, Mellon PL. Cell lines of the pituitary gonadotrope lineage derived by targeted oncogenesis in transgenic mice. Mol Endocrinol 1990;4:597–603.
9. Quarmby VE, Kemppainen JA, Sar M, Lubahn DB, French FS, Wilson EM. Expression of recombinant androgen receptor in cultured mammalian cells. Mol Endocrinol 1990;4:1399–407.

10. Darbre P, Page M, King RJB. Androgen regulation by the long terminal repeat of mouse mammary tumor virus. Mol Cell Biol 1986;6:2847–54.

11. Tan J, Marschke KB, Ho KC, Perry ST, Wilson EM, French FS. Response elements of the androgen regulated C3 gene. J Biol Chem 1992;267:4456–66.

12. Andersen B, Kennedy GC, Nilson JH. A *cis*-acting element located between the cAMP response elements and CCAAT box augments cell-specific expression of the glycoprotein hormone alpha subunit gene. J Biol Chem 1990; 265:21874–80.

13. Kennedy GC, Andersen B, Nilson JH. The human alpha subunit glycoprotein hormone gene utilizes a unique CCAAT binding factor. J Biol Chem 1990; 265:6279–85.

14. Hamernik DL, Keri RA, Clay CM, et al. Gonadotrope- and thyrotrope-specific expression of the human and bovine glycoprotein hormone alpha subunit genes is regulated by distinct *cis*-acting elements. Mol Endocrinol 1992;6:1746–55.

15. Drust DS, Troccoli NM, Jameson JL. Binding specificity of cyclic adenosine 3′,5′-monophosphate responsive element (CRE)-binding proteins and activating transcription factors to naturally occurring CRE sequence variants. Mol Endocrinol 1991;5:1541–51.

16. Burrin JM, Jameson JL. Regulation of transfected glycoprotein hormone alpha-gene expression in primary pituitary cell cultures. Mol Endocrinol 1989;3:1643–51.

17. Ocran KW, Sarapura VD, Wood WM, Gordon DF, Gutierrez-Hartmann A, Ridgway EC. Identification of *cis*-acting promoter elements important for expression of the mouse glycoprotein hormone alpha-subunit gene in thyrotropes. Mol Endocrinol 1990;4:766–72.

18. Kendall SK, Saunders TL, Jin L, et al. Targeted ablation of pituitary gonadotropes in transgenic mice. Mol Endocrinol 1991;5:2025–36.

19. Delegeane AM, Ferland LH, Mellon PL. Tissue specific enhancer of the human glycoprotein hormone alpha-subunit gene: dependent on cyclic AMP-inducible elements. Mol Cell Biol 1987;7:3994–4002.

20. Chatterjee V, Lee JK, Rentoumis A, Jameson JL. Negative regulation of the thyroid-stimulating hormone: receptor interaction adjacent to the TATA box. Proc Nat Acad Sci USA 1989;86:9114–8.

# 5

# Coordinate Regulation of the Gonadotropin Subunit Genes in Pituitary and Placenta

PAMELA L. MELLON, DAVID J. STEGER, FRIEDEMANN HORN,
KERRY M. BARNHART, MARITA BÜSCHER, CLAUDIA STAUBER,
JONATHAN H. HECHT, AND ELAINE T. ALARID

Each of the glycoprotein hormones is composed of two subunits, the common α-subunit and an individual β-subunit (1). The single α-subunit gene is coordinately expressed with an individual β-subunit gene to form each hormone (2). The *chorionic gonadotropin* (CG) β-subunit gene is coexpressed with the α-subunit gene in placental trophoblasts, while the thyroid stimulating hormone β-subunit gene is coexpressed with the α-subunit gene in pituitary thyrotropes. In pituitary gonadotropes, both the *luteinizing hormone* (LH) β-subunit gene and the *follicle stimulating hormone* (FSH) β-subunit gene are expressed with the α-subunit gene, since both hormones are present in this cell type (3). The mechanism whereby the coordinate transcriptional regulation of the genes encoding these subunits is accomplished is a central issue in understanding the control of glycoprotein hormone regulation. Using molecular methods, including DNA transfections, gel retardation assays, and nuclear protein purification in cell culture model systems, several of the transcriptional regulatory proteins involved in controlling expression of the CG genes in placental trophoblast cells (JEG-3 cell line) and the LH genes in pituitary gonadotrope cells (αT3-1 cell line) have been identified. The nuclear proteins involved in specifying expression include both ubiquitous and cell type-specific transcriptional regulatory proteins. Here we present studies that suggest a mechanism for coordinate control of gonadotropin subunit genes. We find that one of the proteins that activates the α-subunit gene in placenta, the *trophoblast-specific element binding protein* (TSEB), is also important for CGβ-subunit gene expression in this tissue, while one of the proteins that activates the α-subunit gene in pituitary gonadotropes, the *gonadotrope-specific element binding protein* (GSEB), also binds to the LHβ-subunit gene. This demonstration of the cor-responding role of individual tissue-specific factors in binding to the regulatory regions of both subunit genes of the CG and LH hormones suggests a model for coordinate tissue-specific regulation.

## Placenta-Specific Expression of the CG-Subunit Genes

CG gene expression is confined to the placental trophoblasts. Expression of the common α-subunit gene is directed to placental cells by three transcriptional regulatory proteins that bind to an enhancer located within 180 base pairs (bp) of the mRNA start site: (i) the *cAMP-responsive element binding protein* (CREB), which is not cell type specific (4, 5) and which is required for basal transcription as well as cAMP responsiveness (6–9); (ii) members of the family of GATA binding proteins (10), which show differential cell-type expression and were first characterized as regulating genes in the erythroid and lymphoid cell lineages (11); and (iii) the *trophoblast-specific element binding protein* (TSEB), which is specifically expressed in placental cells and which acts synergistically with CREB and the GATA factors to activate α-subunit gene expression in placenta (6).

Regulation of the CGβ-subunit genes in placental cells has been studied less extensively than the α-subunit gene. The CGβ genes, which recently evolved from duplication of the LHβ gene, retain more than 90% homology to LHβ at the DNA sequence level, even throughout the 5′ flanking sequence (12, 13). However, the initiation site for transcription differs by 366 bp (14). Thus, the transcription start site for the placenta-specific CGβ gene 5 (the most active gene of the cluster) (15) has evolved from the upstream region of the pituitary-specific LHβ gene without substantial alteration of the local sequences (Fig. 5.1). Transfection experiments into placental cells have demonstrated that the elements important for expression of the CGβ5 gene are spread over several upstream regions (13, 16, 17), including regions as far upstream as −3.7 kb

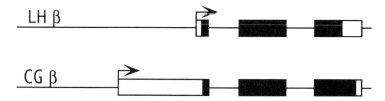

FIGURE 5.1. The structure of the LH and CGβ-subunit genes illustrating the different mRNA initiation sites. The different initiation sites result in promoters located in different regions of the otherwise quite homologous DNA sequence. Adapted from Jameson, Lindell, and Habener (14).

to $-1.7$ kb. Mapping by Albanese et al. (18) localized an important element for basal expression between $-310$ bp upstream of the mRNA start site and $-279$, while others mapped basal expression within $-342$ to $-187$ (16) or $-361$ to $-279$ (17). Furthermore, deletion of the majority of the 5′ untranslated region of the CGβ gene (which evolved from the 5′ flanking region of the LHβ gene) did not effect the basal expression of the transfected CGβ5 gene in placental cells (16).

We have also investigated the tissue-specific elements important for expression of the CGβ-subunit genes using transfection into JEG-3 placental cells. Our fine mapping data indicate that elements important for basal regulation lie within the region downstream of $-305$ (unpublished observations). We performed DNAse I footprinting to detect DNA binding proteins that interact with these important regions of the CGβ5 gene. Footprinting of the region with JEG-3 extract reveals several protected DNA regions (unpublished observations). Some of these binding sites show homology to the binding site for TSEB in the α-subunit gene (Fig. 5.2). To determine whether TSEB binds these elements, we have performed gel retardation assays using DNA affinity-purified TSEB (from JEG-3 cells) and cross-competition with unlabeled oligonucleotides representing these elements. An essential component of tissue-specific expression maps to the sequences containing the element protected by the footprint at $-285$ (FP2). The oligonucleotide representing FP2 binds DNA affinity-purified TSEB with a relative affinity similar to the TSE site from the α-subunit gene (Fig. 5.3). It also strongly competes for binding of TSEB to the TSE from the α-subunit gene and vice versa (Fig. 5.3). Oligonucleotides representing the footprint at $-305$ (FP1) and covering the homologous sites at $-210$ and $-220$ (FP4) also cross-compete with the TSE from the α-subunit gene, but not as strongly (unpublished

```
 ▽▽   ▽▽▽
ACCTAAGGGT   α TSE

TCCTGCGGGC   CGβ -285 (FP2)

CCCCGTGGGC   CGβ -305 (FP1)

GCTTGAGGGT   CGβ -210 (FP4)

CCCGGAGCGG   CGβ -220 (FP4)
```

FIGURE 5.2. Homology between the TSEB binding site in the human α-subunit gene and the footprinted regions of the CGβ-subunit gene 5. The arrows designate nucleotides that are conserved with the human α-subunit gene TSE and that are known to be contacted by TSEB as determined by methylation interference experiments (unpublished observations).

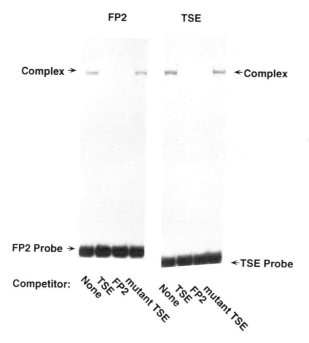

FIGURE 5.3. Gel retardation assay demonstrating the binding of TSEB (DNA affinity-purified by the methods described by Kadonaga et al. [22]) to the FP2 site of the human CGβ-subunit gene 5 (left) and the TSE from the human α-subunit gene (right). Unlabeled competitor oligonucleotides were included at a 100-fold molar excess over the probe concentration to demonstrate specificity (bottom). The oligonucleotides used as probes and competitors were TSE: GATCAAAAATGACCTAAGGGTTGAAACA; mutant TSE: GATCTAAAATGATCTGAGAGTTGAAACAAGATAA; and FP2: GATCCAGGACACACCTCCTGCGGGCCTATTCAA.

observations). Thus, at least three of these footprinted regions bind to the same protein that confers tissue-specific expression on the α-subunit gene, TSEB, while a fourth site binds an as yet unknown protein that may be involved in cAMP activation (unpublished observations).

## Gonadotrope-Specific Expression of the LH-Subunit Genes

In a mouse pituitary gonadotrope cell line, αT3-1, the human α-subunit gene is regulated by CREB (19), the GATA family of transactivators (unpublished observations), and GSEB (19). TSEB is absent, and its site is not bound. Instead, a different sequence, positioned upstream of the TSE at −223 to −197 in the human gene, is required for pituitary

```
                 ∇∇
GGGCTGACCTTGTCGTCACCATCACCT    alpha,   human    -223/-197
AGGCTGACCTTGCAGTCAACACCATCT    alpha,   bovine   -221/-195
AAGCTGTCCTTGAGGTCACCACTACCT    alpha,   murine   -213/-187
AAGCTGTCCTTGAGGTCACCACTACCT    alpha,   rat      -213/-187
AGGCTGACCTTGTGGTCACCACCGCCT    alpha,   equine   -219/-193
TCCCTGGCCATGTGCACCTCTCGCCCC    LH beta, human    -130/-104
CTGGTGGCCTTGCCGCCCCCACAACCC    LH beta, human    -60/-34
    GGCCTTGCCGCCCCCACAGCCC     LH beta, bovine   -67/-36
TTTCTGACCTTGTCTGTCTCGCCCCCA    LH beta, rat      -130/-104
TTAGTGGCCTTGCCACCCCCACAACCC    LH beta, rat      -62/-36
```

FIGURE 5.4. Evolutionary conservation and homology of the human α-subunit gene GSEB binding site with various mammalian α- and LHβ-subunit genes. The arrows designate the two C residues that are contacted by GSEB, as determined by methylation interference (19).

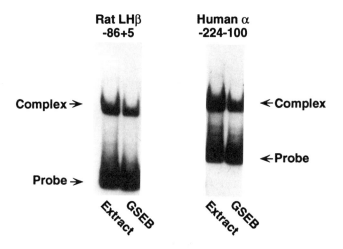

FIGURE 5.5. Gel retardation of DNA fragments containing the region of a GSEB binding site from the rat β-subunit gene (−86 to +5) and the human α-subunit gene (−224 to −100) with αT3-1 nuclear extract (Extract lanes) or DNA affinity-purified GSEB protein (GSEB lanes) (19).

expression. It is bound by GSEB, a transcriptional regulatory protein specifically expressed in cells of the gonadotrope lineage (19). This element is highly conserved across species (Fig. 5.4) and is bound by GSEB in the equine and murine genes in addition to the human gene (19).

The potential role of GSEB in coordinate regulation of the LH genes in the pituitary gonadotrope was investigated using gel retardation assays.

The site bound by GSEB shows homology to two conserved sites in the rat, bovine, and human LHβ-subunit genes (Fig. 5.4). Gel retardation assays with αT3-1 nuclear extract (Fig. 5.5, Extract) or DNA affinity-purified GSEB protein (19) (Fig. 5.5, GSEB) using the region from −86 to +5 of the rat LHβ-subunit gene, which contains one of these sites, demonstrate binding with a relative affinity similar to the site from the human α-subunit gene. In support of the role for GSEB in regulating the LHβ-subunit gene, this protein is also expressed in an immortalized mouse gonadotrope cell line that expresses the endogenous mouse LHβ-subunit gene (unpublished observations).

## Coordinate Expression of Gonadotropin Subunit Genes

These findings suggest that the regulation of the glycoprotein hormone subunit genes may be coordinated within each cell type that expresses them through the participation of common tissue-specific regulatory proteins. Coordinate expression of multiple genes through the use of a common regulatory protein is known in the pituitary in the case of the POU-homeodomain protein, Pit-1 (or GHF-1), which is involved in activation of both the growth hormone and prolactin genes (20, 21). Furthermore, in the erythroid cell lineage, a single transcriptional regulatory protein, GATA-1, activates a variety of erythroid-specific genes, including the globin genes (11). Thus, although the subunit genes of LH and CG require some factors that are not known to be shared, such as CREB or perhaps members of the GATA family, in pituitary, the α- and LHβ-subunit genes may both require the participation of GSEB, and in placenta, the α- and CGβ-subunit genes may both require the participation of TSEB for tissue-specific expression.

*Acknowledgments.* This research was supported by NIH Grants HD-20377 and HD-23818 and March of Dimes Grant 1-FY92-0883 to P.L.M.

## *References*

1. Chin W, Boime I, eds. Glycoprotein hormones. Norwell, MA: Serono Symposia, USA, 1990.
2. Gharib SD, Wierman ME, Shupnik MA, Chin WW. Molecular biology of the pituitary gonadotropins. Endocr Rev 1990;11:177–99.
3. Chin WW, Gharib SD. Organization and expression of gonadotropin genes. Adv Med Biol 1986;205:245–65.
4. Hoeffler JP, Meyer TE, Yun Y, Jameson JL, Habener JF. Cyclic AMP-responsive DNA-binding protein: structure based on a cloned placental cDNA. Science 1988;242:1430–3.

5. Gonzalez GA, Yamamoto KK, Fischer WH, et al. A cluster of phosphory-lation sites on the cyclic AMP-regulated nuclear factor CREB predicted by its sequence. Nature 1989;337:749–52.
6. Delegeane AM, Ferland LH, Mellon PL. Tissue-specific enhancer of the human glycoprotein hormone alpha-subunit gene: dependence on cyclic AMP-inducible elements. Mol Cell Biol 1987;7:3994–4002.
7. Jameson JL, Jaffe RC, Deutsch PJ, Albanese C, Habener JF. The gonado-tropin alpha-gene contains multiple protein binding domains that interact to modulate basal and cAMP-responsive transcription. J Biol Chem 1988;263:9879–86.
8. Jameson JL, Deutsch PJ, Gallagher GD, Jaffe RC, Habener JF. *Trans*-acting factors interact with a cyclic AMP response element to modulate expression of the human gonadotropin alpha gene. Mol Cell Biol 1987;7:3032–40.
9. Silver BJ, Bokar JA, Virgin JB, Vallen EA, Milsted A, Nilson JH. Cyclic AMP regulation of the human glycoprotein hormone alpha-subunit gene is mediated by an 18-base-pair element. Proc Natl Acad Sci USA 1987;84:2198–202.
10. Steger D, Altschmied J, Büscher M, Mellon P. Evolution of placenta-specific gene expression: comparison of the equine and human gonadotropin α-subunit genes. Mol Endocrinol 1991;5:243–55.
11. Orkin S. Globin gene regulation and switching: circa 1990. Cell 1990;63:665–72.
12. Talmadge K, Vamvakopoulos NC, Fiddes JC. Evolution of the genes for the β subunits of human chorionic gonadotropin and luteinizing hormone. Nature 1984;307:37–40.
13. Otani T, Otani F, Krych M, Chaplin D, Boime I. Identification of a promoter region in the CGβ gene cluster. J Biol Chem 1988;263:7322–9.
14. Jameson JL, Lindell CM, Habener JF. Evolution of different transcriptional start sites in the human luteinizing hormone and chorionic gonadotropin β-subunit genes. DNA 1986;5:227–34.
15. Bo M, Boime I. Identification of the transcriptionally active genes of the chorionic gonadotropin β gene cluster in vivo. J Biol Chem 1992;267:3179–84.
16. Jameson J, Lindell C. Isolation and characterization of the human chorionic gonadotropin β subunit (CGβ) gene cluster: regulation of a transcriptionally active CGβ gene by cyclic AMP. Mol Cell Biol 1988;8:5100–7.
17. Otani F, Otani T, Boime I. Effects of adenine nucleotides on choriogonado-tropin α and β subunit synthesis. Biochem Biophys Res Comm 1989;160:6–11.
18. Albanese C, Kay T, Troccoli N, Jameson J. Novel cyclic adenosine 3',5'-monophosphate response element in the human chorionic gonadotropin beta-subunit gene. Mol Endocrinol 1991;5:693–702.
19. Horn F, Windle JJ, Barnhart KM, Mellon PL. Tissue-specific gene expression in the pituitary: the glycoprotein hormone α-subunit gene is regulated by a gonadotrope-specific protein. Mol Cell Biol 1992;12:2143–53.
20. Bodner M, Castrillo JL, Theill LE, Deerinck T, Ellisman M, Karin M. The pituitary-specific transcription factor GHF-1 is a homeobox-containing protein. Cell 1988;55:505–18.

21. Mangalam HJ, Albert VR, Ingraham HA, et al. A pituitary POU domain protein, Pit-1, activates both growth hormone and prolactin promoters transcriptionally. Genes Dev 1989;3:946–58.
22. Kadonaga J, Tjian R. Affinity purification of sequence-specific DNA binding proteins. Proc Natl Acad Sci USA 1986;83:5889–93.

# 6

# Role of Helix-Loop-Helix Proteins in Gonadotropin Gene Expression

STEPHEN M. JACKSON, KERRY M. BARNHART, PAMELA L. MELLON, ARTHUR GUTIERREZ-HARTMANN, AND JAMES P. HOEFFLER

As a cell differentiates into a more specialized cell, it alters the pattern of expression of a subset of genes. Regulation of expression of these genes can occur at almost any point along the pathway from DNA to protein. The most common method that cells use to express specialized genes is to regulate the level of transcription of these genes (1). Combinations of transcription-enhancing and -repressing DNA binding factors can determine the level of expression of certain gene products. These differentiated cell-specific DNA binding factors bind to a defined sequence in the promoter of an expressed gene and somehow interact with the basal transcription machinery to affect the level of expression (2, 3). Knowing this, a simple model for differentiation of a cell may be hypothesized. It may be predicted that each of the genes to be expressed in a differentiated cell contains a single-factor binding site that activates transcription of these genes. The progression from an undifferentiated cell to a differentiated cell would, in theory, require only the activity of a single transcriptional activating protein that would promote the expression of a number of specialized genes. Although cells and tissues behave in a fantastically more complex manner, the above hypothesis has been extremely useful in describing the differentiation of muscle cells and has been extended to many other tissues.

## Muscle Development as a Molecular Paradigm for Tissue Differentiation

Skeletal muscle arises from proliferating myoblasts that fuse to form multinucleated myotubes. Expression of muscle-specific genes, such as myosin heavy chain and muscle creatine kinase, begins as the myoblasts start to fuse. Fortunately, this developmental program can be duplicated in vitro by serum starving proliferating myoblast cells, causing them to fuse into myotubes. It was shown that a single cDNA can convert fibroblasts, which never express muscle-specific genes or fuse, into myotubes (4). This demonstrated that a single protein could act as a "master switch" and initiate the skeletal muscle development program in a nonmuscle cell. Once cloned, this cDNA, named MyoD, was shown to be a sequence-specific DNA binding protein, and it was shown that MyoD binding sites were present in the promoters of the muscle-specific genes (5). When the amino acid sequence of MyoD was closely examined, it was found to have significant homology to the members of the Myc family of proto-oncogenes (4). Curiously, as genes important in the development of *Drosophila* peripheral nervous system (achaete, scute, and other members of the achaete-scute complex) were cloned and sequenced, they also were found to contain similar homologies to c-*myc* (6). When a factor that was responsible for activating the immunoglobulin kappa enhancer (E12/E47) was cloned, sequenced, and found to have the same sequence homology, it was apparent that the homology shared by these factors defined a new class of transcription factors (7) (Fig. 6.1).

## The bHLH Family of Transcriptional Activators

The factors sharing sequence homology to c-*myc* contained regions that could be predicted to form defined secondary structures. First, there is a positively charged region of 10–15 residues that confers DNA binding ability to the protein, possibly by interacting with the negatively charged phosphate backbone of DNA. This is followed by a region of 12 amino acids that have a high probability of forming an alpha helix. Next is a variable 10–25 amino acid domain of poor conserved structure, often containing helix-breaking residues. Finally, there is another region containing conserved amino acids that also has a high probability of forming an alpha helix. Together, these domains define the basic-helix-loop-helix (bHLH) class of proteins (Fig. 6.1) (7). The helix-loop-helix portion of these proteins is responsible for dimerization of these factors (7, 8), and since only dimers bind to DNA, this portion is also necessary for DNA binding. The helix-loop-helix domain appears to form dimers by forming a bundle of four parallel α-helices. Like the leucine zipper proteins, the α-helix content of bHLH proteins increases on binding to DNA (9). The minimal bHLH domain of MyoD alone can induce myogenic conversion in nonmuscle cells (10), suggesting that this domain contains the information necessary for sequence-specific DNA binding and transcriptional activation.

The family of factors containing the bHLH domain is quite large, and interestingly, all appear to have significance in the development of diverse

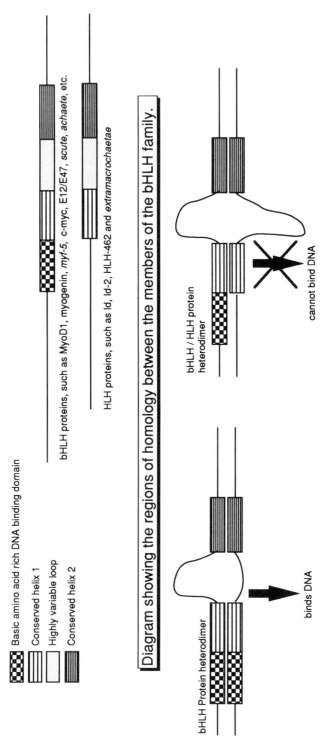

FIGURE 6.1. Interactions between the members of the bHLH family influence DNA binding. Dimerization occurs through the conserved helical regions. Upon dimerization, the DNA binding region assumes the proper configuration for binding to DNA. Monomeric DNA binding regions cannot bind to DNA. Heterodimerization of some bHLH proteins with the widely expressed E12/E47 proteins can enhance DNA binding activity, whereas heterodimerization with proteins that lack a basic region, such as Id, inhibits binding of the complex. Thus, the DNA binding activity, and the transcriptional activation potential, of bHLH proteins can be regulated by heterodimerization.

systems. Examples of other members of the bHLH family include other muscle differentiation factors, (Mrf4, myogenin, and myf) (11–14), factors involved in immunoglobulin expression (E12/E47 and TFEB) (7, 15), insulin-enhancer binding factor (IEBP1) (16), factors important in *Drosophila* for peripheral neural development (members of the achaete-scute gene complex) (6), mesoderm formation (twist) (17), and sex determination (daughterless, da) (18, 19). Homologues of the *Drosophila* proteins have been shown to be conserved in vertebrate cells, further demonstrating their developmental significance. Mammalian homologues of achaete-scute complex, the MASH family, are involved in peripheral and central nervous system development in mice (20, 21). A *Xenopus* homolog of twist, Xtwi, has also been shown to be important for mesoderm formation (22). Furthermore, MyoD homologs have been shown to be involved in muscle development in *Drosophila* and in the nematode *C. elegans* (23). The conservation of these factors across diverse species further illustrates the importance of these proteins in establishing tissue-specific lineages.

The bHLH family of proteins not only share protein structure, but also appear to share sequence-specific DNA binding site preferences. Thus, the MyoD binding site of the *muscle creatine kinase* (MCK) enhancer (5), the μ elements of immunoglobulin enhancers (24), and the *insulin control element* (ICE) of the insulin enhancer (25) contain the sequence CANNTG, demonstrating that a weak sequence homology exists in these protein binding sites. In addition, genes activated by the achaete-scute complex share the same weak homology (5, 24). These sequences can be bound, with varying affinities, by most other members of the bHLH family. Surprisingly, it was discovered that heterodimers can bind to specific sequences tighter than homodimers. For example, it is not a MyoD/MyoD homodimer that binds tightest to the MCK enhancer; rather, it is a heterodimer of MyoD and the ubiquitously expressed protein E12/E47 (E12 and E47 are the products of alternatively spliced transcripts of E2A gene) (7). Although these factors appear to be able to bind to this sequence, there are definite differences in affinity, depending on the heterodimer that binds to the sequence. MyoD/E12 heterodimers bind tightly to a slightly different sequence than E12/E12 homodimers or MyoD/E47 heterodimers (26). The product of the *Drosophila* da gene forms functional heterodimers with members of the achaete-scute complex and has been suggested to be the *Drosophila* homolog of E12/E47 (27). Other members of the bHLH family form heterodimers quite readily (28). Therefore, it is likely that various combinations of heterodimers can modulate the level of expression of the same gene within a single cell.

## bHLH Activity Is Negatively Regulated by Forming Heterodimers with HLH Protein

Many examples of heterodimerization between transcription factors have been discovered (29–31). The importance of heterodimerization within the bHLH family becomes immediately apparent when the regulation of bHLH activity is examined. In addition to being positively regulated by E12/E47, bHLH proteins can be negatively regulated. One of the initial examples suggesting this came from genetic studies of the regulation of the achaete-scute complex in *Drosophila* by the gene *extramacrocheate* (emc). When emc was cloned and sequenced, it was determined to contain a HLH dimerization domain, but lacked a DNA binding basic domain (32, 33). Similarly, in an attempt to identify bHLH proteins in erythroid cells, another protein (an *inhibitor of DNA binding* [Id]) was cloned that contained a HLH domain, but lacked a basic region (34). Thus, these factors make up a special subclass of factors, hereafter referred to as HLH proteins.

Neither of these proteins can bind to DNA, since they lack the basic region. However, they can form heterodimers with other DNA binding bHLH proteins by virtue of the HLH domain they do possess. These heterodimers cannot bind to DNA, thus the activity of the transcriptional activator is blocked (Fig. 6.1) (27, 34). When skeletal myoblasts are grown in normal serum, Id mRNA levels are high and, thus, MyoD cannot bind to its activating sequence. However, when the myoblasts are induced to differentiate, levels of Id drop and, thus, MyoD/E12 heterodimers form and activate muscle-specific genes (34).

Similarly, formation of sensory hairs, called sensilla, in *Drosophila* depends on the activity of the genes of the achaete-scute complex (6). The presence of emc in some cell types allows the formation of nonfunctional heterodimers with achaete or scute and prevents the ectopic appearance of sensilla (35). Id has been found in many cell and tissue types (34) and has been shown to inhibit the cell-specific expression of many different genes. Expression of Id in pancreatic β-cells inhibits the expression of the insulin gene (25), it inhibits the activity of the immunoglobulin enhancer in developing B-lymphoid cells (36), and it inhibits the differentiation of myeloid cells (37), implicating bHLH proteins in expressing the differentiated phenotype in each of those systems.

Two other HLH factors lacking a basic region have been identified and appear to have important developmental roles. Id-2, isolated by *rapid amplification of cDNA ends* (RACE)-PCR from human leukocyte cDNA, is widely expressed in early embryonic development. Interestingly, expression of Id-2 mRNA decreases in neuroblastoma cells when exposed to the morphogen retinoic acid (38). The HLH protein HLH462, another Id analog isolated from 3T3 cells, is induced as part of the early response of mouse 3T3 cells to mitogens (39). Thus, in addition to being important in development, regulation of bHLH function by heterodimerization is also involved in growth control. Although it is known that Id can bind to E12 and to MyoD (albeit with much lower affinity) and HLH462 can inhibit the formation of MyoD/E12 heterodimers, it is not yet known whether Id-2 can bind to the same proteins. Id and Id-2 cDNAs have been shown to be much more abundant in developing and undifferentiated tissue, supporting their roles as developmental regulators. The three HLH proteins Id, Id-2, and HLH462 share 60% amino acid sequence homology in the HLH domain and are less conserved outside this domain. The promiscuous nature of HLH proteins suggests that regulation of bHLH-dependent gene expression by heterodimerization seems to be a common theme used by many different tissues.

## Anterior Pituitary Gland as a Developmental Model

The anterior pituitary gland provides an ideal model system for studying differentiation. The anterior pituitary is populated predominantly by six distinct hormone-producing cell types secreting five major hormones: thyrotropes secrete *thyroid stimulating hormone* (TSH), corticotropes secrete *adrenocorticotropic hormone* (ACTH) and other products of the *proopiomelanocortin* (POMC) gene, lactotropes produce prolactin, somatotropes make growth hormone, and gonadotropes make *leutinizing hormone* (LH) and *follicle stimulating hormone* (FSH). A sixth cell type, a mammosomatotrope, produces both prolactin and growth hormone (Fig. 6.2). The anterior pituitary arises in the developing embryo when a group of undifferentiated cells buds from pharyngeal ectodermal tissue (40). These cells, called Rathke's pouch, migrate until they are met by a protrusion of neural tissue. Cells derived from Rathke's pouch eventually form the anterior pituitary, while the neural-derived cells form the posterior pituitary (41). It is possible, then, that the undifferentiated cells found in Rathke's pouch are stem cells that specialize in response to unknown agents to produce each of the highly differentiated cell types found in the mature anterior pituitary gland.

Examples of factors that are necessary for the restricted expression of pituitary genes exist. The most widely studied of these is Pit-1/GHF-1 (42, 43). This POU-domain factor is expressed in developing rat pituitary tissue at E15.5, before the appearence of lactotropes or somatotropes, but after the appearence of thyrotropes (44). Pit-1/GHF-1 has been shown to be necessary for the expression of growth hormone and also appears to be necessary for the expression of prolactin and TSH (45, 46), but its exact role in expression of the latter two hormones is not yet clear. Pit-1/GHF-1 message is detectable in all pituitary cell types, but Pit-1/GHF-1 protein is found only in somatotropes, lactotropes, and thyrotropes, suggesting a posttranscriptional regulation of Pit-1/GHF-1

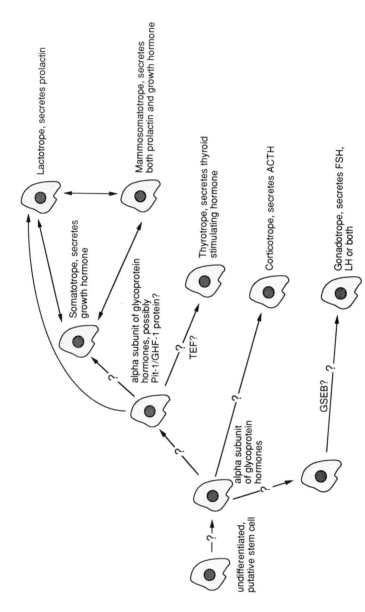

FIGURE 6.2. A simplified diagram showing the various cell types of the anterior pituitary and the hormones they secrete. It is possible that these cells arise from a common precursor cell that, under the influence of certain factors, differentiates into the mature cells. The number of cell types between the undifferentiated cells and hormone-producing cells is not known. The αT3-1 cells used in this study are gonadotrope derived, the αTSH are thyrotrope derived, and the GH3 cells are mammosomatotropes.

protein expression (44). In addition to being a transcriptional activating protein, Pit-1/GHF-1 appears to be a mitogen for pituitary cells (47). Another factor that has been found in developing pituitary tissue is the *thyrotrope embryonic factor* (TEF). Messenger RNA for this factor appears precisely in cells that will eventually form thyrotropes, but appears to be less restricted in more mature tissues (48). The recently described *gonadotrope-specific element binding* (GSEB) protein, purified from αT3-1 cells, restricts exrpession of the glycoprotein hormone α-subunit gene to gonadotrope cells (49), but its ontogeny in developing pituitary remains to be investigated. None of these factors fall into the bHLH class of proteins. However, it is probable that expression of Pit-1/GHF-1, TEF, GSEB and other similar factors requires the activity of other lineage switches.

The α-subunit of glycoprotein hormones is found as a common subunit of the heterodimeric pituitary hormones TSH, FSH, LH, and, in humans, of the placental hormone *chorionic gonadotropin* (hCG). Normal expression of the α-subunit is therefore limited to thyrotropes, gonadotropes, and human placental trophoblast cells (50, 51). Interestingly, many pituitary tumors that do not produce ACTH, PRL, or GH secrete α-subunit (52). Until recently, the only cell lines available that expressed α-subunit were placental choriocarcinomas; no pituitary cell line existed. However, two recently described pituitary cell lines have made possible the study of pituitary-specific expression of α-subunit. Windle et al. were able to produce a mouse gonadotrope cell line that expresses α-subunit (αT3-1) by targeted oncogenesis in transgenic mice (53). Interestingly, although these cells have characteristics of gonadotropes, they do not express the β-subunit of LH or FSH (53). Similarly, Akerblom et al. (54) adapted the α-subunit producing thyrotropic transplantable tumor MGH101A to grow in culture, producing the αTSH cell line. Like the αT3-1 cells, these cells produce the common α-subunit, but do not make TSHβ. It is possible that these cells are precursor cells that, when properly stimulated, can become fully differentiated gonadotropes or thyrotropes and, thus, may be good in vitro models for pituitary differentiation.

## Tissue-Specific Expression of α-Subunit

Normal fetal expression of the α-subunit is detectable before the formation of Rathke's pouch, making it the earliest known cell-specific marker in the developing pituitary (44, 55). Therefore, it is possible that pituitary "master switches," analogous to MyoD, direct the expression of the α-subunit and other specific genes involved in the establishment of the pituitary gland. The promoter of the human α-subunit contains regulatory regions that restrict its expression to the above-described cell types (49, 56, 57). Study of α-subunit expression in placenta has been facilitated by the availability of placental cell lines. Promoter elements responsible for placental expression have been elucidated and are distinct from the elements that control expression in the pituitary. Placenta-specific factor binding sequences are found in the 244 nucleotides 5' to the start site and include an *upstream regulatory element/tissue-specific element* (URE/ TSE), two tandem *cAMP-response elements* (CREs) and a consensus CCAAT box (56, 58, 59). The CREs bind a nuclear transcription factor, CREB, which activates transcription of genes in response to cAMP (60, 61). Optimal utilization of the URE requires the presence of two intact CREs; deletion of a single CRE decreases placental-specific expression, whereas deletion of both CREs abolishes the activity of the URE (57, 59, 62). In addition, a specific CCAAT-box factor (63) and a novel element, the *junctional regulatory element* (JRE) (64), have also been implicated in placental-specific expression of the α-gene.

Recent studies, using the αT3 and αTSH cell lines, have begun to demarcate sequences present on the α-promoter required for pituitary-specific expression. Interestingly, the regulatory sites for pitutary- and placental-specific expression are different. In the αT3-1 cells, a gonadotrope-specific factor has been identified that appears to be required for expression of α-subunit in gonadotrope cells. This factor binds at positions −225 to −208 (the GSE) and has been termed GSEB (49). Deletion analysis of α-subunit gene expression in these cells suggests complex regulation of its expression (49). Two other elements have been demonstrated to be important for pituitary-specific expression of this gene: one at −337 to −330, required for expression in both αT3-1 and αTSH cells, and one at −445 to −438, which appears to be specific for activity in αT3-1 cells (65). A multimerized element, which contained the −337/−330 site, was able to confer pituitary-specific expression to a heterologous promoter (65). Interestingly, a functional CRE is not necessary for pituitary-specific α-subunit gene expression (57, 65–67). The functional elements described so far do not contain CANNTG elements. However, it is conceivable that other lineage-specific switches influence the expression of factors that bind to these sequences. Since the α-subunit gene is the first known marker to be expressed in developing pituitary tissue, it provides an ideal starting point for identifying these switches.

# Pituitary Cells Contain Proteins That Can Bind Specifically to an Oligonucleotide Containing a CANNTG Sequence

Factors of the bHLH family bind to sequences containing the loose consensus sequence CANNTG with varying affinities (28). If bHLH proteins are present in pituitary cells, they would be expected to bind specifically to oligonucleotides containing a CANNTG sequence. To examine if factors are present in pituitary cells that can bind to this consensus, a synthetic oligonucleotide containing the core sequence CACCTG, derived from the MyoD binding site on the MCK enhancer, was used in gel-shift assays with nuclear extracts from adult pituitary tissue, the pituitary-derived GH3, αT3-1, and αTSH cell lines, or HeLa cells (Fig. 6.3, upper panel). Five micrograms of each nuclear extract were incubated with labeled probe in gel-shift buffer (68) for 20 min at room temperature. The binding reaction was then loaded onto a 5% polyacrylamide gel and run in 0.5× TBE. The gel was then dried and exposed to film overnight.

Note the presence of multiple complexes that are pituitary and/or cell-type specific. To examine the specificity of these complexes further, we utilized extracts from the GH3 cell line (Fig. 6.3, lower panel). Excess unlabeled homologous CANNTG containing oligonucleotide (MCK) added to the binding reaction eliminated binding of factors to the labeled oligonucleotide, whereas addition of the same excess of heterologous, CRE-containing oligonucleotide had no effect. Furthermore, addition of Id protein to the binding reaction abolished formation of these complexes. However, addition of Id had no effect on CREB binding to CRE (data not shown).

Since specific complexes were blocked from forming with excess Id protein, these data strongly suggest that these complexes contain bHLH proteins. Identical experiments performed using αT3-1 gonadotrope-derived and αTSH thyrotrope-derived nuclear extracts demonstrate that similar complexes are formed and bind specifically to the MCK oligonu-cleotide, suggesting that common factors exist in these cells. These experiments demonstrate that factors exist in pituitary cells that bind specifically to CANNTG-containing DNA sequences, and since their binding can be inhibited by the addition of excess Id, these complexes probably contain bHLH proteins. Since cell- and tissue-specific differences appear, it will be worthwhile to determine whether factors in these complexes are indeed lineage specific.

# Pituitary Cells Contain Proteins That, When Immobilized on Nitrocellulose, Bind to CANNTG-Containing Oligonucleotides and to Id Protein

In a similar set of experiments, Southwestern assays (70) were performed. Nuclear extracts from pituitary cell lines or HeLa cells were separated on SDS-polyacrylamide gels, transferred to nitrocellulose, and probed with the same oligonucleotide used in the gel-shift experiments described above (Fig. 6.4). The presence of bands on the autoradiogram suggests proteins

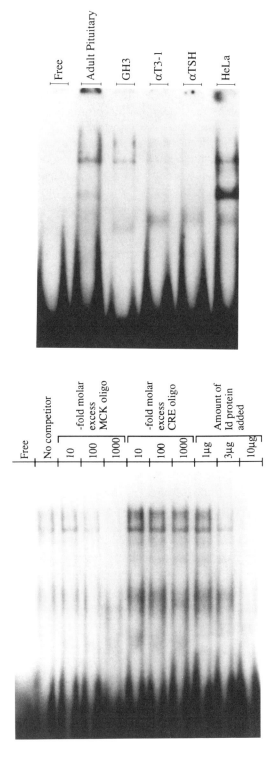

interacting with the DNA probe. Although numerous bands appeared that bound to the MCK oligo sequences (left panel), the specificity of binding of these proteins to the MCK oligonucleotide is demonstrated in the middle and right panels. Pituitary cell line-specific MCK oligo-binding proteins are denoted by the bold arrows. Note that detection of these pituitary cell line-specific proteins is abolished by competition with excess unlabeled MCK oligo, but not in the presence of an excess of an unlabeled CRE-containing oligonucleotide. As described above, bHLH proteins share two properties: the ability to heterodimerize and the ability to bind to CANNTG sequences. It would therefore be expected that if a particular extract contained bHLH proteins, one or more of these proteins could bind to both the CANNTG oligonucleotide probe as well as the Id protein probe, even though the probes are completely different in nature. To further determine whether bHLH proteins exist in the pituitary, far-western assays on pituitary cell nuclear extracts were performed. The far-western assay (68) was used to analyze protein-protein interactions between pituitary cell nuclear extracts and the HLH protein Id directly.

In order to obtain large amounts of Id protein, a prokaryotic expression plasmid was constructed that produced recombinant Id protein. A fragment containing the entire Id cDNA (34) was cloned into the expression plasmid pRSET-B (69). This plasmid contains a number of features that simplify in vitro manipulations of recombinant proteins: a T7 RNA polymerase promoter, an initiator methionine in a good prokaryotic *ribosome binding site* (RBS) context, a polyhistidine sequence to allow purification on nickel chelating columns, the amino terminus from bacteriophage T7 gene 10 that allows detection using anti-gene 10 antibodies, a tyrosine for

---

FIGURE 6.3. *Upper panel*: Pituitary-specific factors interact with an oligonucleotide containing a CANNTG motif. Nuclear extracts from the indicated cell lines or mature adult pituitary tissue were incubated with a radiolabeled probe containing a bHLH consensus binding site, and protein-DNA complexes were separated from free DNA by electrophoresis. It is important to note that each complex present in the extracts from the pituitary cell lines is also present in extracts from the adult pituitary. The slowest migrating complex, absent in HeLa extracts, appears in extracts from the pituitary tissue, GH3, and αT3 1 cells and thus appears to be pituitary specific. *Lower panel*: Id can inhibit the specific interaction of pituitary proteins with the bHLH consensus element. GH3 nuclear extracts were incubated with probe in the presence of either excess unlabeled homologous oligonucleotide (MCK oligo) or the same excess of unrelated oligonucleotide (CRE oligo). A 1000-fold molar excess of homologous DNA completely abolishes binding of the slowest mobility complex, whereas the same excess of CRE had no effect. Incubation of the extracts in the presence of excess Id protein also abolished protein-DNA complex formation, suggesting that Id is forming nonfunctional heterodimers with factors in the extract. Together, these data demonstrate that pituitary cells contain factors that share characteristics of bHLH proteins.

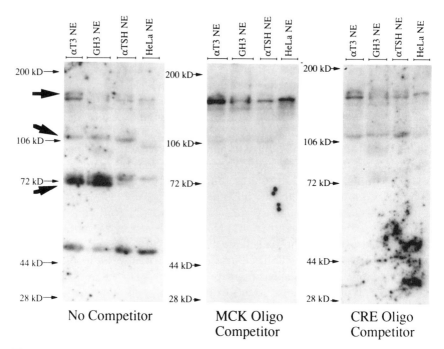

FIGURE 6.4. Southwestern analysis further reveals putative pituitary-specific bHLH proteins. To further examine factors present in pituitary cell extracts that bind to the above-described oligo, Southwestern analysis was performed. Nuclear extracts from the indicated cells were separated by SDS-PAGE, transferred to nitrocellulose, and the blot was then incubated with the same labeled probe as in the gel-shift experiments. A 1000-fold excess of either the homologous (MCK) or heterologous (CRE) oligonucleotide was included in the indicated panels. Although the specificity of the protein-DNA interactions is less obvious in these experiments, it is clear that there are some proteins that bind specifically to this probe. The three arrows indicate proteins that are restricted to pituitary cells and that are not competed by presence of CRE competitor. These results further demonstrate that proteins are present in the pituitary that may be bHLH proteins, based on their DNA binding specificity.

radioiodination, and an enterokinase cleavage site that allows removal of the leader peptide. Together, these sequences make up the YEK/gene 10 epitope tag. We have produced monoclonal antibodies to both the gene 10 epitope and the YEK epitope, facilitating detection of recombinant proteins containing epitope tags. A multiple cloning site containing common restriction enzyme recognition sites in all three reading frames allows introduction of desired cDNAs into this vector. Transformation of these plasmids into bacteria containing the T7 RNA polymerase under the control of the lac promoter will produce large amounts of recombinant protein when induced with IPTG. An Id cDNA, generously provided by

H. Weintraub, was cloned into this vector, and the encoded protein was overexpressed and purified on nickel-chelating columns. The availability of large amounts of Id allowed the previous and subsequent studies to be done.

Far-western assays were performed essentially as described originally (68). Nuclear extracts from cultured cells were separated on SDS-polyacrylamide gels and transferred to nitrocellulose. Id protein was radioiodinated and used to probe the nitrocellulose blot containing immobilized nuclear extracts. Bands appearing on the autoradiograph correspond to proteins that interact with the labeled Id protein probe. It should be noted that a coomassie-stained gel of the same extracts would show many more bands than appear in the autoradiogram. Simultaneous comparison of far- (Fig. 6.5, right panel) and Southwestern (Fig. 6.5, left panel) blots reveals that bands at 70 kd, 120 kd, and 180 kd are common to both probes. In addition, comparison between the αT3-1 and HeLa extracts demonstrates that there are pituitary-specific proteins that bind to each probe. It was later determined that the L6 myoblast nuclear extracts were degraded and not suitable for use in these experiments. Utilizing far-western assays, similar proteins were detected when radio-iodinated MyoD was used as a probe (data not shown), suggesting that these nuclear proteins bind to at least two separate members of the bHLH family. Comparison of bands seen with GH3, αT3-1, αTSH, HeLa, and JEG-3 nuclear extracts (data not shown) reveals that size and abundance differences exist. It is tempting to speculate that these reflect cell-and tissue-specific differences, but this remains to be rigorously demonstrated. Regardless, the results from these experiments further demonstrate that there are pituitary proteins that share properties of bHLH factors.

## Id Inhibits Expression of a Glycoprotein Hormone α-Subunit Reporter Plasmid When Transiently Coexpressed in Gonadotropic or Thyrotropic Pituitary Cells, but not Mammosomatotropic Pituitary Cells or Placental Choriocarcinoma Cells

As described, the α-subunit promoter contains regulatory sites responsible for its expression in the anterior pituitary. To test whether Id can influence the activity of the α-promoter in gonadotropes, αT3-1 and αTSH cells were transiently cotransfected with a human α-promoter/luciferase reporter and an Id overexpression plasmid. We would expect that, if bHLH proteins were involved in the expression of α-subunit, overexpression of Id should influence expression of this gene. Thus, 1 μg of reporter DNA was electroporated into cells with a 19-fold mass excess of either an

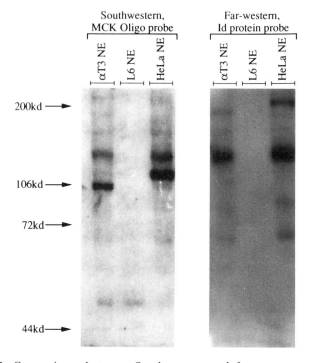

FIGURE 6.5. Comparisons between Southwestern and far-western assays demonstrate similar proteins can interact with dissimilar probes. Side-by-side far- and Southwestern experiments were performed in order to directly compare proteins which bound to the two probes. We may expect that putative bHLH proteins may bind to both probes. Several bands in the αT3-1 and HeLa lanes clearly interact with both probes. Comparison between the αT3-1 and HeLa extracts reveals that the bands at about 106 and 180 kd are specific for αT3-1 cells; the two others at 120 kd and 70 kd appear in both cell extracts. Other bands may also be present which can interact with both probes; however, the intensity of these bands makes comparisons difficult. The L6 myoblast nuclear extract was later found to be extensively degraded; thus, it is not surprising that interactive proteins do not appear. Results from these experiments again demonstrate that proteins are present in pituitary cells which share characteristics of bHLH proteins.

expression plasmid driving expression of Id (pE:Id(S)) (34) or the identical expression plasmid lacking Id to control for nonspecific promoter effects. In preliminary experiments, we found that maximal inhibition of expression is seen at 24 h after transfection. This suggests that Id does indeed inhibit the expression of α-subunit/luciferase reporters. The specific nature of this effect was elucidated by performing the same cotransfection experiments on both the α-subunit/luciferase constructions and a strong viral promoter, derived from the CMV LTR, driving expression of luciferase. After electroporation, the cells were allowed to incubate for

24 h for maximal Id activity. The total amount of DNA was kept constant by using the Id-less expression vector. These experiments were repeated in triplicate and performed on at least three different days. The data, summarized in Table 6.1, demonstrate that overexpression of Id has no effect on the strong viral promoter derived from the CMV LTR. However, the maximal dose of Id expression plasmid decreases α-subunit-dependent expression of luciferase to one-half to one-third of control levels in αTSH and αT3-1 cells, respectively, but the expression of the α-luciferase reporter is unaffected by Id in the human choriocarcinoma cell line, JEG-3. These results are consistent with the apparent lack of bHLH proteins detected using gel-shift, South-, and far-western analyses.

The specificity of the effect of Id to inhibit pituitary hormone gene expression is demonstrated in Table 6.2. Overexpression of Id in GH3 cells had no effect on prolactin or growth hormone reporter gene activity. In addition, the expression of the α-subunit reporter in these cells was completely unaffected by overexpression of Id (data not shown). These data further demonstrate that Id specifically inhibits α-subunit promoter activity in restricted cell types of the anterior pituitary and that the

TABLE 6.1. Overexpression of Id decreases expression of α-glycoprotein hormone subunit gene in pituitary but not placental cells.

| | 0.5 μg pCMVLuc | | 1 μg (−242)alphaLuc | |
|---|---|---|---|---|
| Cell type | No Id | 20 μg Id | No Id | 20 μg Id |
| αT3-1 | 100 | 100 | 100 | 38 |
| αTSH | 100 | 100 | 100 | 55 |
| JEG-3 | 100 | 100 | 100 | 94 |

*Note:* The indicated reporter and expression plasmids were electroporated into the indicated cell lines, extracts prepared and luciferase activity measured according to standard protocols. The data presented here represent multiple experiments performed in triplicate on at least three different days. In some cases, the overexpression of Id slightly decreased the activity of the nonspecific CMV promoter; therefore, the data are expressed as percent of control. The p(−1763)alphaLuc construct behaved identically to the p(−224)alphaLuc in these experiments. In the JEG-3 cells, expression of α-subunit was unaffected by Id overexpression, suggesting that bHLH proteins do not have a major role in α-subunit gene expression in these cells.

TABLE 6.2. Overexpression of Id has no specific effect in GH3 cells.

| | 0.5 μg pCMVLuc | | 1 μg pPRLLuc | | 1 μg prGHLuc | |
|---|---|---|---|---|---|---|
| Cell type | No Id | 20 μg Id | No Id | 20 μg Id | No Id | 20 μg Id |
| GH3 | 100 | 100 | 100 | 143 | 100 | 117 |

*Note:* Experiments were performed and data expressed as in Table 6.1. In experiments not shown, Id had no effect on the (meager) expression of α-subunit in these cells. Thus, expression of growth hormone and prolactin do not appear to be directly dependent on bHLH proteins.

element responsible for this inhibition is contained within the 242 bp 5' to the transcriptional start site. It cannot be determined from these experiments whether Id has a direct effect on the α-promoter or whether Id effects expression of a different gene required for the subsequent expression of the α-subunit. In any event, it is evident that Id has a reproducible inhibitory effect on α-subunit gene expression in two different pituitary cell types. As α-subunit is the first detectable marker in developing pituitary tissue, it is tempting to speculate that its expression depends on the activity of lineage-specific factors. Since Id affects the expression of α-subunit in two different cell types, an intriguing possibility is that these factors are members of the bHLH family. These results further demonstrate the specific effect that Id has on expression of α-subunit and indicate the need to further examine the effect of Id on pituitary-specific gene expression.

Nothern blotting data (not shown) demonstrates that Id message is present in αT3-1 and αTSH cells. In combination with the gene transfer data presented above, it could be hypothesized that levels of expression of the α-subunit gene are already repressed in these cells. This proves to be true when the levels of α-subunit produced by αT3-1 cells and αTSH cells are compared. We know that Id levels in αT3-1 cells are higher than in αTSH cells, and indeed, the levels of α-subunit secreted by αT3-1 cells $(140 \pm 15.5 \, \text{ng}/2 \times 10^6 \, \text{cells})$ (53) are about 73 times lower than the expression of α-subunit in αTSH cells $(5161 \pm 791 \, \text{ng}/10^6 \, \text{cells})$ (54). Thus, it appears that Id already present in these cells may be having an effect on regulating the endogenous expression of the α-subunit gene in these cells. High levels of transient expression further decrease the activity of the α-subunit promoter. Further studies on the nature and involvement of bHLH proteins in the pituitary will help us better understand the role of these proteins in pituitary gene expression and differentiation programs.

*Acknowledgments.* We would like to thank the other members of the Hoeffler laboratory for many helpful discussions and suggestions regarding this work and the chapter. We would also like to thank Harold Weintraub for the Id cDNA and eukaryotic expression plasmid pE:Id(S). These studies were supported by grants from the CU Cancer Research Foundation, the Cancer League of Colorado, and NIH Grant GM-45872 to J.P.H. These studies were also supported in part by ACS Grant BE-532 and NIH Grant DK-37667 to A.G.H., and NIH Grant HD-20377 to P.L.M.

## References

1. Sawadogo M, Sentenac A. RNA polymerase II and general transcription factors. Annu Rev Biochem 1990;59:711–54.
2. Manitatis T, Goodbourn S, Fischer JA. Regulation of inducible and tissue-specific gene expression. Science 1987;236:1237–45.

3. Lewin B. Commitment and activation at pol II promoters: a tail of protein-protein interactions. Cell 1990;61:1161–4.
4. Davis RL, Weintraub H, Lassar AB. Expression of a single transfected cDNA converts fibroblasts to myoblasts. Cell 1987;51:987–1000.
5. Lassar AB, et al. MyoD is a sequence-specific DNA binding protein requiring a region of myc homology to bind to the muscle creatine kinase enhancer. Cell 1989;58:823–31.
6. Villares R, Cabrera CV. The achaete-scute gene complex of *D. melanogaster*: conserved domains in a subset of genes required for neurogenesis and their homology to myc. Cell 1987;50:415–24.
7. Murre C, McCaw PS, Baltimore D. A new DNA binding and dimerization motif in immunoglobulin enhancer binding, daughterless, MyoD and myc proteins. Cell 1989;56:777–83.
8. Voronova A, Baltimore D. Mutations that disrupt DNA binding and dimer formation in the E47 helix-loop-helix protein map to distinct domains. Proc Natl Acad Sci USA 1990;87:4722–6.
9. Anthony-Cahill SJ, et al. Molecular characterization of helix-loop-helix peptides. Science 1992;255:979–83.
10. Tapscott SJ, et al. MyoD1: a nuclear phosphoprotein requiring a myc homology region to convert fibroblasts to myoblasts. Science 1988;242:405–11.
11. Rhodes SJ, Konieczny SF. Identification of MRF4: a new member of the muscle regulatory gene family. Genes Dev 1989;3:2050–61.
12. Edmondson DG, Olson EN. A gene with homology to the myc similarity region of MyoD1 is expressed during myogenesis and is sufficient to activate the muscle development program. Genes Dev 1989;3:628–40.
13. Wright WE, Sassoon DA, Lin VK. Myogenin, a factor regulating myogenesis, has a domain homologous to MyoD. Cell 1989;56:607–17.
14. Braun T, et al. Differential expression of myogenic determination genes in muscle cells: possible autoactivation by the myf gene products. EMBO J 1989;8:3617–25.
15. Carr CS, Sharp PA. A helix-loop-helix protein related to the immunoglobulin E box-binding proteins. Mol Cell Biol 1990;10:4384–8.
16. Shibasaki Y, Sakura H, Takaku F, Kasuga M. Insulin enhancer binding protein has a helix-loop-helix structure. Biochem Biophys Res Comm 1990; 170:314–21.
17. Leptin M. Twist and snail as positive and negative regulators during *Drosophila* mesoderm development. Genes Dev 1991;5:1568–76.
18. Cronmiller C, Schedl P, Cline TW. Molecular characterization of daughterless, a *Drosophila* sex determination gene with multiple roles in development. Genes Dev 1988;2:1666–76.
19. Caudy M, et al. The maternal sex determination gene daughterless has zygotic activity necessary for the formation of peripheral neurons in *Drosophila*. Genes Dev 1988;2:843–52.
20. Johnson JE, Birren SJ, Anderson DJ. Two rat homologs of *Drosophila* achaete-scute specifically expressed in neuronal precursors. Nature 1990; 346:858–61.
21. Lo L-C, Johnson JE, Wuenschell CW, Saito T, Anderson DJ. Mammalian achaete-scute homolog 1 is transiently expressed by spacially restricted subsets of early neuroepithelial and neural crest cells. Genes Dev 1991;5:1524–37.

22. Hopwood ND, Pluck A, Gurdon JB. A *Xenopus* mRNA related to *Drosophila* twist is expressed in response to induction in the mesoderm and the neural crest. Cell 1989;59:893–903.
23. Weintraub H, et al. The MyoD gene family: nodal point during specification of the muscle cell lineage. Science 1991;251:761–6.
24. Church GM, Ephrussi A, Gilbert W, Tonegawa S. Cell type specific to immunoglobulin enhancers in nuclei. Nature 1985;313:798–801.
25. Cordle SR, Henderson E, Masuoka H, Weil PA, Stein R. Pancreatic β-cell-type-specific transcription of the insulin gene is mediated by basic helix-loop-helix DNA binding proteins. Mol Cell Biol 1991;11:1734–8.
26. Blackwell KT, Weintraub H. Differences and similarities in DNA-binding preferences of MyoD and E2A proteins revealed by binding site selection. Science 1990;250:1104–10.
27. Van Doren M, Ellis HM, Posakony JW. The *Drosophila* extramacrochaetae protein antagonizes sequence-specific DNA binding by daughterless/achate-scute protein complexes. Development 1991;113:245–55.
28. Murre C, et al. Interactions between heterologous helix-loop-helix proteins generate complexes that bind specifically to a common DNA sequence. Cell 1989;58:537–44.
29. Jones N. Transcriptional regulation by dimerization: two sides to an incestuous relationship. Cell 1990;61:9–11.
30. Hai T, Liu F, Coukos WJ, Green MR. Transcription factor ATF cDNA clones: an extensive family of leucine zipper proteins able to selectively form DNA binding heterodimers. Genes Dev 1989;3:2083–90.
31. Curran T, Franza JR. Fos and Jun: the AP-1 connection. Cell 1988;55:395–7.
32. Ellis HM, Spann DR, Posakony JW. Extramacrochaetae, a negative regulator of sensory organ development in *Drosophila*, defines a new class of helix-loop-helix proteins. Cell 1990;61:27–38.
33. Garell J, Modolell J. The *Drosophila* extramacrochaetae locus, an antagonist of proneural genes that, like these genes, encodes a helix-loop-helix gene. Cell 1990;61:39–48.
34. Benezra R, Davis RL, Lockshon D, Turner DL, Weintraub H. The protein Id: a negative regulator of helix-loop-helix DNA binding proteins. Cell 1990;61:49–59.
35. Skeath JB, Carroll SB. Regulation of achaete-scute gene expression and sensory pattern formation in the *Drosophila* wing. Genes Dev 1991;5:984–95.
36. Wilson RB, et al. Repression of immunoglobulin enhancers by the helix-loop-helix protein Id: implications for B-lymphoid-cell development. Mol Cell Biol 1991;11:6185–91.
37. Kreider BL, Benezra R, Rovera G, Kadesch T. Inhibition of myeloid differentiation by the helix-loop-helix protein Id. Science 1992;255:1700–2.
38. Biggs J, Murphy EV, Israel MA. A human Id-like helix-loop-helix protein expressed during early development. Proc Natl Acad Sci USA 1992;89:1512–6.
39. Christy BA, et al. An Id-related helix-loop-helix protein encoded by a growth factor-inducible gene. Proc Natl Acad Sci USA 1991;88:1815–9.
40. Voss JW, Rosenfeld MG. Anterior pituitary development: short tales from dwarf mice. Cell 1992;70:527–30.

41. McCann SM. In: Griffen JE, Ojeda SR, eds. Textbook of endocrine physiology. New York: Oxford University Press, 1988:70–100.
42. Ingraham HA, et al. A tissue-specific transcription factor containing a homeodomain specifies a pituitary phenotype. Cell 1988;55:519–29.
43. Bodner M, et al. The pituitary-specific transcription factor GHF-1 is a homeobox-containing protein. Cell 1988;55:505–18.
44. Simmons DM, et al. Pituitary cell phenotypes involve cell-specific Pit-1 mRNA translation and synergistic interactions with other transcription factors. Genes Dev 1990;4:695–711.
45. Crenshaw EB III, et al. Cell-specific expression of the prolactin gene in transgenic mice is controlled by synergistic interactions between Pit-1 recognition elements. Genes Dev 1989;3:959–72.
46. Mangalam HJ, et al. A pituitary POU domain protein, Pit-1, activates both growth hormone and prolactin promoters transcriptionally. Genes Dev 1989; 3:946–58.
47. Castrillo JL, Theill LE, Karin M. Function of the homeodomain protein GHF-1 in pituitary cell proliferation. Science 1991;253:197–9.
48. Drolet DW, et al. TEF, a transcription factor expressed specifically in the anterior pituitary during embryogenesis, defines a new class of leucine zipper proteins. Genes Dev 1991;5:1739–53.
49. Horn F, Windle JJ, Barnhart KM, Mellon PL. Tissue-specific gene expression in the pituitary: the glycoprotein hormone α-subunit gene is regulated by a gonadotrope-specific protein. Mol Cell Biol 1992;12:2143–53.
50. Pierce JG, Parsons TF. Glycoprotein hormones: structure and function. Annu Rev Biochem 1981;50:465.
51. Gharib SD, Wierman ME, Shupnik MA, Chin WW. Molecular biology of the pituitary gonadotropins. Endocr Rev 1990;11:177–99.
52. Ridgeway EC, et al. Pure alpha-secreting adenomas. N Engl J Med 1981; 304:1254–9.
53. Windle JJ, Weiner RI, Mellon PL. Cell lines of the pituitary gonadotrope lineage derived by targeted oncogenesis in transgenic mice. Mol Endocrinol 1990;4:597–603.
54. Akerblom IE, Ridgeway EC, Mellon PL. An alpha-subunit secreting cell line derived from a mouse thyrotrope tumor. Mol Endocrinol 1990;4:589–96.
55. Mulchahey JJ, DiBlasio AM, Martin MC, Blumenfeld Z, Jaffe RB. Hormone production and peptide regulation of the human fetal pituitary gland. Endocr Rev 1987;8:406–25.
56. Jameson JL, Deutsch PJ, Gallagher GD, Jaffe RC, Habener JF. Trans-acting factors interact with a cyclic AMP response element to modulate expression of the human gonadotropin alpha gene. Mol Cell Biol 1987;7:3032–40.
57. Bokar JA, et al. Expression of the glycoprotein hormone alpha subunit gene in the placenta requires a functional cyclic AMP response element, whereas a different cis-acting element mediates pituitary-specific expression. Mol Cell Biol 1989;9:5113–22.
58. Delegeane AM, Ferland LH, Mellon PM. Tissue-specific enhancer of the glycoprotein hormone alpha gene: dependence on cyclic AMP-inducible elements. Mol Cell Biol 1987;7:3994–4002.
59. Jameson JL, Jaffe RC, Deutsch PJ, Albanese C, Habener JF. The gonadotropin alpha gene contains multiple protein binding domains that interact to

modulate basal and cAMP-responsive transcription. J Biol Chem 1988;263: 9879–86.

60. Silver BJ, et al. Cyclic AMP regulation of the human glycoprotein hormone alpha subunit gene is mediated by an 18-base-pair element. Proc Natl Acad Sci USA 1987;84:2198–202.

61. Hoeffler JP, Meyer TE, Yun Y, Jameson JL, Habener JF. Cyclic AMP-responsive DNA binding protein: structure based on a cloned placental cDNA. Science 1988; 242:1430–3.

62. Jameson JL, Powers AC, Gallagher GD, Habener JF. Enhancer and promoter element interactions dictate cyclic adenosine monophosphate mediated and cell-specific expression of the glycoprotein hormone alpha gene. Mol Endocrinol 1989;3:763–72.

63. Kennedy GC, Andersen B, Nilson JH. The human alpha glycoprotein hormone gene utilizes a unique CCAAT binding factor. J Biol Chem 1990; 265:6279–85.

64. Andersen B, Kennedy GC, Nilson JH. A *cis*-acting element located between the cAMP response elements and CCAAT box augments cell-specific expression of the glycoprotein hormone alpha subunit gene. J Biol Chem 1990; 265:21874–80.

65. Schoderbek WE, Kim KY, Ridgeway EC, Mellon PL, Maurer RA. Analysis of DNA sequences required for pituitary-specific expression of the glycoprotein hormone α-subunit gene. Mol Endocrinol 1992;6:893–903.

66. Ocran KW, et al. Identification of *cis*-acting promoter elements important for expression of the mouse glycoprotein hormone alpha subunit gene in thyrotropes. Mol Endocrinol 1990;4:766–72.

67. Fox N, Solter D. Expression and regulation of the pituitary and placenta-specific human glycoprotein hormone alpha-subunit gene is restricted to the pituitary in transgenic mice. Mol Cell Biol 1988;8:5470–6.

68. Hoeffler JP, Lustbader JW, Chen C-Y. Identification of multiple nuclear factors that interact with cyclic adenosine 3′,5′-monophosphate response element binding protein and activating transcription factor-2 by protein-protein interactions. Mol Endocrinol 1991;5:256–66.

69. Kroll DJ, et al. A multifunctional prokaryotic expression system: overproduction, affinity purification and selective detection. DNA Cell Biol 1993.

70. Chen C-Y, Bessesen DH, Jackson SM, Hoeffler JP. Overexpression and purification of transcriptionally competent CREB from a recombinant baculovirus. Protein Expression and Purification 1991;2:402–11.

# 7

# Cyclic AMP-Mediated Hormonal Regulation of Gene Transcription

JOEL F. HABENER AND TERRY E. MEYER

The complex metabolic activities of cells respond to environmental cues by sensing transmitter substances, such as ligands or hormones. Peptide hormones cannot enter the cell directly and transmit their information by recognizing and binding to receptors located on the surface of the cell. The receptor serves as the communicative link between the outside and the inside of the cell. In response to the binding of hormone, receptors activate one or more signal transduction pathways, resulting in the generation of the important second messengers, such as cAMP and *diacylglycerol* (DAG), that in turn activate the protein kinases A and C, respectively (1–5).

One major role of the protein kinases and of phosphatases is to regulate the activities of DNA binding proteins (transcription factors) that interact with specific DNA control elements located in the promoter regions of genes and to thereby activate or repress transcription (4–7). One group of transcription factors responsive to the cAMP and DAG signaling pathways are members of a subfamily of a large superfamily of DNA binding proteins known collectively as the bZIP proteins, so named because of the similar structures of their DNA binding domains that consist of a *basic region* (b) involved in recognition and binding to DNA and a *leucine zipper* (ZIP), a coiled-coil structure with heptad repeats of leucines responsible for dimerization (8–10) (Fig. 7.1).

The transcriptional transactivational activities of CREB and the proteins that most closely resemble it in their structures are intensely regulated by the cAMP-dependent signaling pathway and appear to be unique in this regard. Given that the bZIP transcriptional proteins now identified constitute a large family, members of which compete for binding to their enhancer DNAs, it is understandable that nature would evolve only a limited set of these proteins for activation by cAMP-dependent protein kinase A, with the other proteins responsive to other non-cAMP-directed

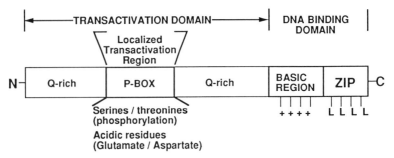

FIGURE 7.1. Diagram of cAMP-response element binding protein CREB. The structure of the protein is typical of the family of bZIP transcription factors inasmuch as it consists of two distinct domains: (i) a carboxyl-terminal DNA binding domain consisting of a basic region (BR) and a leucine zipper (ZIP) involved in DNA recognition and in dimerization, respectively, and (ii) an amino-terminal transcriptional transactivation domain. The transactivation domain of CREB consists of a localized highly phosphorylated region (P-box) flanked by sequences enriched in glutamine residues (Q-rich). The P-box, otherwise known as the kinase-inducible domain (KID) is initially phosphorylated on a single serine by the cAMP-dependent protein kinase A followed by multiple phosphorylations by secondary, processive kinases (See Fig. 7.3). The combined phosphorylations appear to allosterically alter the conformation of the transactivation domain so as to switch the domain from an inactive to active conformation. Both the glutamine-rich and acidic phosphorylated regions appear to be important for transactivational activity.

pathways. Thus, as far as is known, CREB, CREM, and ATF-1 represent the final communicative link in the regulation of gene expression in response to the activation of cAMP-dependent signaling systems.

## Role of cAMP-Dependent Phosphorylation in the Regulation of the Transactivation Functions of CREB

The primary amino acid sequences of the bZIP proteins diverge considerably outside their DNA binding and dimerization domains—that is, their transcription transactivation domains have evolved so as to mediate signal transduction, each factor in response to distinct stimuli. For example, of the known bZIP transcription factors, only CREB, CREM, and ATF-1 are activated upon phosphorylation by cAMP-dependent *protein kinase A* (PKA) (9), and also in some circumstances CREB is activated by calcium-calmodulin kinase (11, 12).

Remarkably, the regulation of the transactivational activity of CREB in the promotion of gene transcription by cAMP is regulated at the level of the reversible translocation of the active catalytic subunit of cAMP-

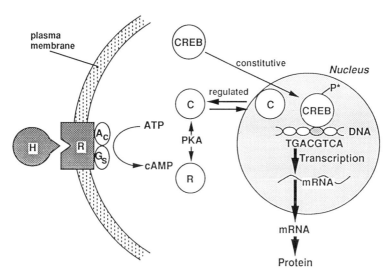

FIGURE 7.2. Model describing CREB-mediated cAMP-dependent transcription. Hormone (H) bound to receptor (R) activates G-protein (G$_s$) and adenylyl cyclase (AC), resulting in the formation of cAMP. The regulatory subunit (R) of protein kinase A binds cAMP, thereby dissociating the catalytic subunit (C). By mechanisms that are unknown, the catalytic subunit enters the nucleus, where it phosphorylates the cAMP-response element binding protein (CREB) and activates gene transcription. CREB is believed to be constitutively translocated to the nucleus. The translocation of C from the cytoplasm to the nucleus is reversible and is regulated by cAMP. When the cAMP signal decays, C is transported from the nucleus back to the cytoplasm.

dependent PKA from the cytoplasm to the nucleus (13, 14) (Fig. 7.2). Newly-synthesized CREB appears to be constitutively imported into the nucleus, where it presumably binds to the *cAMP-response elements* (CREs) located in the promoters of cAMP-responsive genes. The activation of CREB awaits the nuclear import of PKA catalytic subunits, which phosphorylates and thereby activates the transcriptional transactivation functions of CREB.

Consistent with the role of CREB in the cAMP pathway, there is a consensus site, RRPSY (A-kinase box), for phosphorylation by PKA at serine-119 in CREB327 and at the corresponding serine-133 in CREB341 (15–18), an isoform of CREB formed by an alternatively spliced exon (exon D) in the protein coding domain of the mRNA (19, 20) (Fig. 7.3). The identical A-kinase box, RRPSY, is present in the corresponding location in CREM (21) and ATF-1 (22). The PKA site in CREB resides in a *kinase-inducible domain* (KID) (23). In addition to PKA, consensus sites for potential phosphorylations by *protein kinase C* (PKC), *casein kinase II* (CKII), and *glycogen synthase kinase-3* (GSK-3) reside in

FIGURE 7.3. Amino acid sequence of the phosphorylated, localized transactivation domain of CREB. The circled serines are phosphorylated by protein kinases, resulting in the generation of transcriptional transactivation functions. The initial and key phosphorylation is a serine-119 by cAMP-dependent protein kinase A. Additional serines are then phosphorylated by secondary processive kinases.

the kinase-inducible domain. The protein kinases CKII and GSK-3 are processive (secondary) kinases inasmuch as phosphorylation of one site facilitates the successive phosphorylation of adjacent sites (24) (Fig. 7.3).

Several studies indicate that phosphorylation of the PKA site (serine-119) is essential for the activation of CREB. Phosphorylation of the serine-119 of CREB327, however, is necessary, but not sufficient to generate the transactivation functions of CREB. Sequences located carboxy-proximal to serine-119 appear to be required to generate the transcriptional transactivation functions of CREB (23). Phosphorylation of the serine-119 may allosterically alter CREB to reveal the protein conformational structure required for transcriptional transactivation (23). The regions involved in transactivation are believed to be the glutamine-rich regions corresponding to exons C and/or G of CREB and CREM (21, 23), and the activity of these regions may require negatively charged phosphorylated regions that flank the A-kinase box (Fig. 7.3).

## Characterization of the CREB Gene and Its Expression

The CREB and CREM genes consist of multiple exons (Fig. 7.4). At least 11 exons have been identified in the human CREB gene, across a span of more than 80 kb (19, 25).

## Alternative Exon Splicing

Alternative splicing of several exons accounts for the synthesis of CREB327, CREB341, and several less abundant isoforms. The mRNAs for the less abundant CREB isoforms are found only in rat testes, and not in other rat tissues, and have alternatively spliced exons that contain stop codons that interrupt the usual translation reading frame (26, 27). The resulting truncated CREB proteins lack the bZIP domain and the nuclear translocation signal located in the basic region and are unable to bind to

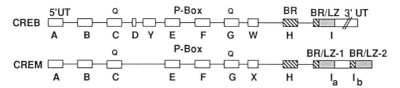

FIGURE 7.4. The organization of the CREB and CREM genes. Shown are diagrams of the exonic structures of the genes encoding the cAMP-response element binding protein (CREB) and the cAMP-response element modulator protein (CREM). The phosphorylation box (P-box) consists of exons E and F, the flanking glutamine-rich exons are C and G, and the DNA binding domains consist of exons H and I, encoding the basic regions and leucine zippers, respectively. The CREM gene contains two alternatively spliced exon I's, $I_a$ and $I_b$. Exons D, Y, and W in CREB are alternatively spliced. Exons Y and W contain trans-locational stop codons in all three reading frames, resulting in the synthesis of truncated proteins.

CREs (28). The functions of these truncated CREBs remains unknown. It has been proposed, however, that the alternative splicing in of an exon that results in the synthesis of a truncated CREB may serve to interrupt a positive feedback loop in CREB gene transcription (see below).

The mouse CREB gene spans at least 100 kbs and encodes two pre-viously unidentified exons, $\psi$ and H' (25). Like several of the alternatively used exons in the rat and human CREB genes (19), these two new mouse CREB exons contain stop codons that encode truncated CREB proteins (25). CREB327 and CREB341 are expressed ubiquitously and are the primary isoforms, except in testes (human, rats, and mice), where a multitude of CREB proteins are observed as a consequence of translations from alternatively spliced mRNAs (25–27). With the exception of exon D, which retains the translational reading frame, all of the other four alternatively spliced exons of CREB identified so far contain blocked translational reading frames. Further, when exons are deleted, the resultant reading frame is often blocked. These splicing events lead to the formation of CREB isoforms devoid of the DNA binding domain. In contrast to CREB, all of the alternatively spliced forms of CREM retain the coding sequence CREBs. CREM also uniquely has two alternatively spliced exons encoding the DNA binding and dimerization domains (29).

## The CREB Gene Promoter

To better understand the mechanisms governing CREB gene expression, the 5' flanking sequence containing the promoter region of CREB was isolated from a human genomic library (30). A 1.2 GREs clone containing an exon for part of the 5' untranslated tract and the promoter region was sequenced (30) (Fig. 7.5). By inspection, the DNA sequence is highly

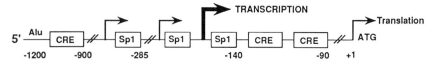

FIGURE 7.5. Diagram of the 5′ flanking promoter region (1200 bp) of the CREB gene. The major transcription start site is shown in heavy arrow and two minor start sites by light arrows to the left. The start of translation at the AUG codon is designated base +1. Potential binding sites for CREB (CRE) and Sp1 are designated by the boxes.

enriched in guanines and cytosines and contains canonical binding sites for the transcription factors CREB and Sp1, among other sites. Transfection-expression studies indicate that CREs participate in regulating transcription from the CREB gene (30). Many of these binding sites for proteins are also present in the mouse CREB promoter region, a sequence of which was reported recently (25, 31). The human and mouse CREB promoter sequences are nearly identical within the 303 bp of the mouse sequence reported (25). The similarity of the nucleotide sequences between mouse and human CREB DNAs diminishes considerably elsewhere in the genes, especially in the introns, and even in the exons (despite the high protein sequence identity).

The degree to which the CREB promoter has been conserved during evolution suggests that much of this region of the gene is functionally critical for the regulation of expression. There is a notable absence of TATA and CCAAT sequence motifs, a finding consistent with primer extension data that indicate the existence of multiple transcription start sites for the creb gene. A major start site occurs within an *initiator* (InR) sequence, CCTCA (32). These features are reminiscent of many of the housekeeping genes that recently have been described (33). Typically these genes are expressed ubiquitously in cells of diverse phenotypes. Many of the genes that have GC-rich promoter sequences devoid of TATA and CCAAT boxes encode components of signal transduction pathways—for example, plasma membrane bound receptors, protein kinases, GTP binding proteins, and cell adhesion molecules (33 and references therein).

Characteristically, CREB has been detected in all cells and tissues tested thus far. Whereas the CREB protein is easily detected by either antibodies or DNA probes, the CREB mRNA is particularly difficult to detect by standard Northern blotting techniques. This circumstance suggests that the CREB mRNA is expressed at a relatively low, basal level and/or is rapidly degraded. The CREB protein, however, must be relatively stable. A constant supply of stable CREB protein would be consistent with the observation that cAMP activation of the transcriptional transactivation functions of CREB, mediated through the protein kinase

A pathway, occurs rapidly, reaching a peak within 30 min of stimulation, even when cycloheximide is used to block new protein synthesis (34, 35). Thus, posttranslational modifications of CREB by way of phosphorylations play a major and critical role in the regulation of its transactivational functions.

The expression of the CREB gene is modulated in certain tissues. In the rat testes, CREB mRNA and protein levels show striking increases at specific developmental stages in specific cell types. For example, the intensity of CREB mRNA in Sertoli cells of the testis from a three-day autoradiographic exposure of in situ hybridized CREB mRNA from adult rat testes is comparable to that for a six-week exposure of a similarly probed rat brain section. This cyclical increase in expression of the CREB gene during the spermatogenic cycle is also interesting because it exhibits developmental regulation (26, 27), suggesting that these marked swings in CREB expression in the testes may serve an important role in spermatogenesis. It has been proposed that the cyclical splicing in and out of exons, such as exons W (26) or Y (27), during spermatogenesis in the adult rat serves to modulate autopositive feedback control of CREB gene transcription. Inasmuch as transcripts containing exons W or Y encode truncated CREB isoforms missing the DNA binding domain with its nuclear translocation signal (28), the proposed stimulation of CREB gene transcription by CREB itself would be interrupted by the splicing in of these exons. Conversely, splicing out of the exons would reestablish the autopositive feedback loop (26).

# Evidence for Positive Autoregulation of CREB Gene Expression

Results from transient transfections of the human CREB promoter show that basal transcription driven by the 1.2-kb CREB promoter fragment is very high (30). The human CREB promoter region contains three nonpalindromic CREs, labeled CRE1, CRE2, and CRE3. Two of these, CRE1 and CRE2, are conserved in the smaller promoter fragment reported for mouse CREB (25), but not enough of the mouse CREB promoter is known to determine whether it has a sequence corresponding to CRE3 of the human CREB gene. These three CREs from the human promoter, as well as the entire 1.2 kb of cloned promoter region, have been tested for potential autoregulation by CREB (or at least participation of CREB or CREB-like proteins) (30). Most of the constructs (both the 1.2-kb promoter and subfragments derived from it) exhibited 2- to 4-fold cAMP responses, 3- to 5-fold TPA responses, and 5- to 7-fold responses for cAMP + TPA, relative to basal activity. Creation of a point mutation in CRE3 of the promoter also reduced basal and cAMP-stimulated transcription relative to the wild-type clone, indicating functional importance of CRE3 (30).

Further testing of CRE1, CRE2, and CRE3 by DNAse I footprinting and by gel electrophoretic mobility shift assays shows that isolated recombinant CREB and proteins in extracts of cell nuclei indeed bind these CREs (30). Antibodies to the CREB P-box perturb binding of the proteins to these CREs, suggesting that CREB or a related protein (e.g., CREM or ATF-1) participates in regulating the expression of the CREB gene.

*Acknowledgments.* We thank Gérard Waeber and Yungdae Yun for their participation in some of the studies and Townley Budde for preparation of the manuscript. J.F.H. is an Investigator with the Howard Hughes Medical Institute.

## *References*

1. Scott JK. Cyclic nucleotide-dependent protein kinases. Pharmacol Ther 1991; 50:123–45.
2. Nishizuka Y. The role of protein kinase C in cell surface signal transduction and tumour promotion. Nature 1984;308:693–8.
3. Berridge MF. Inositol trisphosphate and calcium signalling. Nature 1993; 361:315–25.
4. Karin M. Signal transduction from cell surface to nucleus in development and disease. FASEB J 1992;6:2581–90.
5. Hunter T, Karin M. The regulation of transcription by phosphorylation. Cell 1992;70:375–87.
6. Ptashne M. How eukaryotic transcriptional activators work. Nature 1988; 335:683–9.
7. Mitchell PH, Tjian R. Transcriptional regulation in mammalian cells by sequence-specific DNA binding proteins. Science 1989;245:371–8.
8. Johnson PF, McKnight SL. Eukaryotic transcriptional regulatory proteins. Annu Rev Biochem 1989;58:799–839.
9. Meyer TE, Habener JF. Cyclic AMP-dependent transactivation of gene transcription mediated by CREB phosphoprotein. In: Mond JJ, Cambier JC, Weiss A, eds. Advances in regulation of cell growth; vol 2. Cell activation: genetic approaches. New York: Raven Press, 1991:61–82.
10. Habener JF. Cyclic AMP response element binding proteins: a cornucopia of transcription factors. Mol Endocrinol 1990;4:1087–94.
11. Dash PK, Karl KA, Colicos MA, Prywes R, Kandel ER. cAMP response element-binding protein is activated by $Ca^{2+}$/calmodulin as well as cAMP-dependent protein kinase. Proc Natl Acad Sci USA 1991;88:5061–5.
12. Sheng M, Thompson MA, Greenberg ME. A $Ca^{2+}$-regulated transcription factor phosphorylated by calmodulin-dependent kinases. Science 1991;252: 1427–30.
13. Nigg EA, Hilz H, Eppenberger HM, Dutly F. Rapid and reversible translocation of the catalytic subunit of cAMP-dependent protein kinase type II from the Golgi complex to the nucleus. EMBO J 1985;4:2801–6.

14. Meinkoth JL, Ji Y, Taylor SS, Feramisco JR. Dynamics of the distribution of cyclic AMP-dependent protein kinase in living cells. Proc Natl Acad Sci USA 1990;87:9595–9.

15. Hoeffler JP, Meyer TE, Yun Y, Jameson JL, Habener JF. Cyclic-AMP responsive DNA-binding protein: structure based on a cloned placental cDNA. Science 1988;242:1430–3.

16. Gonzalez GA, Yamamoto KK, Fischer WH, et al. A cluster of phosphorylation sites on the cyclic AMP-regulated nuclear factor CREB predicted by its sequence. Nature 1989;337:749–52.

17. Lamph WW, Dwarki VJ, Ofir R, Montminy M, Verma IM. Negative and positive regulations by transcription factor cAMP response element-binding protein is modulated by phosphorylation. Proc Natl Acad Sci USA 1990; 37:4320–4.

18. Gonzalez GA, Montminy MR. Cyclic AMP stimulates somatostatin gene transcription by phosphorylation of CREB at serine 133. Cell 1989;59:675–80.

19. Hoeffler JP, Meyer TE, Waeber GW, Habener JF. Multiple adenosine 3',5'-cyclic monophosphate response element DNA-binding proteins generated by gene diversification and alternative exon splicing. Mol Endocrinol 1990;4: 920–30.

20. Yamamoto KK, Gonzalez GA, Menzel P, Rivier J, Montminy MR. Characterization of a bipartite activator domain in transcription factor CREB. Cell 1990;60:611–7.

21. Foulkes NS, Mellstrom B, Benusiglio E, Sassone-Corsi P. Developmental switch of CREM function during spermatogenesis: from antagonist to activator. Nature 1992;355:80–4.

22. Rehfuss RP, Walton KM, Loriaux MM, Goodman RH. The cAMP-regulated enhancer binding protein ATF-1 activates transcription in response to cAMP-dependent protein kinase A. J Biol Chem 1991;266:18431–4.

23. Gonzalez GA, Menzel P, Leonard J, Fischer WH, Montminy MR. Characterization of motifs which are critical for activity of the cyclic AMP-responsive transcription factor CREB. Mol Cell Biol 1991;11:1306–12.

24. Roach PJ. Multi-site and hierarchal protein phosphorylation. J Biol Chem 1991;266:14139–42.

25. Ruppert S, Cole TJ, Boshart M, Schmid E, Schutz G. Multiple mRNA isoforms of the transcription activator protein CREB: generation by alternative splicing and specific expression in primary spermatocytes. EMBO J 1992;11:1503–12.

26. Waeber G, Meyer TE, LeSieur M, Hermann H, Gérard N, Habener JF. Developmental stage-specific expression of the cyclic AMP response element binding protein CREB during spermatogenesis involves alternative exon splicing. Mol Endocrinol 1991;5:1418–30.

27. Waeber G, Habener JF. Novel testis germ cell-specific transcript of the CREB gene contains alternatively spliced exon with multiple in frame stop codons. Endocrinology 1992;131:2010–5.

28. Waeber G, Habener JF. Nuclear translocation and DNA recognition signals colocalized within the bZIP domain of cAMP response element-binding protein CREB. Mol Endocrinol 1991;5:1431–8.

29. Foulkes NS, Borrelli E, Sassone-Corsi P. CREM gene: use of alternative DNA-binding domains generates multiple antagonists of cAMP-induced transcription. Cell 1991;64:739–49.

30. Meyer TE, Waeber G, Beckmann W, Lin J, Habener JF. The promoter of the gene encoding cAMP-response element binding protein CREB contains cAMP response elements: evidence for positive autoregulation of gene transcription. Endocrinology 1993.

31. Cole TJ, Copeland NG, Gilbert DJ, Jenkins NA, Schutz G, Ruppert G. The mouse CREB (cAMP responsive element binding protein) gene: structure, promoter analysis, and chromosomal location. Genomics 1992;13:974–82.

32. O'Shea-Greenfield A, Smale ST. Roles of TATA and initiator elements in determining the start site location and direction of DNA polymerase II transcription. J Biol Chem 1992;267:1391–407.

33. Blake MC, Jambou RC, Swick AG, Kahn JW, Azizkhan JC. Transcriptional initiation is controlled by upstream GC-box interactions in a TATAA-less promoter. Mol Cell Biol 1990;10:6632–41.

34. Hagiwara M, Alberts A, Brindle P, et al. Transcriptional attenuation following cAMP induction requires PP-1-mediated dephosphorylation of CREB. Cell 1992;70:105–13.

35. Lewis EJ, Harrington CA, Chikarshi DM. Transcriptional regulation of the tyrosine hydroxylase gene by glucocorticoid and cyclic AMP. Proc Natl Acad Sci USA 1987;84:3550–4.

# 8

# Regulation and Expression of Thyroid Stimulating Hormone

BRUCE D. WEINTRAUB, MYUNG K. KIM, DONALD L. BODENNER,
N. RAO THOTAKURA, MARIUSZ W. SZKUDLINSKI, LATA JOSHI, AND
YOKO MURATA

*Thyroid stimulating hormone* (TSH) is an anterior pituitary hormone that stimulates thyroid hormone production (1). It is composed of two subunits: the alpha subunit, which is in common with the gonadotropins, luteinizing hormone, and follicle stimulating hormone; and the unique beta subunit, which confers biological specificity to the intact hormone and is expressed exclusively in the thyrotropes of the anterior pituitary (2). Previous studies have indicated that the transcription of the TSHβ gene is stimulated by *thyrotropin releasing hormone* (TRH) (3–6), phorbol esters (5, 7), and the adenylyl cyclase activator, forskolin, or the cAMP analog 8-bromo-cAMP (4–6, 8), and is inhibited by thyroid hormone (9–15).

The anterior pituitary contains a mixed population of cells producing various pituitary hormones, including *growth hormone* (GH), *prolactin* (PRL), and TSH (16). The tissue-specific expression of the GH and PRL genes is regulated by the pituitary-specific homeobox protein Pit-1 (or GHF-1) (17–20), which is present only in the three cell types of the anterior pituitary: GH-producing somatotropes, PRL-producing lactotropes, and a subset of TSH-producing thyrotropes (18). Furthermore, mutations in the Pit-1 gene, as shown in dwarf mice (21) and in patients carrying combined pituitary deficiency (22–24), led to a decrease in pituitary gland size and/or a complete loss of expression of GH, PRL, and TSH. Thus, in the anterior pituitary, Pit-1 appears to serve as a tissue-specific transcription factor and a developmental regulator.

Unlike the activation of the GH and PRL genes, where Pit-1 alone is sufficient to activate these promoters (19, 20), Pit-1 appears to be necessary, but not sufficient to induce the transfected human or mouse TSHβ gene (8, 25). Previous studies demonstrated that TRH or forskolin induction of the hTSHβ gene is mediated by a sequence containing Pit-1 binding sites (6). Furthermore, in cells that do not produce Pit-1, such as human embryonal kidney 293 cells, the hTSHβ gene was induced only when cells were cotransfected with the Pit-1 expression vector and stimulated by forskolin. However, Pit-1 alone or forskolin treatment in the absence of Pit-1 did not stimulate expression (8). Thus, these studies revealed that, in addition to Pit-1, activating signals from effectors such as TRH, forskolin or, perhaps, phorbol esters are also essential for the induction of the hTSHβ promoter.

Although little is known about the mechanism by which TRH, phorbol esters, or forskolin activate the hTSHβ gene, these inducing agents might exert their positive regulatory effects via induction or phosphorylation of Pit-1. Indeed, Pit-1 was shown to be induced by cAMP (26, 27) and phosphorylated after treatment with phorbol esters or cAMP (28) in pituitary cells. Furthermore, recent studies showed that Pit-1 phosphorylated by either protein kinase C or A recognizes Pit-1 binding sites in the hTSHβ gene more efficiently than does nonphosphorylated Pit-1 (29). Alternatively, TRH, phorbol esters, or cAMP might activate the hTSHβ gene by inducing and/or modifying factors involved in protein kinase C or A pathways that, in turn, might interact with Pit-1 to lead to the cell-specific activation of the hTSHβ gene. In this regard, it is interesting that the $-1/+6$ sequence (TGGGTCA) of the hTSHβ gene differs at only one nucleotide from the consensus binding sequence for transcription factor AP-1 (a heterodimer of members of the Fos and Jun families of proto-oncogenes) or the consensus *phorbol ester-response element* (TRE, TGAG/CTCA) (30). The $-1/+6$ sequence is also similar to the consensus *cAMP-response element* (CRE, TGACGTCA) (31–33). Thus, the presence of the TGGGTCA element in the hTSHβ gene raised the possibility that it could be involved in the induction of the hTSHβ gene by interacting with factors that mediate second messenger pathways.

Recently, we have found that an AP-1 (c-*fos*/c-*jun*)-like factor that interacts with the TGGGTCA element and that, together with Pit-1, apparently mediates both the PMA and forskolin induction of the hTSHβ gene (34). In transient transfection assays, the mutations of this element alone completely abolished induction by PMA and forskolin and impaired transactivation by c-*fos* and c-*jun* even in the presence of Pit-1. In addition, we report that the TGGGTCA element does not recognize the cAMP-responsive transcription factor CREB, although this element functions as a CRE upon forskolin stimulation. Moreover, we found that in the context of the hTSHβ promoter, the TGGGTCA element is a more potent TRE and CRE than the corresponding consensus sequences, suggesting that the sequences flanking the TGGGTCA element in the hTSHβ gene may influence responsiveness to phorbol esters or cAMP.

The biological significance of glycosylation variants of pituitary glycoprotein hormones remains controversial because of indirect methods usually employed to determine carbohydrate composition or structure, as well as the use of unreliable biologic/immunologic ratio to determine bioactivity. We have previously characterized *recombinant human TSH* (rhTSH) secreted by Chinese hamster ovary cells attached to microcarrier beads in a large-scale bioreactor after stable transfection of hCGα and hTSH minigenes (35). Recently rhTSH has been used as a model to determine structure-function relationships of different isoforms of glycoprotein hormones. We have now produced greater than 200 mg of rhTSH using a hollow-fiber bioreactor. The highly purified rhTSH produced in the hollow-fiber bioreactor (rhTSH-N), as well as rhTSH commercially produced in a large-scale bioreactor (rhTSH-G), was quantitated by immunoassays, receptor binding assay, and amino acid analysis and further characterized by a variety of physico-biochemical methods, including chromatofocusing and carbohydrate analysis. rhTSH-G, rhTSH-N, and *pituitary human TSH* (phTSH) have been separated by chromatofocusing on a Mono P column into several isoforms with different pI values. Compositional analysis of the fractions showed higher sialic acid content in the more acidic rhTSH-G fractions. phTSH acidic isoforms showed higher total sulfate and sialic acid content than the more basic fractions. The bioactivities of various TSH isoforms based on rigorous quantitation of mass by amino acid analysis determined in three different FRTL-5 cell bioassays showed that the more basic and less sialylated fractions of rhTSH-G were more active than the more acidic fractions. In contrast to the in vitro data, highly sialylated and acidic rhTSH-G isoforms showed longer plasma half-life and a higher in vivo bioactivity than the basic forms. These results indicate that the secreted rhTSH, similar to the intrapituitary phTSH, exists as a mixture of charge isoforms that are related, at least in part, to the degree of sialylation. The degree of sialylation, highly dependent on the bioreactor production conditions, appears to be the major factor affecting the charge heterogeneity, metabolic clearance rate, and bioactivity of rhTSH.

## References

1. Condliffe PG, Weintraub BD. In: Gray CH, James VHT, eds. Hormones in blood. London: Academic Press, 1979:499 574.
2. Pierce JG, Parsons TF. Annu Rev Biochem 1981;50:465–95.
3. Shupnik MA, Greenspan SL, Ridgway EC. J Biol Chem 1986;261:981–7.
4. Franklyn JA, Wilson M, Davis JR, Ramsden K, Docherty K, Sheppard MC. J Endocrinol 1986;111:R1–2.
5. Shupnik MA, Rosenzweig BA, Showers MO. Mol Endocrinol 1990;4:829–36.
6. Steinfelder HJ, Hauser P, Nakayama Y, et al. Proc Natl Acad Sci USA 1991;88:3130–4.
7. Carr FE, Galloway RJ, Reid AH, et al. Biochemistry 1991;30:3721–8.
8. Steinfelder HJ, Radovick S, Mroczynski MA, et al. J Clin Invest 1992;89:409–19.

9. Shupnik MA, Chin WW, Habener JF, Ridgway EC. J Biol Chem 1985;260: 2900–3.
10. Carr FE, Burnside J, Chin WW. Mol Endocrinol 1989;3:709–16.
11. Burnside J, Darling DS, Carr FE, Chin WW. J Biol Chem 1989;264:6886–91.
12. Wood WM, Kao MY, Gordon DF, Ridgway EC. J Biol Chem 1989;264: 14840–7.
13. Chatterjee VKK, Lee J-K, Rentoumis A, Jameson JL. Proc Natl Acad Sci USA 1989;86:9114–8.
14. Wondisford FE, Farr EA, Radovick S, et al. J Biol Chem 1989;264:14601–4.
15. Bodenner DL, Mroczynski MA, Weintraub BD, Radovick S, Wondisford FE. J Biol Chem 1991;266:21666–73.
16. Voss JW, Rosenfeld MG. Cell 1992;70:527–30.
17. Bodner M, Castrillo J-L, Theill LE, Deerinck T, Ellisman M, Karin M. Cell 1988;55:505–18.
18. Ingraham HA, Chen R, Mangalam HJ, et al. Cell 1988;55:519–29.
19. Nelson C, Albert VR, Elsholtz HP, Lu LI-W, Rosenfeld MG. Science 1988;239:1400–5.
20. Mangalam HJ, Albert VR, Ingraham HA, et al. Genes Dev 1989;3:946–58.
21. Li S, Crenshaw EB, Rawson EJ, Simmons DM, Swanson LW, Rosenfeld MG. Nature 1990;347:528–33.
22. Pfaffle RW, DiMattia GE, Parks JS, et al. Science 1992;257:1118–21.
23. Radovick S, Nations M, Du Y, Berg LA, Weintraub BD, Wondisford FE. Science 1992;257:1115–8.
24. Tatsumi K, Miyai K, Notomi T, et al. Nature Genetics 1992;1:56–8.
25. Gordon DF, Haugen BR, Sarapura VD, Wood WM, Ridgway EC. [Abstract]. Endocr Soc Annu Meet, 1992.
26. McCormick A, Brady H, Theill LE, Karin M. Nature 1990;345:829–32.
27. Chen R, Ingraham HA, Treacy MN, Albert VR, Wilson L, Rosenfeld MG. Nature 1990;346:583–6.
28. Kapiloff MS, Farkash Y, Wegner M, Rosenfeld MG. Science 1991;253:786–9.
29. Steinfelder HJ, Radovick S, Wondisford FE. Proc Natl Acad Sci USA 1992;89:5942–5.
30. Angel P, Imagawa M, Chiu R, et al. Cell 1987;49:729–39.
31. Montminy MR, Sevarino KA, Wagner JA, Mandel G, Goodman RH. Proc Natl Acad Sci USA 1986;83:6682–6.
32. Deutsch PJ, Jameson JL, Habener JF. J Biol Chem 1987;262:12169–74.
33. Silver BJ, Bokar JA, Virgin JB, Vallen EA, Milsted A, Nilson JH. Proc Natl Acad Sci USA 1987;84:2198–202.
34. Kim MK, McClaskey JH, Bodenner DL, Weintraub BD. An AP-1-like factor and the pituitary-specific factor Pit-1 are both necessary to mediate hormonal induction of human thyrotropin beta gene expression. J Biol Chem, 1993.
35. Thotakura NR, Desai RK, Bates LG, Cole ES, Pratt BM, Weintraub BD. Biological activity and metabolic clearance of a recombinant human thyrotropin produced in Chinese hamster ovary cells. Endocrinology 1991; 128:341–8.

# Part III

## Hormonal Protein Structure/Function

# 9

# Human Chorionic Gonadotropin: Progress in Determining Its Tertiary Structure

Joyce W. Lustbader, Chuan Wang, Xiaolu Zhang,
Steven Birken, Hao Wu, Jonathan Miles Brown,
David Yarmush, Susan Pollak, and Robert E. Canfield

In humans, maintenance of pregnancy requires extending the lifetime of the corpus luteum, which produces essential steroids. The signal for this maintenance is provided by *human chorionic gonadotropin* (hCG), a dimeric glycoprotein hormone produced by the trophoblast cells of the early embryo. The absence of hCG would result in rapid termination of early pregnancy.

Solution of the three-dimensional structure of hCG, the hormone of pregnancy, could have a profound impact on society. Uncontrolled population growth is a major obstacle to improving the quality of life in many nations. One potential route to controlling population growth is the use of antagonists to hCG, since pregnancy cannot be sustained in its absence. Furthermore, solution of the structure of hCG would also permit a reasonable prediction of the structure of follicle stimulating hormone, the homologous pituitary hormone controlling ovulation.

Understanding the action of hCG may also be important in alleviating infertility due to short or poor luteal phase, the condition in which hCG secretion is insufficient to maintain adequate formation of progestational compounds. Development of hCG agonists may improve therapeutic prospects for patients with this form of infertility. A rational approach to the design of both agonists and antagonists of hCG presupposes knowledge of the hormone's three-dimensional structure.

For more than twenty years laboratory groups have been attempting to determine the three-dimensional structure of the glycoprotein hormones, the family to which hCG belongs. Knowledge of the tertiary structure of the hormones would aid in understanding their structure-function relationships. Detailed knowledge of a given glycoprotein hormone's three-dimensional structure would allow the chemist to identify a specific amino acid's contribution to the function of the hormone and then to change those properties through site-specific mutagenesis. The chemist could also, for example, "build a better hormone" by synthesizing molecules that had higher affinities for the receptor than the native hormone. Alternatively, a synthetic peptide or small molecule could be developed that would block the binding of the hormone to the receptor. To make such compounds requires an intimate knowledge of the hormone's surface, and that understanding cannot be derived from knowledge of the primary amino acid structure of the protein. With regard to hCG in particular, two groups, one of which is ours, obtained large diffractable crystals of hCG, and yet little new structural information has been presented since those initial reports (1, 2).

This chapter describes the progress in efforts to determine the structure of hCG. hCG is one of four glycoprotein hormones, and the only one secreted by the trophoblast. The other three glycoproteins secreted by the pituitary are *follicle stimulating hormone* (FSH), *thyroid stimulating hormone* (TSH), and *luteinizing hormone* (LH). The hormones are dimeric in structure, and within a species they share an identical α-subunit, with the biological specificity of the hormone conferred by the β-subunit; the primary action of hCG is to signal the corpus luteum of pregnancy to maintain steroid production. The intact hormone's subunits are held together by noncovalent interactions. Dissociation of the dimer into its monomeric subunits can be achieved by adding chaotropic reagents or acid pH. Both subunits possess a large number of disulfide bonds, and this greatly restricts the number of different conformations the hormone can assume. The degree of homology in the sequences in the α-subunit from different species can be determined by aligning the positions of the cysteines; the α has 10 cysteines (all conserved in species that have been protein-sequenced to date), resulting in 5 disulfide bridges in the folded subunit. Similarly the β-subunits can be aligned by their 12 cysteines, resulting in 6 disulfide bridges. Since more than 70% of the β-subunit associated with any given glycoprotein hormone is homologous with the β-subunits of the other members of the family, one could reasonably extrapolate from knowledge of the structure of one subunit to the structures of the other subunits. Since CG and its most homologous analog LH, both bind to the same receptor, indicating that their conformations are similar, either one could have been chosen for the studies to be described.

A three-dimensional structure of hCG can be generated in one of three ways: (i) model building, (ii) crystallography, and (iii) *nuclear magnetic resonance* (NMR) imaging. The first technique is an indirect method that capitalizes on other known structures that have significant sequence homologies to the protein under study. The latter two methods rely on direct physical data.

## Model Building

Obstacles to hCG model building are threefold: (i) No sequence homology to any known structure has been identified, (ii) little in the way of interatomic distance data is available, and (iii) the disulfide bridge pairings are still unresolved. Therefore, traditional model building, primarily based on known homologous structures, cannot be accomplished until the structure of at least one member of the glycoprotein hormones is solved.

Unfortunately, a method to derive the three-dimensional structure of a protein from its primary amino acid sequence has not yet been determined, although testable hypothetical models can be generated from the many computer-based modeling programs. Such models are probably incorrect in their detail, but they aid the investigator in visualizing the surface topology of the protein. Molecular biology experiments aimed at defining epitopes, biologic efficacy, and dimeric contacts have greatly increased our knowledge of the surface residues, providing additional inputs to a modeling program. Because of the considerable degree of homology between the α- and β-subunits from different species, initial modeling paradigms relied heavily on the placement of conserved regions. Regions of conserved sequences probably exist internalized in the center of the molecule and may not participate directly in the function of the hormone, but may aid in the proper folding of the molecule and, consequently, be crucial in relation to function. Those positions in the sequences with variable amino acids are probably located on the surface and are subject to a greater degree of evolutionary change.

For the studies to be discussed in this chapter, the modeling program PAKGGRAF was used. A brief summary of the modeling studies is presented here, while an extensive report appears elsewhere (3). The precursor of PAKGGRAF was originally developed by Levinthal more than thirty years ago (4). It has been continually revised over the ensuing years. Its advantage over other modeling programs is that it performs both torsional and Cartesian minimization operations and can alternate between the two interactively. This allows the researcher to start with unrealistic coordinates and refine the structure by moving atoms over several angstroms. A number of models have been presented by other groups (5, 6). These models, although informative, do not assign coordinates to atoms and therefore cannot be stringently compared. The model presented in this paper relies heavily on the Mise and Bahl assignments for the disulfide pairings (7, 8). Other experimental data incorporated into this model were derived from the reactivities of glycoprotein hormones of various species with a selected panel of monoclonal anti-bodies (3). These comparisons allowed for the placement of certain amino acids on the surface of the hormone.

## α-Subunit

The α-subunit appears to have a structure with three dominant loops. As has been indicated by *circular dichroism* (CD) spectra analyses, the α-subunit is predominantly β sheet with little α helix. Figure 9.1 depicts the α-subunit indicating the loops: loop 1, residues 7–31; loop 2, residues 32–59; and loop 3, residues 60–84. The amino terminal of the first loop is exposed in the dimer and consequently can be cleaved by enzymes. The remainder of the first loop is hidden in the dimer. Residues 27–30 of this loop are likely to be the nucleating site for the folding of this subunit. The greatest divergence in sequences between species appears in the first part of loop 1; this observation is consistent with the previously reported data indicating that this part of the molecule is not involved in the subunit interface. The second loop is more strongly conserved, except at four positions, 45, 48, 54, and 57. The data suggest that the region preceding Asn 52 is involved in subunit association; that Tyr 37 is not iodinated in the dimer; that glycosylation of Thr 39 prevents subunit association; and that Lys 45 can be cross-linked to Asn 111 on β. Evidence from site-specific mutagenesis studies indicates that a mutation or a deletion of residues 38, 39, or 40 reduces or destroys the ability of α to combine with β. The amino terminal of the third loop is also involved in the subunit

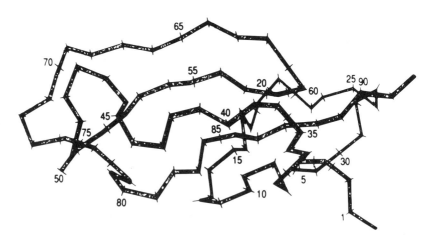

FIGURE 9.1. The α-carbon backbone of the α-subunit of hCG with every fifth residue labeled. Reprinted with permission from Lustbader, Yarmush, Birken, Puett, and Canfield, © The Endocrine Society, 1993.

interface (Fig. 9.2). The carboxyterminal region (73–92) of this loop is antigenic and may play a role in receptor binding.

## β-Subunit

Our model includes residues 1–112 of the β-subunit of hCG only, omitting the carboxyterminal peptide, which includes residues 113–145. The β-subunit of our proposed model contains five antiparallel strands of β sheet, with turns proximal to residues 20, 38, 58, 70, 83, 90, and 96 (Fig. 9.3). The CD spectrum of hCGβ depicts a preponderance of β structure with only limited, if any, helicity (3). Puett and Birken demonstrated that reduced, carboxymethylated hCGβ and some tryptic fragments become helical in aqueous solution and that the helix index increases in the presence of trifluoroethanol (9).

The initial portion of the first cysteine loop (9–90) projects away from the subunit interface, influenced probably by the large bulky carbohydrate group at Asn 13. Additionally, residues 8–10 have been described as part

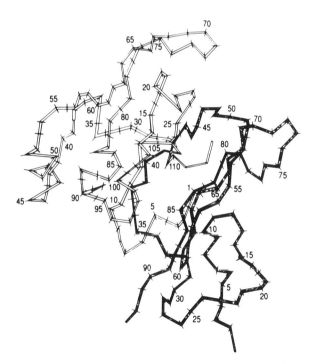

FIGURE 9.2. The α-carbon backbone placement of the α-subunit (shaded) of hCG relative to the β-subunit (residues 1–112) (outlined) with every fifth residue labeled. Reprinted with permission from Lustbader, Yarmush, Birken, Puett, and Canfield, © The Endocrine Society, 1993.

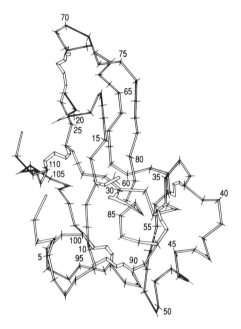

FIGURE 9.3. The α-carbon backbone of the β-subunit of hCG with every fifth residue labeled. Reprinted with permission from Lustbader, Yarmush, Birken, Puett, and Canfield, © The Endocrine Society, 1993.

of the B204 epitope. An area of high sequence conservation is proximal to Glu 19. A mutation of Lys 20 decreases receptor binding significantly, indicating that this region has a role in the receptor interface. Because of carbohydrate at Asn 30, the region between Cys 26 and Cys 34 is likely to project away from any contact. The CAGY region is highly conserved among species and is therefore likely to be involved in receptor or subunit interaction. The conserved region around Asp 111 is likely to be located near the subunit contact since this residue can be cross-linked to the α-subunit (Fig. 9.2).

## X-Ray Crystallography

Crystallography is so far the most powerful technique for determining structures of proteins of moderate molecular weight. Since the solution of the hemoglobin structure, developments in methodology and computers have brought about an exponential increase in the number of solved three-dimensional protein structures. The first step toward a structure determination is the purification and crystallization of the target protein.

The history of the successful crystallizations of hCG is a story of international cooperation. Although our laboratory has worked with hCG for over two decades and has prepared all of the reference preparations distributed by the NIH, we had never seriously pursued crystal growth in our own laboratory. We lacked a crystallographer and had been discouraged in such attempts by the numerous anecdotal tales from experts in the field who had spent a great deal of time in fruitless efforts to obtain crystals. These stories had discouraged many investigators from such efforts, and it was widely accepted that hCG and the other glycoprotein hormones could not be crystallized into useful, diffractable crystals by current technology and that the solution of the structures of these hormones would probably not be by crystal diffraction.

In 1982 a crystallographer joined the group (J.W.L.) to pursue a postdoctoral project of bacterial expression of hCG under the tutelage of James Roberts, who was then at Columbia University. Since bacteria do not glycosylate their proteins, this was an avenue of obtaining a naturally carbohydrate-free form of hCG to be used for crystallographic studies. It was known that carbohydrate interfered with crystal growth, and this seemed to be a reasonable study. Unfortunately, we could never obtain intact hCG from the bacteria and only observed some denatured material. The sugar moieties must somehow assist in folding of the molecule. It was at this time that we embarked on a collaboration with Integrated Genetics, who had also tried bacterial expression with hCG and had the same result, but had gone on to express intact hCG in Cos cells. This was the first demonstration of an expression of a biologically active dimeric hormone in recombinant cells (10). These studies also led to an expression system in C127 cells that could produce significant quantities of this recombinant hCG that could be used for crystallographic experiments (11).

Late in 1988 our lab was visited by Francis Morgan from St. Vincent's Institute of Medical Research in Melbourne, Australia. Dr. Morgan was a long-time friend of the laboratory, having spent several years working with us (R.E.C.) in the early 70s on the primary structure of hCG and, in fact, was the postdoctoral mentor of one of us (S.B.). Dr. Morgan indicated that there was some successful crystallization work with hCG by a member of his laboratory, Dr. Neil Isaacs, and a student, D. Harris. Recognizing this as a very major accomplishment, we sought and received approval from the NIH to send Dr. Isaacs several hundred milligrams of the most recently purified hCG reference preparation to facilitate his progress.

In early 1989 Dr. Isaacs sent us a preprint of his manuscript describing the hCG crystallization, which soon appeared in the *Journal of Biological Chemistry*. This manuscript seemed quite remarkable since the hCG employed was the Organon material, which is only 1/3 hCG by weight. This crude material was simply treated with *hydrogen fluoride* (HF) to remove most of the sugar groups and then gel filtered. The hCG crystals appeared to have grown from material that may have been 75% or less of HF-treated hCG in the mother liquor. No characterization of the content of the crystals was presented. It was not clear that these crystals were composed of hCG. However, future studies proved that they were pure hCG. Fortuitously, at that time, the crystallographer (J.W.L.) in our laboratory was able to pursue crystallization work in our own facility. We proceeded to repeat Neil Isaacs's studies, but with purified hCG in our own laboratories. Nick Pileggi in the University's Protein Core laboratory was quite adept at the HF deglycosylation procedure, and we tailored as careful a treatment scheme as possible. We also made modifications in an attempt to treat the protein in as gentle a manner as possible after HF deglycosylation, such as adding a high-strength neutral Tris buffer instead of sodium hydroxide to neutralize residual HF. We also did a very careful size selection of the molecules for the crystallization work by running an SDS-gel profile on each fraction from a gel-permeation column separation of the HF-treated hCG. Crystals of hCG were successfully grown and completely characterized.

Through the efforts of Mary Ann Gawinowicz, it was found that the HF-treated hCG seemed to accumulate peptide bond cleavages during the long period of crystal growth in acid pH. Since examination of the SDS-gel patterns of HF-treated hCG showed that the heterogeneity of the molecules had increased, not decreased, after HF treatment and that the total carbohydrate content was only reduced by about half, we reasoned that simple removal of the negatively charged sugars may provide the only necessary pretreatment needed to allow the formation of orderly crystals. This proved to be the case, and neuraminidase-treated hCG crystals diffracted better than HF-treated materials. They were found to undergo less peptide bond cleavage during crystal growth, indicating less damage from the HF-treatment. In the light of Neil Isaacs's findings that hCG crystallized from impure mixtures, we also investigated whether the crystals selected particular species of hCG molecules and found that this was not the case and that the crystals reflected the content of the mother liquor. In fact, the crystals seemed to be even more heterogeneous than the mother liquor because of accumulated additional peptide bond cleavages over time.

We have crystallized urinary HF-hCG, urinary asialo-hCG, and r-HF-hCG (Fig. 9.4) using the hanging drop vapor diffusion method (2) with a range of ammonium sulfate at pH 4.2. Whereas large crystals of HF-hCG tended to form within 4 to 6 weeks, the neuraminidase-treated hCG required 8 to 12 weeks to grow (2). The crystals are morphologically identical and appear as stout bipyramids whose faces are usually somewhat rounded. These crystals typically reach a size of $0.4 \times 0.4 \times 0.6$ mm. The HF-treated hCG crystals diffract to a limit spacing of $2.8$ Å, whereas the asialo-hCG crystals diffract to a slightly better resolution

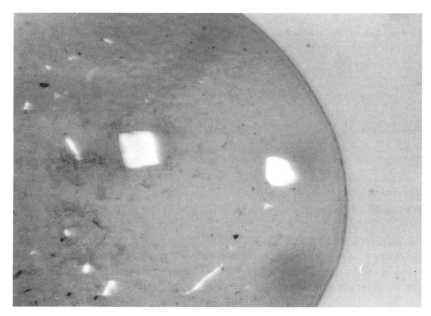

FIGURE 9.4. Hydrogen-fluoride treated recombinant hCG (HF-r-hCG) crystals. The crystals appear as stout bipyramids.

of 2.6 Å. Crystals from neuraminidase-treated hCG, however, have not been regularly reproducible. Each of these crystals possesses the pre-requisites for a successful structural determination at intermediate resolution.

Diffraction patterns from these crystals could be indexed in a hexagonal lattice that belongs to space group $P6_122$, or its enantiomorph $P6_522$ (2). Unit cell dimensions are roughly a = b = 88 Å and c = 177 Å, with slight variation among the three forms of hCG. The solvent content of the crystal is estimated to be 59%, assuming one protein molecule per crystal-lographic asymmetric unit. Most protein crystals have solvent contents of between 27% and 65%, with 43% occurring most frequently (12). The patterns of diffracted intensities from the HF-hCG crystals and the asialo-hCG crystals are not visually distinguishable, suggesting that additional sugars removed by the HF-treatment might be rather disordered in the neuraminidase-treated hCG crystals (2).

Chemical characterization of the various forms of hCG showed that HF treatment increased size heterogeneity of the β-subunit, which was augmented even further by the crystallization process (2). Although the starting HF-hCG contained 1%–3% of β-subunit beginning at Leu 45, N-terminal sequencing revealed the presence of peptide bond cleavages at the acid-labile Asp-Pro (112–113) position and to a lesser degree at Asp-Ser (117–118) in the HF-hCG crystals (2). Although the resolution limits of urinary and recombinant hCG crystals appear to be similar, the expressed HF-hCG had considerably less nicking than the urinary source material and is therefore a better starting material for crystallization. The neuraminidase-treated hCG starting material or crystal had the least cleavages and was the most homogeneous of the three forms of hCG (2).

In the meantime, we have continued to collaborate with Dr. Isaacs toward the solution of the structure of urinary forms of hCG by the crystal diffraction method. Conventionally, the most commonly used method for determining a nonhomologous protein structure is *multiple isomorphous replacement* (MIR) (13). In this method, differences in diffraction intensities due to incorporation of heavy atoms into the crystals provide information about the detailed structural arrangement of the protein. In collaboration with Dr. Isaacs and his group, we are searching heavy atom derivatives using urinary HF-hCG crystals. Such searches are usually empirical, and some proteins turn out to be more difficult to derivatize than others. Heavy atom-labeled crystals may also have different crystal parameters, since they are nonisomorphous with the native crystal. Both problems necessitate trials of various heavy atom compounds and different derivatizing methods. We (J.W.L.) have provided over 50 high-quality crystals to Dr. Isaacs for the heavy atom replacement procedure. During this period of time we learned of the importance of very careful HF treatment if crystals were to be obtained. Dr. Isaacs informed us of problems that he had with a number of batches of the purified hCG (sent to him earlier), which did not crystallize well presumably due to the harshness of the HF treatment or problems in the selection of the proper fractions to pool for crystallization. We also ran into problems in the capability of different batches of neuraminidase-treated hCG to crystallize. This may have been due to protease contamination. In any event, the effort to solve the structure of urinary hCG remains in the hands of Dr. Isaacs by collaborative agreement.

We have pursued a parallel, but different, approach, since we do not know if it will prove possible to obtain suitable heavy atom replacements for the urinary HF-treated hCG crystals. This work is being pursued collaboratively with Wayne Hendrickson and is dependent on use of recombinant expressed hCG. These studies generally require incorporation of selenomethionine into hCG and are described below.

We obtained high levels of expression of *recombinant hCG* (rhCG) from *Chinese hamster ovary* (CHO) cells subcloned from cells provided by Irving Boime. We have also formed a collaboration with Ares Advanced Technology (a division of Serono) for high-level expression of hCG. Serono had acquired the glycoprotein hormone section of Integrated Genetics, which first expressed hCG and with whom we had worked earlier. A method developed by Dr. Wayne Hendrickson, *multiwavelength anomalous diffraction* (MAD) (14, 15), has resulted in many successful applications (16–20). With the presence of strong diffractors, in this case selenium atoms introduced by substituting methionines in hCG with selenomethionines, diffraction intensities from a single-type selenomethionine-containing crystal would vary between two sets of otherwise equivalent reflections (Bijvoet difference[1]) and wavelength (dispersive difference[2]). This phenomenon is a result of the so-called anomalous scattering, which is inappreciable for lighter atoms in proteins at typically achievable wavelengths. These variations provide phasing information of the structure. Since only one type of crystal is required, there is perfect isomorphism and no derivatization is required. MAD experiments require a wavelength-tunable synchrotron radiation source. The strong synchrotron radiation should provide a more accurate measurement of diffraction data, and hence, generally small signals from different wavelengths could be extracted. There are a total of four methionines in the primary sequence of hCG—that is, one methionine in every 59 residues. Assuming 100% of the methionines are substituted by selenomethionines, the phasing power is 7.8% from Bijvoet difference and 4.7% from dispersive difference. Previous studies have shown that signals of this magnitude should be ample for a successful structure determination by MAD phasing (21). Theoretically, phasing information from MIR and MAD experiments are complementary to each other and thus can be combined in the structure solution.

The difficulty associated with MAD is obtaining sufficient quantities of hCG with a high incorporation level of selenomethionine. Although some *Escherichia coli* strains are able to grow for 100 generations (22) in selenomethionine, the slight difference in chemistry and metabolism between selenomethionine and methionine makes it toxic to cells, especially mammalian cells as a result of their greater complexity. The starting cell culture must grow in methionine-containing medium since the cells do not tolerate selenomethionine at this stage (21). The expression of selenomethionine protein therefore mandates careful arrangement of timing and concentration levels of the selenomethionine for the preincubation and incubation period to replace the native methionine. We have recently successfully expressed and purified sufficient quantities of selenomethionine HF-rhCG. Amino acid analysis and mass spectroscopy showed complete selenomethionine substitution within the limits of the sensitivity of our analysis.

Small crystals have been obtained from these selenomethionine HF-hCG preparations using the same crystallization condition that has produced native crystals. In many cases, selenomethionine proteins crystal-

---

[1] Bijvoet difference is the difference in structure factor amplitudes of Friedel pairs (structure factor hkl and hkl).
[2] Dispersive difference is the difference in structure factor amplitudes at different wavelengths.

lize isomorphously under conditions similar to those in which the native materials crystallize (21). Small adjustments to the crystallization conditions are often necessary to allow for possible differences in solubility and hydrophobicity. Here, the selenomethionine rhCG appears to differ from the wild-type rhCG to a somewhat greater extent, and this may necessitate a search for completely new crystallization conditions. In collaboration with Dr. Wayne Hendrickson and coworkers, we have recently initiated a joint effort toward this end. Once diffraction quality crystals are obtained, we should be able to determine the three-dimensional structure of hCG. Such a structure would also facilitate further structural studies on individual subunits of hCG and on the interactions of hCG with its receptor and antibodies.

## Nuclear Magnetic Resonance

Although X-ray crystallography is still the technique of choice for large proteins, both NMR (23, 24) and X-ray crystallography can be used to determine the *three-dimensional* (3D) structure of small- to medium-sized proteins. NMR is particularly attractive for intact hCG, given the difficulties of obtaining intact hCG crystals (1, 2). In addition, NMR allows the determination of the structure of intact hCG in solution, with its full complement of carbohydrate, which most closely resembles its physiologic condition. In the past, structures of many small proteins (less than approximately 10 kd) have been successfully determined by using the *two-dimensional* (2D) proton NMR method (24–27). For larger proteins (up to approximately 40 kd), however, it is necessary to apply the multidimensional heteronuclear NMR method (28–32).

Intact hCG is too large to be studied by the 2D proton NMR method. The problems are threefold: First, the large molecular weight (38.6 kd) severely broadens proton NMR resonances, preventing the investigator from obtaining high-quality NMR spectra. Second, the very large number of proton resonances leads to extremely complex NMR spectra for analysis. Third, the presence of a large amount of carbohydrate (comprising 1/3 of the molecular weight) causes further signal overlaps in NMR spectra, which also makes a complete analysis of NMR spectra difficult.

In order to probe the structure of hCG, we have applied the 2D proton NMR method to the α-subunit, which is common to all members of the glycoprotein hormone family. To facilitate proton resonance assignments, the α-subunit was treated with HF to partially remove carbohydrate. In this case, it was found necessary to treat the α-subunit under harsher conditions with HF in order to remove enough carbohydrate to permit gathering some interpretable NMR data. In one case, the α-subunit was treated with HF twice, and a pool was made of the smallest molecular weight alpha species. NMR studies of the HF-treated α-subunit were carried out in our laboratories, as well as those of Dinshaw Patel. To obtain structural information, the majority of proton resonances, at least those of backbone protons, must be first identified. Therefore, the study starts with proton resonance assignments. This usually involves two steps (24). Amino acid spin systems[3] can be identified according to their characteristic covalent connectivity[4] as observed in various correlation spectra. This is accomplished by relying mainly on three-bond proton-proton J-couplings[5] (24). The order of individual amino acids along the peptide chain is then determined by observation of *nuclear Overhauser effects* (NOE)[6] (33) that result in NOE spectra from protons that are close in space (24).

To identify amino acids, a series of homonuclear correlation spectra of the HF-treated α-subunit, including *correlation* (COSY)[7], *relayed correlation* (RELAYED-COSY), *total correlation* (TOCSY), and *double-quantum correlation* (DQ-COSY) experiments, were recorded in both $H_2O$ and $D_2O$. Figure 9.5 shows the NH-CαH fingerprint region of the correlation spectrum of the α-subunit of hCG in $H_2O$. In this spectral region, each amino acid shows a cross-peak at the chemical shifts of its amide and α protons, except that each proline shows none and each glycine shows two. The same region of the relayed-correlation spectrum (Fig. 9.6) shows not only the direct NH-CαH J-coupling correlation, but also the indirect NH-CβH J-coupling correlation. With careful analysis of the data, the NH, CαH, and/or CβH protons of all 85 amino acids other than proline are identified. Among these 85 amino acids, all alanines, glycines, threonines, and most valines are identified by their characteristic patterns in the correlation spectra. The assignments are summarized in Table 9.1. At present, sequential proton resonance assignments have not been completed because the extensive overlap of the proton resonances precludes reliable identification of some signals. As shown in Table 9.1, approximately 62% of the amide protons and more than 90% of the CαH and CβH protons possess indistinct chemical shifts from one another. Degeneracy of the proton resonances causes great difficulties in identification of sequential NOE connectivities.

---

[3] Spin systems refers to the type and number of equivalent nuclear spins in a molecule and their relationships.
[4] Covalent connectivity indicates a covalent bond connection between atoms in a molecule.
[5] J-coupling arises from a specific interaction of one nucleus with another that produces a resonance splitting in an ordinary spectra.
[6] Nuclear Overhauser effect refers to the change in intensity of a nuclear spin when another spin is perturbed by spin saturation or spin inversion (through-space experiments).
[7] COSY, RELAYED COSY, TOCSY, and DQ-COSY refer to types of homonuclear 2D NMR experiments (through-bond experiments).

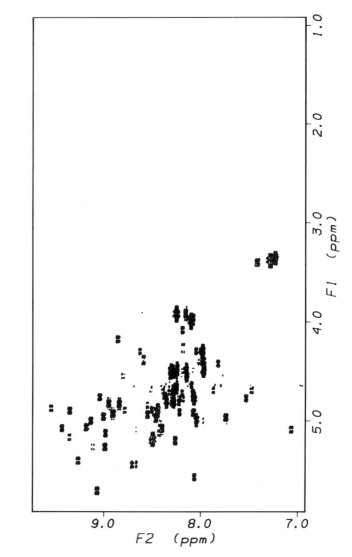

FIGURE 9.5. The NH-CαH fingerprint region of COSY spectrum of HF-treated α-hCG at pH 3.5. The spectrum is recorded on a Bruker AMX500 spectrometer at 500 MHz frequency and 60.0 °C. Each cross peak identifies the NH and CαH of individual amino acids.

Multidimensional triple-resonance ($^1$H, $^{13}$C, and $^{15}$N) NMR has proved to be a powerful tool for determining the structures of larger proteins (28–39). As in the 2D proton NMR method (24), proton resonance assignments are crucial in structure elucidation. This new technique provides significant improvement in spectral resolution by increasing the

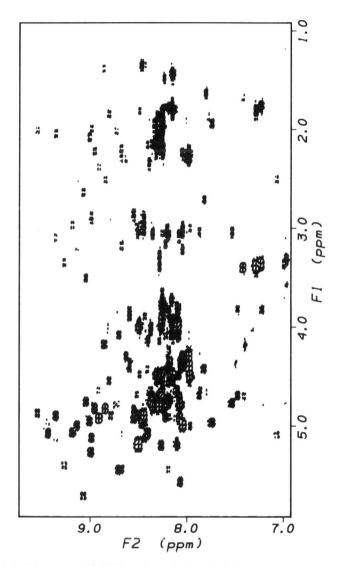

FIGURE 9.6   The NH-CαH-CβH region of relayed-COSY spectrum of HF-treated α-hCG at pH 3.5 and 60.0 °C. This spectrum is used to identify the NII-CαII-CβII fragments of individual amino acids.

spectral dimensionality to spread proton resonances according to the chemical shifts of the directly attached $^{13}C$ or $^{15}N$. In addition, this method provides high sensitivity because it utilizes large one-bond heteronuclear J-couplings and is thus much less sensitive to the broad line width of proton resonances. The applications of multidimensional triple-resonance NMR have been successfully demonstrated by deter-

TABLE 9.1. Proton chemical shifts (ppm) of HF-deglycosylated α-hCG[a].

| Residue[b] | NH | CαH | CβH | CγH | |
|---|---|---|---|---|---|
| 1 | 9.54 | 4.87 | 2.01 | | |
| 2 | 9.43 | 5.07 | 3.65/3.52 | | |
| 3 | 9.35 | 5.16 | 3.09/2.62 | | |
| 4 | 9.35 | 4.89 | 2.04 | | |
| 5 | 9.27 | 5.40 | 3.35/2.48 | | |
| 6 | 9.18 | 5.06 | 2.97 | | |
| 7 | 9.13 | 4.99 | 3.19 | | |
| 8 | 9.11 | 5.25 | 2.55 | | |
| 9 | 9.07 | 5.70 | 2.64/3.18 | | |
| 10 | 9.04 | 4.76 | 3.50 | | |
| 11 | 8.99 | 4.95 | 2.08 | | |
| 12 | 8.98 | 5.26 | 2.88 | | |
| 13 | 8.98 | 5.12 | 1.99 | | |
| 14 | 8.95 | 4.82 | 2.22 | 0.90/1.03 | Val |
| 15 | 8.91 | 4.63 | 2.13 | | |
| 16 | 8.90 | 4.92 | 2.38 | | |
| 17 | 8.85 | 4.17 | 1.39 | | Ala |
| 18 | 8.84 | 4.82 | 2.51 | 0.98/1.02 | Val |
| 19 | 8.80 | 4.54 | 1.84 | | |
| 20 | 8.78 | 4.89 | c | | |
| 21 | 8.72 | 4.78 | 2.01 | | |
| 22 | 8.70 | 5.46 | 4.08 | | |
| 23 | 8.68 | 4.62 | 2.19/2.30 | | |
| 24 | 8.67 | 5.44 | 3.18 | | |
| 25 | 8.65 | 5.08 | c | | |
| 26 | 8.62 | 4.29 | 2.29 | | |
| 27 | 8.55 | 4.83 | c | | |
| 28 | 8.55 | 4.94 | 2.85 | | |
| 29 | 8.52 | 4.68 | 2.99 | | |
| 30 | 8.51 | 4.89 | 3.02 | | |
| 31 | 8.49 | 5.18 | 3.99 | | |
| 32 | 8.48 | 4.47 | 1.81 | 1.09/1.05 | Val |
| 33 | 8.46 | 4.97 | 1.35 | | Ala |
| 34 | 8.44 | 4.89 | 2.92/3.02 | | |
| 35 | 8.43 | 4.92 | 3.89 | | |
| 36 | 8.42 | 5.12 | 0.75 | | Ala |
| 37 | 8.42 | 5.01 | 4.14 | 1.28 | Thr |
| 38 | 8.41 | 4.42 | 2.15 | | |
| 39 | 8.40 | 5.07 | 4.12 | 1.25 | Thr |
| 40 | 8.39 | 4.10 | 2.39 | 1.19 | Val |
| 41 | 8.37 | 4.74 | 4.00 | | |
| 42 | 8.35 | 4.80 | 3.05 | | |
| 43 | 8.30 | 4.04 | 2.40 | | |
| 44 | 8.28 | 4.47 | 4.14 | 1.31 | Thr |
| 45 | 8.26 | 5.19 | 3.76 | | |
| 46 | 8.25 | 3.91 | 2.27/2.16 | | |
| 47 | 8.22 | 4.81 | 3.05 | | |
| 48 | 8.21 | 4.91 | 3.16 | | |
| 49 | 8.19 | 4.71 | c | | |
| 50 | 8.19 | 4.07 | 5.44 | | |

TABLE 9.1. *Continued.*

| | | | | | |
|---|---|---|---|---|---|
| 51 | 8.18 | 4.49 | 1.78 | | |
| 52 | 8.16 | 3.91 | 3.72 | | |
| 53 | 8.15 | 4.48 | 1.44 | | Ala |
| 54 | 8.15 | 4.55 | 1.79 | | |
| 55 | 8.10 | 3.99 | 5.18 | | |
| 56 | 8.09 | 4.64 | 3.79 | | |
| 57 | 8.08 | 4.48 | 3.05 | | |
| 58 | 8.08 | 4.71 | 4.47 | 1.29 | Thr |
| 59 | 8.07 | 4.77 | 4.36 | 1.43 | Thr |
| 60 | 8.07 | 4.91 | 4.10 | 1.21 | Thr |
| 61 | 8.06 | 5.57 | 3.37/3.01 | | |
| 62 | 8.05 | 4.29 | 2.23 | 1.08/1.10 | Val |
| 63 | 8.04 | 4.99 | 3.23/3.01 | | |
| 64 | 7.98 | 4.07 | c | | |
| 65 | 7.98 | 4.40 | 2.29 | 1.08 | Val |
| 66 | 7.98 | 4.29 | 2.25 | | |
| 67 | 7.97 | 4.53 | 4.38 | 1.37 | Thr |
| 68 | 7.97 | 4.48 | 4.38 | 1.36 | Thr |
| 69 | 7.97 | 5.00 | 3.00 | | |
| 70 | 7.87 | 4.69 | 3.03 | | |
| 71 | 7.82 | 4.42 | 2.71 | | |
| 72 | 7.80 | 4.67 | 1.63 | | Ala |
| 73 | 7.78 | 4.52 | 1.88 | | |
| 74 | 7.74 | 4.96 | 1.94 | | |
| 75 | 7.53 | 4.77 | 3.03 | | |
| 76 | 7.47 | 4.69 | 3.82 | | |
| 77 | 7.42 | 3.39 | 1.67/1.87 | | |
| 78 | 7.41 | 4.70 | c | | |
| 79 | 7.28 | 3.38 | 1.81 | | |
| 80 | 7.23 | 3.35 | 1.75 | | |
| 81 | 7.07 | 5.09 | 2.06 | | |
| 82 | 8.80 | 4.31/3.99 | | | Gly |
| 83 | 8.59 | 4.38/3.86 | | | Gly |
| 84 | 8.18 | 4.25/3.81 | | | Gly |
| 85 | 8.07 | 4.77/3.93 | | | Gly |

[a] Proton chemical shifts of HF deglycosylated α-hCG at pH 3.5 and 60 °C.

[b] The ordering of residues is purely arbitrary.

[c] Not identified due to signal overlap.

*Source:* Reprinted with permission from Lustbader, Yarmush, Birken, Puett, and Canfield (3). © The Endocrine Society, 1993.

mination of the structures of several medium-sized (about 20 kd) $^{13}$C,$^{15}$N-labeled proteins (31, 32).

Determination of the entire structure of intact hCG presents a challenge for today's NMR technology because of its large molecular weight, as well as its large carbohydrate content. Multidimensional triple-resonance NMR will be applied to intact hCG. This approach requires the hormone to be uniformly enriched with $^{13}$C and $^{15}$N isotopes. Therefore, we are pursuing expression of uniformly $^{13}$C,$^{15}$N-labeled hCG in mammalian cells.

In 1992 we were searching for more feasible methods for performing NMR studies on hCG. It was obvious that chemical deglycosylation, even the harsh double HF treatment that we had performed with the α-subunit, was inadequate because of residual sugar content. Incorporation of the NMR active isotopes, $^{13}$C and $^{15}$N, was the only feasible response. The problem was that whereas $^{13}$C and $^{15}$N can easily be incorporated into bacterial proteins by growing the bacteria on $^{13}$C glucose and $^{15}$N ammonium salts, mammalian cells require a medium containing all the amino acids preformed. For the production of $^{13}$C- and $^{15}$N-labeled mammalian proteins, these amino acids must all be present and all universally labeled with $^{13}$C and $^{15}$N. Unfortunately, not all amino acids are commercially available in $^{13}$C- and $^{15}$N-labeled form, and moreover, some cannot be prepared from published procedures.

Fortuitously, we saw an advertisement from Martek that stated that dual-labeled amino acids would soon be available. Martek was contacted at an apparently critical time in their efforts to produce such dual-labeled materials, and one key member of their research team, Jonathan Miles Brown, was already active in producing NMR-labeled materials for use by the general research community. He was at the point of designing universally labeled mammalian cell growth media and needed a collaborator to help in assessing various formulations of these media. We entered into such a collaboration using an ongoing expression system for hCG that had been refined to produce hCG for use in the crystallization work. Several members of our group visited Martek and found an innovative company producing products from algae. $^{13}$C- and $^{15}$N-labeled proteins were produced by supplying the algae with $^{13}$C- and $^{15}$N-labeled carbon dioxide and nitrogen salts, respectively. Jonathan Miles Brown had developed a scheme to hydrolyze the algal proteins in a fashion that preserves all the labeled amino acids, remarkably, including the amides, and that produces them in high purity. Much exploration of mixtures of concentrations of amino acids was necessary, but in a period of months, a suitable media formulation was developed that would sustain mammalian cells that were producing NMR-labeled hCG. This represents a great advance in the potential to solve structures of the mammalian proteins, especially glycoproteins, which cannot be correctly expressed with sugars intact in any cells other than mammalian.

In our upcoming studies, the α- and β-subunits will first be dissociated and their structures determined separately. This enables the analysis of a smaller molecular weight molecule, increasing the probability of high-quality NMR spectra and facilitating data analysis. A variety of triple-resonance experiments have been developed for resonance assignments of polypeptides (28–30, 34–43). These experiments are also essential for glycoproteins. Alternatively, it will be neceessary for us to develop more effective resonance assignment procedures for the carbohydrate moieties of the glycoprotein. To determine the entire structure of hCG, two intact hCG molecules will be prepared, each consisting of a $^{13}C,^{15}N$-labeled subunit and its unlabeled complementary subunit. This approach will facilitate resonance assignments, as well as measurements of structural information. Complexity of heteronuclear NMR spectra of such heterodimeric hormones is the same as these of their subunits alone because the heteronuclear NMR permits the selective observation of those protons bonded to $^{13}C$ and/or $^{15}N$ instead of the protons bonded to $^{12}C$ or $^{14}N$, or vice versa. However, the greater molecular weight of intact hCG will severely broaden NMR resonances, especially those carbons bonded to protons, such as α carbons. For intact hCG, optimization experiments are needed to overcome this problem. Generally, this problem is a major obstacle to the application of NMR technology applied to larger molecules, including the glycoprotein hormones.

The application of computer modeling, X-ray crystallography, and NMR methods is bringing about encouraging progress toward the solution of the three-dimensional structure of hCG. Modeling has yielded a representation of the structure with defined atomic coordinates that can be tested against experimental data. However, the chief benefit of modeling will come after the hCG structure is solved by one or both of the methods described in this chapter. The structures of the other homologous glycoprotein hormones will then be predicted by modeling techniques. Most importantly, hormone analogs will be precisely designed from the known geometry of the hormone surface. Agonists and antagonists can then be constructed by the techniques of molecular biology and tested in vitro and in vivo. From these studies, small molecules may be designed that will serve as effective regulators of human fertility and permit a simple and humane control of human population growth.

*Acknowledgment.* This work was supported by NIH Grant HD-15454.

## References

1. Harris DC, Machin KJ, Evin GM, Morgan FJ, Isaacs NW. Preliminary X-ray diffraction analysis of human chorionic gonadotropin. J Biol Chem 1989;264: 6705–6.
2. Lustbader JW, Birken S, Pileggi NF, et al. Crystallization and characterization of human chorionic gonadotropin in chemically deglycosylated and enzymatically desialylated states. Biochem 1989;28:9239–43.
3. Lustbader JW, Yarmush DL, Birken S, Puett D, Canfield RE. The application of chemical studies of human chorionic gonadotropin to visualize its three-dimensional structure. Endocr Rev 1993;14:1–21.

4. Katz L, Levinthal C. Interactive computer graphics and representation of complex biological structures. Annu Rev Biophys Bioeng 1972;1:465–504.
5. Moyle WR, Matzuk MM, Campbell RK, et al. Localization of residues that confer antibody binding specificity using human chorionic gonadotropin/luteinizing hormone β subunit chimeras and mutants. J Biol Chem 1990;265: 8511–8.
6. Willey KP, Leidenberger F. Functionally distinct agonist and receptor-binding regions in human chorionic gonadotropin. J Biol Chem 1989;264:19716–29.
7. Mise T, Bahl OP. Assignment of disulfide bonds in the beta subunit of human chorionic gonadotropin. J Biol Chem 1981;256:6587–92.
8. Mise T, Bahl OP. Assignment of disulfide bonds in the alpha subunit of human chorionic gonadotropin. J Biol Chem 1980;255:8516–22.
9. Puett D, Birken S. Helix formation in reduced S-carboxymethylated human choriogonadotropin β subunit and tryptic peptides. J Protein Chem 1989;8: 779–94.
10. Reddy VB, Beck AK, Garramone AJ, Vellucci V, Lustbader J, Bernstine EG. Expression of human choriogonadotropin in monkey cells using a single simian virus 40 vector. Proc Natl Acad Sci USA 1985;82:3644–8.
11. Lustbader J, Birken S, Pollak S, et al. Characterization of the expression products of recombinant human choriogonadotropin and subunits. J Biol Chem 1987;262:14204–12.
12. Matthews BW. Solvent content of protein crystals. J Mol Biol 1968;33: 491–7.
13. Watenpaugh KD. Overview of phasing by isomorphous replacement. Methods Enzymol 1985;115:3–14.
14. Hendrickson WA. Determination of macromolecular structures from anomalous diffraction of synchrotron radiation. Science 1991;254:51–8.
15. Hendrickson WA, Smith JL, Sheriff S. Direct phase determination based on anomalous scattering. Methods Enzymol 1985;115:41–54.
16. Leahy DJ, Hendrickson WA, Aukhil I, Erickson HP. Structure of a fibronectin type III domain from Tenascin phased by MAD analysis of the selenomethionyl protein. Science 1992;258:987–91.
17. Weis WI, Kahn R, Fourme R, Drickamer K, Hendrickson WA. Structure of the calcium-dependent lectin domain from a rat mannose-binding protein determined by MAD phasing. Science 1991;254:1608–15.
18. Yang W, Hendrickson WAS, Crouch RJ, Satow Y. Structure of ribonuclease H phased at 2 A resolution by MAD analysis of the selenomethionyl protein. Science 1990;249:1398–405.
19. Graves BJ, Hatada MH, Hendrickson WA, Miller JK, Madison VS, Satow Y. Structure of interleukin 1 A at 2.7 A resolution. Biochemistry 1990;29:2679–84.
20. Guss JM, Merritt EA, Phizackerley RP, et al. Phase determination by multiple-wavelength x-ray diffraction: crystal structure of a basic "blue" copper protein from cucumbers. Science 1988;241:806–11.
21. Hendrickson WA, Horton JR, LeMaster DM. Selenomethionyl proteins produced for analysis by multiwavelength anomalous diffraction (MAD): a vehicle for direct determination of three-dimensional structure. EMBO J 1990;9:1665–72.
22. Cowie DB, Cohen GN. Biochim Biophys Acta 1957;26:252–61.

23. Ernst RR, Bodenhausen G, Wokaun A. Principles of nuclear magnetic resonance in one and two dimensions. Oxford: Oxford Science Publications, 1986.
24. Wuthrich K. NMR of proteins and nucleic acids. New York: Wiley, 1986.
25. Markley JL. Two-dimensional nuclear magnetic resonance spectroscopy of proteins: an overview. Methods Enzymol 1989;176:12–34.
26. Clore GM, Gronenborn AM. Determination of three-dimensional structures of proteins and nucleic acids in solution by nuclear magnetic resonance spectroscopy. CRC Crit Rev Biochem Mol Biol 1989;24:479–564.
27. Wagner G. NMR investigations of protein structure. Prog NMR Spectroscopy 1990;22:101–39.
28. Kay L, Ikura M, Tschudin R, Bax A. Three-dimensional triple-resonance NMR spectroscopy of isotopically enriched proteins. J Magn Reson 1990;89: 496–514.
29. Ikura M, Kay M, Bax A. Novel approach for sequential assignment of $^1$H, $^{13}$C and $^{15}$N spectra of larger proteins: heteronuclear triple-resonance three-dimensional NMR spectroscopy: application to calmodulin. Biochemistry 1990;29:4659–67.
30. Montelione G, Wagner G. Conformation-independent sequential NMR connections in isotope-enriched polypeptides by $^1$H-$^{13}$C-$^{15}$N triple-resonance experiments. J Magn Reson 1990;87:183–8.
31. Ikura M, Clore GM, Gronenborn AM, Zhu G, Klee G, Bax A. Solution structure of a Calmodulin-target peptide complex by multidimensional NMR. Science 1992;256:632–7.
32. Clore GM, Wingfield P, Gronenborn AM. High-resolution three-dimensional structure of Interleukin β in solution by three- and four-dimensional nuclear magnetic resonance spectroscopy. Biochemistry 1991;30:2315–23.
33. Noggle JH, Schirmer RE. The nuclear Overhauser effect: chemical applications. New York: Academic Press, 1971.
34. Boucher W, Laue E, Campbell S, Domaille P. Four-dimensional heteronuclear triple-resonance NMR methods for the assignment of backbone nuclei in proteins. J Am Chem Soc 1992;114:2262–4.
35. Kay L, Wittekind M, McCoy M, Friedrichs M, Miller L. 4D NMR triple-resonance experiments for assignment of protein backbone nuclei using shared constant-time evolution periods. J Magn Reson 1992;98:443–50.
36. Olejniczak ET, Xu RX, Petros AM, Fesik SW. Optimized constant-time 4D HNCαHα and HN(CO)CαHα experiments: applications to the backbone assignments of the FKBP/Asomycin complex. J Magn Reson 1992;100:444–50.
37. Grzesiek S, Bax AJ. Improved 3D triple-resonance NMR techniques applied to a 31 kD protein. Magn Reson 1992;96:432–40.
38. Palmer AG III, Fairbrother WJ, Cavangh J, Wrught PE, Rance M. Improved resolution in three-dimensional constant-time triple resonance NMR spectroscopy of proteins. J Biomol NMR 1992;2:103–8.
39. Farmer BT II, Venters RA, Spicer LD, Wittekind MG, Mueller L. Refocused optimized HNCα: increased sensitivity and resolution in large macromolecules. J Biomol NMR 1992;2:195–8.
40. Clubb RT, Thanabal V, Wagner G. A constant-time three-dimensional triple-resonance pulse scheme to correlate intraresidue $^1$H, $^{15}$N and $^{13}$C: chemical shifts in $^{15}$N-$^{13}$C-labeled proteins. J Magn Reson 1992;97:213–7.

41. Bax A, Clore GM, Driscoll PC, Gronenborn AM, Ikura M, Kay L. Practical aspects of proton-carbon-carbon-proton three-dimensional correlation spectroscopy of $^{13}$C-labeled proteins. J Magn Reson 1990;87:620–7.
42. Bax A, Clore GM, Gronenborn AM. $^1$H-$^1$H correlation via isotopic mixing of $^{13}$C magnetization, a new three-dimensional approach for assigning $^1$H and $^{13}$C spectra of $^{13}$C-enriched proteins. J Magn Reson 1990;88:425–31.
43. Fesik SW, Eaton HL, Olejniczak ET, Zuiderweg ERP, McIntosh LP, Dahlquist FW. Proton-proton correlation via carbon-carbon couplings: a three-dimensional NMR approach for the assignment of aliphatic resonances in proteins labeled with carbon-13. J Am Chem Soc 1990;112:886–8.

# 10

# Receptor Binding Regions of hLH and hCGβ-Subunit: Structural and Functional Properties

Henry T. Keutmann

For the study of structure-activity relations, we encounter in the glyco-protein hormones a group of molecules with characteristics of both a typical peptide hormone and a larger globular protein. As noted by Dyson and Wright (1), the active conformation of a small hormone or bioactive peptide may be imposed predominantly by the shape of the receptor, while regions of interest within a protein are inevitably involved in interactions within the molecule itself. In the absence of a detailed three-dimensional structure, the extent and nature of these influences within the glycoprotein hormone molecule remain largely uncharted. They must, however, be considered an important factor in the alignment and relative contributions of the multiple receptor binding sites revealed in all four glycoprotein hormones by assays of individual synthetic peptides (2–8).

The concept of multiple receptor binding sites is paralleled by similar observations in other small proteins, such as growth hormone (9) and inflammatory factor C5a (10). In the glycoprotein hormones, this pattern is reinforced substantially—if not universally—by study of the products of site-directed mutagenesis (11–16), together with a wealth of earlier results using chemically modified subunits and hormones (reviewed in 17–21). Since each of these experimental approaches has its own par-ticular assets and drawbacks, integration of the results in concerted fashion continues to be warranted in developing an accurate functional map of the hormone. Indeed, for example, some of the differences observed between assays of individual peptides and mutagenized hormone might not be a contradiction, but rather a clue to the nature of the peptide's environment within the whole molecule.

Our own assays of synthetic peptides representing intercysteine loops and overlapping linear sequences have revealed five distinct binding sequences within the β-subunit of LH and hCG (2, 3, 22–24) (Fig. 10.1). Despite the lack of more knowledge about their interactions within the subunit, can common features or unique properties be discerned that might enable us to predict the contribution of each to a common receptor binding domain?

The binding affinities for rat ovarian cell membranes fall within a relatively narrow range, as summarized in Table 10.1. The active sites found in hLH coincide closely with those in hCG, as might be expected from their generally close structural similarity; in some peptides the potencies differ somewhat, reflecting the modest sequence differences between the hormones. The unique 30-residue extension in hCG has no

```
        ┌─1──────────────────────────────────────10──────────────────┐
hLH:    │ SER - ARG - GLU - PRO - LEU - ARG - PRO - TRP - CYS - HIS - PRO - ILE -
hCG:    │       LYS                                 ARG         ARG
        └────────────────────────────────────────────────────────────
                                        20
 - ASN - ALA - ILE - LEU - ALA - VAL - GLU - LYS - GLU - GLY - CYS - PRO - VAL -
     *           THR
                              *
                              30
 - CYS - ILE - THR - VAL - ASN - THR - THR - ILE - CYS - ALA - GLY - TYR - CYS -
                              *
        ┌──40────────────────────────────────────────────────50──────┐
 - PRO - THR - MET - MET - ARG - VAL - LEU - GLN - ALA - VAL - LEU - PRO - PRO -
                      THR                       GLY                  ALA
        ────────────────────────────────────────────────────────────
                                        60
 - LEU - PRO - GLN - VAL - VAL - CYS - THR - TYR - ARG - ASP - VAL - ARG - PHE -
                                      ASN
                                  ──70──
 - GLU - SER - ILE - ARG - LEU - PRO - GLY - CYS - PRO - ARG - GLY - VAL - ASP -
                                                                          ASN
          80                                                        ──90──
 - PRO - VAL - VAL - SER - PHE - PRO - VAL - ALA - LEU - SER - CYS - ARG - CYS -
                           TYR   ALA                               GLN
        ────────────────────────────────────100────────────────────
 - GLY - PRO - CYS - ARG - ARG - SER - THR - SER - ASP - CYS - GLY - GLY - PRO -
   ALA   LEU                                THR
                                   110                      114
 - LYS - ASP - HIS - PRO - LEU - THR - CYS - ASP - HIS - PRO - GLN - PHE - GLN -
   LEU                                           ASP         ARG
                       120
 - GLY - SER - SER - SER - SER - LYS - ALA - PRO - PRO - PRO - SER - LEU - PRO -
                     *                                       *
   130                                          140
 - SER - PRO - SER - ARG - LEU - PRO - GLY - PRO - SER - ASP - THR - PRO - ILE -
               *                                 *
         145
 - LEU - PRO - GLN
```

FIGURE 10.1. Linear amino acid sequences of human LH and hCGβ-subunits. Substitutions found in hCG are indicated, including the unique (115–145) C-terminal extension. Shaded boxes denote ovarian membrane receptor binding regions identified by use of synthetic peptides. Asterisks above and below denote locations of side-chain carbohydrate in LH and hCG, respectively. Modified with permission from Keutmann (3).

TABLE 10.1. Binding activity of β-subunit peptides.

| Peptide | ID$_{50}$ (mM)* | |
| --- | --- | --- |
| | hLH | hCG |
| 1–15 | 260.0 | 42.0 |
| 38–57 | 32.0 | 150.0 |
| 57–72 | 110.0 | 140.0 |
| 85–97 | 9.3 | 8.3 |
| 93–101 | 260.0 | 220.0 |

*Concentration effecting half-maximal inhibition of [$^{125}$I]-labeled hCG binding.

binding regions, consistent with recent data that mutant hCG lacking this sequence retains activity (25, 26). Three of the active sites, (38–57), (62–72), and (93–100), fall within intercysteine loops, while the other two—including the most active and recently identified fragment (85–97)—contain internal cysteines involved in disulfide bridging to other subunit regions. The close association of receptor binding sites with subunit contact sites is increasingly apparent, as discussed later in this chapter and in Dr. Puett's chapter (see Chapter 12).

Basic residues are prominent binding elements among peptide hormones in general (27), and their requirement for the activity of intact gonadotropins has been appreciated for some time (17–19, 28). It is thus not surprising that these are frequent (Fig. 10.2) and important for activity in most, if not all, of the hLH/hCGβ binding sites. A classical example is the "determinant loop" (93–100) region, in which the critical nature of Arg-93 and -94 in LH/hCG binding and specificity was first proposed by Ward and colleagues (19) and substantiated in assays using both synthetic analogs (23) and hormone modified by site-directed muta-

(1—15):    S **K** E P L **R** P **R** C **R** P T N A T

(38—57):    C P T M T **R** V L Q G V L P A L P Q V V C

(62—72):    V **R** F E S I **R** L P G C

(85—97):    A L S C Q C A L C **R** **R** S T

(93—100):    C **R** **R** S T T D C

FIGURE 10.2. Amino acid sequences of hCGβ-subunit binding regions, with basic residues (K = Lys, R = Arg) highlighted in bold type. The segment (85–101) is divided into two separate subdomains, each dependent for activity on essential arginines at positions 94 and 95.

genesis (11, 12). Arg-93 also promotes the activity of the overlapping (85–97) region (Morbeck, Roche, Keutmann, McCormick, see Appendix, this volume), in which it represents a component of the thioredoxin active-center homology (29) that may be capable of covalently linking the receptor, as described further below.

The singular importance of Arg-43 within the (38–57) intercysteine loop segment is emphasized by results from both peptide synthesis and site-directed mutagenesis. In the synthetic peptide, replacement by either a neutral (Ala) or negative (Asp) residue abolishes receptor binding (2). Point mutation of Arg-43 to Leu in expressed whole hCG reduces binding by 10-fold (11); retention of partial activity can be attributed to other binding sites, as the "double mutant" with the additional replacement of Arg-94 by Leu was devoid of activity. As discussed below, Arg-43 is a component of the hydrophilic face of the N-terminal amphipathic helix contained within the (38–57) region (30). An essential Arg appears in a similar configuration in the C-terminal helix of the G-protein α-chain (31). In the case of the hLH/hCGβ-(1–15) sequence, multiple basic residues (Fig. 10.1) appear critical for activity; their replacement by substitution or elimination through truncation results in an incremental decrease in binding activity (24). The lower activity of the native hLHβ-(1–15) peptide appears mainly attributable to substitution of Arg-10 by His. In fact, these residues are so influential that a skeleton hCGβ-(1–15) fragment, comprised of a glycine backbone with only the four basic residues ($Lys_2, Arg_{6,8,10}$) and the Cys at position 9, was comparable in binding activity to the native peptide (Keutmann, Zschunke, unpublished observations).

The mechanism by which basic residues influence hormone action is not certain, but charge interaction with one or more negatively charged cell-surface elements has been proposed (27), and there are numerous potential sites for salt-bridge formation in the LH/hCG receptor (32, 33). Basic residues are less pervasive among the binding peptides reported from FSH and TSH, but appear important in at least one region from each hormone (34, 35). Thus, while affinity for LH/hCG receptors is clearly promoted by these residues, specificity among the hormones undoubtedly also depends on additional primary and secondary structural features.

Cysteine residues form the extensive subunit tertiary structure through disulfide formation, but it has become evident that they may also play a role in binding as primary structural elements. This is suggested by the decrease in activity consequent to isosteric replacement of Cys in most of the active peptides. Most striking is the complete loss of activity upon substitution of Ala or Ser for the single Cys at position 9 in hCGβ-(1–15), summarized in Figure 10.3. Restoring Cys to any position within the peptide is sufficient to reestablish partial activity. Decreased activity in the loop peptides (93–101) and (38–57) after substitution for Cys may

| PEPTIDE/ANALOG | BINDING CONSTANT ($M \times 10^{-5}$) | RELATIVE POTENCY |
|---|---|---|
| 1                                8  9                    14<br>SER---------------------ARG CYS--------ALA----<br>(NATIVE SEQUENCE) | 4.21 | 1.0 |
| ----------------------------SER--------------- | >500.* | <.01 |
| ----------------------------ALA--------------- | >500.* | <.01 |
| ---------------------------CYS ALA--------------- | 115. | .03 |
| CYS----------------------ALA--------------- | 75. | .06 |
| ----------------------------ALA--------CYS---- | 28. | .15 |

*NO INHIBITION AT $5 \times 10^{-3}$ M PEPTIDE.

FIGURE 10.3. Analogs of hCGβ-(1–15) showing importance of Cys-9 to binding activity. Partial activity is restored by insertion of Cys at other locations in the peptide.

represent distortion or disruption of loop formation, but also may represent a primary structural requirement for this residue. Replacement of one or more cysteines in the peptide β-(81–95) also diminishes activity (Morbeck, Roche, Keutmann, McCormick, see Appendix, this volume). Cysteine, grouped together with a basic residue as seen in Figure 10.2, is an essential component of the thioredoxin-like motif described by Boniface and Reichert (29) and Grasso et al. (36) in peptides from this region of both LH and FSH. These investigators have suggested the possibility that peptide-receptor disulfide formation or disulfide rearrangement within the receptor, mediated by disulfide isomerase activity in the thioredoxin-like sequence, may accompany binding and/or post-receptor activation. Cysteines are numerous, and fairly well conserved, among glycoprotein hormone receptors, although their relative contribution to ligand binding versus membrane expression has yet to be defined (37).

Hydrophobic residues, typically envisioned as components of the protein interior, appear far less prominent as binding determinants. Their potential role in activity has also been more difficult to assess, however, as (with the exception of tyrosine) the hydrophobic side chains are not readily amenable to chemical modification. Synthetic peptides have thus proven especially useful in probing their importance. Influential hydrophobic residues are found in two of the β-subunit binding peptides. They are abundant in the (38–57) loop sequence, including several as components of an amphipathic helix, and may be more involved in interactions within the loop or with regions elsewhere in the subunit than in direct receptor contact, as described further below. One important hydrophobic residue appears to be Leu-86 in the (85–97) region. Its replacement by Ala in the peptide hCGβ-(81–95) diminishes activity 3-fold (Morbeck, Roche, Keutmann, McCormick, see Appendix, this volume), and we have found the truncated peptide (88–97) to have only 5% of the activity of (85–97). Among the other hormones, the active hTSHβ peptide (71–85) is rich in hydrophobic residues, and the paired tyrosines at the C-terminus of alpha have long been recognized to be essential to activity (16–21).

Secondary structure is another characteristic that varies widely among peptides. In general, ordered structure has not been considered prominent in glycoprotein hormone subunits due to constraints imposed by the multiple disulfide bridges. Indeed, removal of the disulfides by reduction and carboxymethylation enhances ordered structure significantly (38), and its true extent in the intact subunit will be an important objective for future studies of three-dimensional structure. Among the active regions, the (1–15) fragment is strongly predicted by several algorithms to have a β-turn about Pro-7, but its significance to receptor binding is questionable since replacement of this proline by glycine in the generic peptide mentioned above is tolerated. The (93–100) fragment is entirely random. Other fragments (such as alpha 26–46) show evidence for inducible ordered structure, but their ability to form it in the whole hormone may be inhibited by extensive disulfide linking. Secondary structure, including a helical segment, is predicted (39) for the intercysteine region (9–23), but peptides replicating this region are inactive.

Among the active regions of hLH/hCG, secondary structure is most prominent in the intercysteine (38–57) loop sequence. The properties of this region are of special interest, with its biological activity in the natural hormone (40–42). Our earlier studies of peptides representing this sequence using circular dichroism and molecular modeling (2, 43) had predicted an amphipathic helix in the N-terminal portion of the peptide, flanked by turns about several proline residues. We have recently confirmed this by *two-dimensional nuclear magnetic resonance* (2D NMR) spectroscopy (30), using an analog with hydrophilic extensions at either terminus to enhance solubility at the higher concentrations needed for NMR (Fig. 10.4). The peptide is disordered in aqueous solution, but when introduced into the lipophilic solvent *trifluoroethanol* (TFE), inter-residue contacts characteristic of helix are observed between residues 41 and 48, as shown schematically in Figure 10.5.

The involvement of induced helical structure in binding by β-(38–57) was suggested by several structure-activity studies using peptide analogs (2). We have documented formally the role of this structure using circular-dichroic and 2D NMR spectroscopy of analogs substituting a hydrophilic residue (Glu) at positions adjacent in the linear sequence, but falling on opposite surfaces of the amphipathic helix (Fig. 10.3). The analog sub-stituted on the hydrophilic face ($Gln_{46} \rightarrow Glu$) was identical to the native sequence in binding activity and in its content of ordered structure by

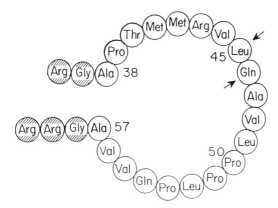

FIGURE 10.4. Structure of hLHβ-(38–57) analog used for NMR spectroscopy. Arginine extensions (shaded residues) were provided to enhance solubility, and Ala replaced Cys to eliminate dimer formation at the 5 mM concentrations employed. Arrows denote locations of Glu substitution (positions 45 and 46) in analogs designed to test importance of amphipathic-helical structure (see text). Reprinted with permission from Keutmann, Hua, and Weiss (30), © The Endocrine Society, 1992.

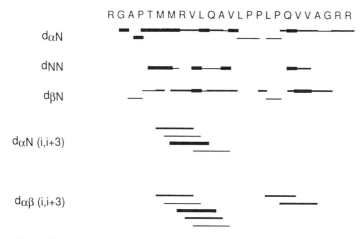

FIGURE 10.5. Common sequential and helical interresidue contacts (NOEs) found in NMR spectra of hLHβ-(38–57), designated by horizontal bars. Sequence of the peptide appears at top. Sequential amide/amide connectivities (second row) and medium-range (i, i + 3) contacts (bottom two rows) are characteristic of helical structure. Line thickness indicates relative intensity of the contacts. Reprinted with permission from Keutmann, Hua, and Weiss (30), © The Endocrine Society, 1992.

circular dichroism and NMR. The analog (Leu$_{45}$ → Glu) substituted on the hydrophobic face was devoid of binding activity, and we observed markedly diminished helical interactions in both the circular dichroic and NMR spectra. This is especially striking since Glu itself is highly prevalent in helical structures.

Conversely, enhancement of helical structure may be associated with increased binding activity. Among analogs with higher affinity than the native (38–57) peptide is that replacing Pro-53 with the nonnatural amino acid, *amino-isobutyric acid* (AIB); rotational motion about the peptide backbone at this position is restricted by the extra alpha-methyl group in AIB. Increased helix content was evident in the circular dichroic spectra (43). By 2D NMR, the boundaries of the N-terminal helix were similar to the wild-type peptide, but the number and intensity of the N-terminal helix were similar to the wild-type peptide, but the number and intensity of characteristic contacts was increased (Hua, Weiss, Keutmann, unpublished observations). Additional helical interactions were also observed in the C-terminal region. Thus, the AIB residue appears to stabilize the helix and may confer broader effects on ordered structure that provide a more favorable conformation for receptor binding.

The contrasting structural consequences of these two mutations confirm the importance of the helix in (38–57) for receptor binding. It may also contribute to specificity; the loop sequences corresponding to (38–57) in hFSHβ and hTSHβ are also active in their homologous assay systems, but lack significant induced helical structure (8, 35). The sequence in FSHβ (residues 33–53) has been shown by NMR to contain extensive β sheet structure instead (35).

The correlation between TFE-induced helical structure found in peptides with that in corresponding sequences in the holoprotein is now well established (44). It is likely, therefore, that the helix in hLHβ-(38–57) is found in the native subunit and hormone as well. This type of structure has been implicated in the action of many smaller hormones and bioactive peptides, in which inducible helix occupies a significant portion of the sequence. This may facilitate approach to the lipophilic membrane environment as an initial step in receptor binding. Amphipathic structure in hLH/hCGβ is perhaps more appropriately compared with a larger protein, such as the G-protein α-subunit (31) or interleukin-6, in which a short amphipathic-helical sequence at the C-terminus of the 173-residue molecule is essential for binding to hepatic cells (45). In this context, a question yet to be answered is whether the β-(38–57) helix is induced only on contact with membrane, as proposed for the shorter peptide hormones, or whether it is stabilized within the subunit or hormone as a preformed recognition element. The latter would occur through interaction of the hydrophobic surface of the helix with hydrophobic regions elsewhere on either subunit. Either mechanism could be envisioned to orient the hydrophilic face, including the essential Arg-43, for interaction with receptor. However, this issue illustrates the dilemma posed by LH/hCG as peptide versus protein, outlined at the outset of this chapter.

Differences in conformation and intrasubunit interactions may present different portions of the linear sequence to the surface in each of the hormones, as implied by the alignment of binding regions shown in Figure 10.6. Given the homologous location of cysteine residues and the need to accommodate the common α-subunit, it is not surprising that certain active regions coincide, such as the large intercysteine loop represented by (38–57) in LH/hCGβ, which has active counterparts in FSHβ and TSHβ (6–8, 46). The "determinant loop" corresponding to (93–100) in LH/hCGβ is also active in FSH (5), but not in TSH (7). On the other hand, the active C-terminal (101–112) sequence in hTSHβ has no counterpart in LH/hCG. Its place may be taken in hLH/hCGβ by (1–15), with which it shows some structural homology. Thus, specificity may be influenced not only by the primary and, probably, secondary structure within each binding region, but also by the orientation of binding regions along the linear subunit. In the α-subunit, TSH and LH receptor binding are mediated by the same sequences, but there are differences in the relative importance of certain residues within them (47). This may reflect more subtle differences in orientation of the alpha regions in the hormone binding domain imposed by the marked structural differences in the component beta regions.

Important to the organization of these sites is undoubtedly the location of contacts between α and β. Although these sites have proven less easily determined than receptor-binding sites, a consensus regarding many

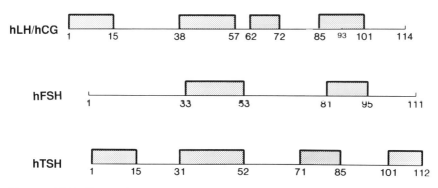

FIGURE 10.6. Comparison of receptor binding sequences (shaded regions) in linear glycoprotein hormone β-subunits, based on cumulative data from assays of synthetic peptides. Subunits are oriented with conserved cysteines in alignment. In hCGβ, the sequence (85–101) includes the two regions (85–97) and (93–101), as shown in Figure 10.2. The C-terminal (115–145) extension in hCG, devoid of binding activity, is omitted for clarity. Modified with permission from Keutmann (3).

is gradually being attained. Insights into potential regions have been provided by use of immunological probes (48, 49), inhibition of association by synthetic fragments (50, 51), and by studies comparing the results of chemical modification of the whole hormone versus separate subunits. Recently, effects of mutations to subunits have helped define more precisely the importance of specific positions to subunit association, as reported by the laboratories of Boime (15) and Puett (26) and described further in succeeding chapters. Noteworthy to us is the role of residues (37–40) in the α-subunit (15). These appear to occupy an inert segment between two active domains found in the α-(30–45) binding region (47) and may "tether" this region of α to other binding sites in β.

Chemical cross-linking has the advantage of identifying complementary sites on the two subunits, as shown initially in bovine LH by Weare and Reichert (52) and more recently in porcine LH by van Dijk and Ward (53). These studies showed coupling of the lysine at 49 in α (corresponding to Lys-45 in human α) with Asp-111 in β. We have used the photoreactive cross-linking reagent *l-benzoylphenylalanine* (Bpa) to investigate the association of the β-(1–15) sequence with α (54). A peptide from this region had previously been shown by Salesse et al. (50) to inhibit subunit association. The aromatic tryptophan residue at position 8 in hLHβ-(1–15) provided us with an optimal location for substitution by [$^3$H]-Bpa during solid-phase synthesis. After photoreaction in presence of α-subunit, the labeled subunit was isolated and the site of cross-linking determined by mapping of proteolytic fragments. A single [$^3$H]-labeled chymotryptic peptide (18–33) was identified and the labeling site localized by sequence analysis to residues (29–30) (Gly-Met), again adjacent to the major α receptor binding site (30–45).

Other evidence (50, 54) suggests that the contact site in β-(1–15) resides in the C-terminal half of the peptide, but it clearly overlaps determinants for receptor binding as well. A conformational change in presence of the receptor may expose sites concealed in the unbound hormone, or a transitional state may exist between two conformations. Another β region that has frequently been implicated in subunit contact is the *CAGY* (Cys-Ala-Gly-Tyr) region, comprising residues (34–37) in hLH/hCG. It is conserved among all β-subunits, and mutant forms of hCGβ (55) and TSHβ (56) with Arg replacing Gly in the CAGY sequence have been found incapable of assembly into an active hormone. However, attempts to inhibit association using synthetic peptides containing this sequence have not been successful (5, 20, 50), and a Bpa-substituted CAGY peptide does not cross-link to α (54). It is therefore possible that modifications in the CAGY region affect assocation by inducing conformational changes elsewhere in the subunit.

Whereas contact and binding sites had often been visualized as quite separate regions, closer coincidence or even overlap between the two is evident from the studies just described. This is reasonable if the multiple binding sites on both subunits are to be brought into close approximation for cooperative interaction with the receptor. An outstanding challenge remains to assemble these regions into a simplified "synthetic hormone" with full bioactivity. Results to date suggest that this may not be easily achieved. Typically, dimers or oligomers of different component binding regions have proven, at most, additive in binding activity. For example, the peptides hCGβ-(1–15) + β-(38–57) or α-(26–46) + β-(38–57), joined by varying numbers of glycines or carbons (e.g., ε-aminocaproic acid), bind with $ID_{50}$s slightly higher than the most active peptide alone. In a detailed study of coupled FSH peptides by Santa Coloma et al. (46), contributions of each sequence to binding was documented, but the affinities of the heterodimers were well below theoretical predictions based on the binding constants of the individual peptides. One recent report (57) describes the assembly of three binding regions from both FSH subunits into a molecule with significantly enhanced affinity, using a homologous protein (thioredoxin) as a form of template. Clearly, then, the correct orientation of the respective binding components presents a major obstacle, the solution for which will require a model that accounts for interactions within the subunit as well as surfaces facing the receptor. This may become possible only as efforts progress to provide a complete three-dimensional structure of the glycoprotein hormone molecule.

*Acknowledgments.* This work represents a collaborative effort among members of the Endocrine Unit of Massachusetts General Hospital, the Department of Biochemistry and Molecular Phamacology of Harvard Medical School, and the Department of Biochemistry and Molecular Biology of Mayo Medical School. Participating colleagues are recognized with appreciation: Dr. Robert J. Ryan, Dr. Patrick C. Roche, Dr. Cris Charlesworth, Dr. D.J. McCormick, Michael Zschunke, Kathleen Kitzmann, and Elizabeth Bergert at Mayo Medical School, Drs. Qing-Xin Hua and Michael Weiss at Harvard Medical School, and Kathleen A. Mason, Teofila Ostrea, Leslie Johnson, Xiang-Chen Huang, and David Rubin at Massachusetts General Hospital. This work was supported by Grants HD-09140, HD-12851, and HD-28138 from the National Institutes of Health.

# References

1. Dyson HJ, Wright PE. Defining solution conformations of small linear peptides. Annu Rev Biophys Biophys Chem 1991;20:519–38.
2. Keutmann HT, Charlesworth MC, Kitzmann K, Mason KA, Johnson L, Ryan RJ. Primary and secondary structural determinants in the receptor binding sequence β-(38–57) from human luteinizing hormone. Biochemistry 1988;27:8939–44.

3. Keutmann HT. Receptor-binding regions in human glycoprotein hormones. Mol Cell Endocrinol 1992;86:C1–6.
4. Charlesworth MC, McCormack DJ, Madden B, Ryan RJ. Inhibition of human choriogonadotropin binding to receptor by human choriogonadotropin alpha peptides: a comprehensive synthetic approach. J Biol Chem 1987; 262:13409–16.
5. Santa Coloma TA, Reichert LE Jr. Identification of a follicle-stimulating hormone receptor-binding region in hFSH-beta-(81–95) using synthetic peptides. J Biol Chem 1990;265:5037–42.
6. Santa Coloma TA, Crabb JW, Reichert LE Jr. A synthetic peptide encompassing two discontinuous regions of hFSH-β subunit mimics the receptor binding surface of the hormone. Mol Cell Endocrinol 1991;78:197–204.
7. Morris JC, McCormick DJ, Ryan RJ. Inhibition of thyrotropin binding to receptor by synthetic human thyrotropin beta peptides. J Biol Chem 1990; 265:1881–4.
8. Freeman SL, McCormick DJ, Ryan RJ, Morris JC. Inhibition of TSH bioactivity by synthetic beta TSH peptides. Endocr Res 1992;18:1–17.
9. Cunningham BC, Jhurani P, Ng P, Wells JA. Receptor and antibody epitopes in human growth hormone identified by homolog-scanning mutagenesis. Science 1989;243:1330–6.
10. Mollison KW, Mandecki W, Zuiderweg ERP, et al. Identification of receptor-binding residues in the inflammatory complement protein C5a by site-directed mutagenesis. Proc Natl Acad Sci USA 1989;86:292–6.
11. Chen F, Puett D. Contributions of arginines-43 and -94 of human choriogonadotropin-beta to receptor binding and activation as determined by oligonucleotide-based mutagenesis. Biochemistry 1991;30:10171–5.
12. Chen F, Wang Y, Puett D. Role of the invariant aspartic acid 99 of human choriogonadotropin β in receptor binding and biological activity. J Biol Chem 1991;266:19357–61.
13. Campbell RK, Dean Emig DM, Moyle WR. Conversion of human choriogonadotropin into a follitropin by protein engineering. Proc Natl Acad Sci USA 1991;88:760–4.
14. Yoo J, Ji I, Ji T. Conversion of lysine-91 to methionine or glutamic acid in human choriogonadotropin alpha results in the loss of cAMP inducibility. J Biol Chem 1991;266:17741–3.
15. Bielinska M, Boime I. Site-directed mutagenesis defines a domain in the gonadotropin alpha-subunit required for assembly with the chorionic gonadotropin B-subunit. Mol Endocrinol 1992;6:267–71.
16. Chen F, Wang Y, Puett D. The carboxy-terminal region of the glycoprotein hormone alpha subunit: contributions to receptor binding and signaling in human chorionic gonadotropin. Mol Endocrinol 1992;6:914–9.
17. Pierce JG, Parsons TF. Glycoprotein hormones: structure and function. Annu Rev Biochem 1981;50:465–95.
18. Sairam MR. Gonadotropic hormones: relationship between structure and function with emphasis on antagonists. In: Li CH, ed. Hormonal proteins and peptides. New York: Academic Press, 1983:1–79.
19. Gordon WL, Ward DN. Structural aspects of luteinizing hormone actions. In: Ascoli M, ed. Luteinizing hormone receptors and action. Paris: CRC Press, 1985:173–98.

20. Ryan RJ, Keutmann HT, Charlesworth MC, et al. Structure and function of the glycoprotein hormones. Recent Prog Horm Res 1987;43:383–429.
21. Combarnous Y. Molecular basis of the specificity of binding of glycoprotein hormones to their receptors. Endocr Rev 1992;13:670–91.
22. Keutmann HT, Charlesworth MC, Mason KA, Ostrea T, Johnson L, Ryan RJ. A receptor-binding region in human choriogonadotropin/lutropin beta subunit. Proc Natl Acad Sci USA 1987;84:2038–41.
23. Keutmann HT, Mason KA, Kitzmann K, Ryan RJ. Role of the beta-(93–100) determinant loop sequence in receptor binding and biological activity of human luteinizing hormone and chorionic gonadotropin. Mol Endocrinol 1989;3:526–31.
24. Keutmann HT, Rubin DA, Mason KA, Kitzmann K, Zschunke M, Ryan RJ. Receptor binding and subunit interaction by the N-terminal (1–15) region of LH/hCG β subunit shown by use of synthetic peptides. In: Smith JA, Rivier JE, eds. Peptides: chemistry and biology. Leiden: ESCOM, 1992: 74–6.
25. Matzuk MM, Hsueh AJW, LaPolt P, Tsafriri A, Keene JL, Boime I. The biological role of the carboxyl-terminal extension of human chorionic gonadotropin beta subunit. Endocrinology 1990;126:376–83.
26. Chen F, Puett D. Delineation via site-directed mutagenesis of the carboxyl-terminal region of human choriogonadotropin beta required for subunit assembly and biological activity. J Biol Chem 1991;266:6904–8.
27. Dohlman JG, Loff HD, Segrest JP. Charge distributions and amphipathicity of receptor-binding alpha-helices. Mol Immunol 1990;27:1009–20.
28. Sairam M. Role of arginine residues in ovine lutropin: reversible modification by 1,2-cyclohexanedione. Arch Biochem Biophys 1976;176:197–205.
29. Boniface JJ, Reichert LE Jr. Evidence for a novel thioredoxin-like catalytic property of gonadotropic hormones. Science 1990;247:61–4.
30. Keutmann HT, Hua QX, Weiss MA. Structure of a receptor-binding fragment from human luteinizing hormone β-subunit determined by [$^1$H]- and [$^{15}$N]-nuclear magnetic resonance spectroscopy. Mol Endocrinol 1992;6:904–13.
31. Sullivan KA, Miller RT, Masters SB, Beiderman B, Heideman W, Bourne HR. Identification of receptor contact site involved in receptor-G protein coupling. Nature 1987;330:758–60.
32. Koo YB, Ji I, Slaughter RG, Ji T. Structure of the luteinizing hormone receptor gene and multiple exons of the coding sequence. Endocrinology 1991;128:2297–308.
33. Roche PC, Ryan RJ, McCormick DJ. Identification of hormone-binding regions of the luteinizing hormone/human chorionic gonadotropin receptor using synthetic peptides. Endocrinology 1992;131:268–74.
34. Leinung MC, Bergert ER, McCormick DJ, Morris JC. Synthetic analogs of the carboxyl-terminus of β-thyrotropin: the importance of basic amino acids in receptor binding activity. Biochemistry 1992;31:10094–8.
35. Agris PF, Guenther RH, Sierzputowska GH, et al. Solution structure of a synthetic peptide corresponding to a receptor binding region of FSH (hFSH-beta 33–53). J Protein Chem 1992;11:495–507.
36. Grasso P, Santa Coloma TA, Reichert LE Jr. Synthetic peptides corresponding to human follicle-stimulating hormone (hFSH)-β-(1–15) and hFSH-β-(51–65) induce uptake of $^{45}Ca^{++}$ by liposomes: evidence for calcium-

conducting transmembrane channel formation. Endocrinology 1991;128: 2745–51.

37. Wadsworth HL, Russo D, Nagayama Y, Chazenbalk GD, Rapoport B. Studies on the role of amino acids 38–45 in the expression of a functional thyrotropin receptor. Mol Endocrinol 1992;6:394–8.

38. Birken S, Gawinowicz-Kolks MA, Amr S, Nisula B, Puett D. Tryptic digestion of the alpha subunit of human chorionic gonadotropin. J Biol Chem 1986; 261:10719–27.

39. Garnier J. Molecular aspects of the subunit assembly of glycoprotein hormones. In: McKerns KW, ed. Structure and function of the gonadotropins. New York: Plenum Press, 1978:381–414.

40. Bousfield GR, Ward DN. Selective proteolysis of ovine lutropin or its beta subunit by endoproteinase Arg-C. J Biol Chem 1988;263:12602–7.

41. Cole LA, Kardana A, Andrade-Gordon P, et al. The heterogeneity of human chorionic gonadotropin (hCG), III. The occurrence and biological and immunological activities of nicked hCG. Endocrinology 1991;129:1559–67.

42. Weiss J, Axelrod L, Whitcomb RW, Harris PE, Crowley WF, Jameson JL. Hypogonadism caused by a single amino acid substitution in the β-subunit of luteinizing hormone. N Engl J Med 1992;326:179–83.

43. Milius RP, Keutmann HT, Ryan RJ. Molecular modeling of residues 38–57 of the beta-subunit of human lutropin. Mol Endocrinol 1990;4:859–68.

44. Dyson HJ, Merutka G, Waltho JP, Lerner RA, Wright PE. Folding of peptide fragments comprising the complete sequence of proteins: models for initiation of protein folding, I. Myohemerythrin. J Mol Biol 1992; 226:795–817.

45. Leebeek FWG, Kariya K, Schwabe M, Fowlkes DM. Identification of a receptor binding site in the carboxyl terminus of human interleukin-6. J Biol Chem 1992;267:14832–8.

46. Santa Coloma TA, Dattareyamurty B, Reichert LE Jr. A synthetic peptide corresponding to human FSH-beta subunit 33–53 binds to FSH receptor, stimulates basal estradiol biosynthesis, and is a partial antagonist of FSH. Biochemistry 1990;29:1194–200.

47. Reed DK, Ryan RJ, McCormick DJ. Residues in the alpha subunit of human choriogonadotropin that are important for interaction with the lutropin receptor. J Biol Chem 1991;266:14251–5.

48. Bidart JM, Troalen F, Bousfield GR, Birken S, Bellet DH. Antigenic determinants on human choriogonadotropin alpha-subunit, I. Characterization of topographic sites recognized by monoclonal antibodies. J Biol Chem 1988;263:10364–9.

49. Weiner RS, Andersen TT, Dias JA. Topographic analysis of the alpha-subunit of human follicle-stimulating hormone using site-specific antipeptide antisera. Endocrinology 1990;127:573–9.

50. Salesse R, Bidart JM, Troalen F, Bellet D, Garnier J. Peptide mapping of intersubunit and receptor interactions in human choriogonadotropin. Mol Cell Endocrinol 1990;68:113–9.

51. Krystek SR Jr, Dias JA, Andersen TT. Identification of subunit contact sites on the alpha-subunit of lutropin. Biochemistry 1991;30:1858–64.

52. Weare JA, Reichert LE Jr. Studies with carbodiimide-cross-linked derivatives of bovine lutropin, II. Location of the cross-link and implication for interaction with the receptors in testes. J Biol Chem 1979;254:6972–9.

53. van Dijk S, Ward DN. Chemical cross-linking of porcine luteinizing hormone: location of the cross-link and consequences for stability and biological activity. Endocrinology 1993;132:534–8.
54. Keutmann HT, Rubin DA. A subunit interaction site in human luteinizing hormone: identification by photoaffinity cross-linking. Endocrinology 1993; 132:1305–12.
55. Azuma C, Miyai K, Saji F, et al. Site-specific metagenesis of human chorionic gonadotrophin (hCG)-β subunit: influence of mutation on hCG production. J Mol Endocrinol 1990;5:97–102.
56. Hayashizaki Y, Hiroaka Y, Endo Y, Matsubara K. Thyroid-stimulating hormone (TSH) deficiency caused by a single base substitution in the CAGYC region of the beta subunit. EMBO J 1989;8:2291–6.
57. Hage van Noort M, Puijk WC, Plasman HH, et al. Synthetic peptides based upon a three-dimensional model for the receptor recognition site of follicle-stimulating hormone exhibit antagonistic or agonistic activity at low concentrations. Proc Natl Acad Sci USA 1992;89:3922–6.

# 11

# Evidence for Altered Splicing in Two Members of the Chorionic Gonadotropin β Gene Cluster

Masaki Bo, Brian L. Strauss, and Irving Boime

Human *chorionic gonadotropin* (CG) is a placental hormone that is required for the maintenance of pregnancy. It stimulates the corpus luteum to produce steroid hormones during the *first trimester* (FT). CG is a member of the glycoprotein hormone family that includes the pituitary hormones *lutropin* (LH), follitropin, and thyrotropin and consists of an α-β heterodimer. The α-subunit is encoded by a single gene (1) and the β-subunit is encoded by a cluster of six genes (or pseudogenes [2, 3]) (see Fig. 11.1A). Linked to this cluster is the single gene encoding the related LHβ-subunit.

Although the general structure and sequences of the CGβ genes have been reported, it is unclear which CGβ genes are expressed in vivo and whether there is a preferential utilization of one or more unique genes during gestation and tumorigenesis. Talmadge et al. (4) transfected heterologous cells with genomic clones linked to an SV40 promoter and also compared restriction polymorphisms of cDNA clones from placental RNA. They concluded that potentially only three genes (CGβ genes 1, 3, and 5) could be expressed in the placenta. Otani et al. (5) showed that genomic subclones of four of the genes (CGβ genes 3, 5, 7, and 8) with their natural promoters could be expressed in a mouse adrenal cell line. No expression was detected for CGβ genes 1 and 2, which presumably reflected the presence of a noncanonical splice site at the exon I/intron I boundary. In the same study CAT constructs bearing either the CGβ5 or CGβ7 promoters were transfected into choriocarcinoma cells, and only the CGβ5 construct was active. These studies however, did not address which of the genes were expressed in vivo. It was not possible to distinguish the expression of particular genes using conventional techniques because the translated sequences and predicted mRNA sizes are virtually

**A**

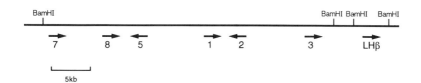

**B**

**C**

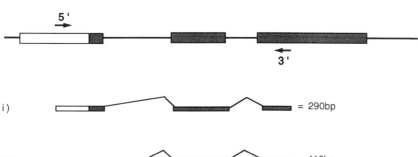

FIGURE 11.1. Splicing of CGβ1/2 RNAs. *A:* Map of the CGβ gene cluster. Number assignments are based on Boorstein, Vamvakopoulos, and Fiddes (3). *B.* Diagram of the general organization of CGβ genes. The open and stippled bars indicate the untranslated and translated exons, respectively; CAP and ATG correspond to the transcription and translation start sites, respectively. The splicing donor site for the first intron is represented as GAA. This site is mutated in β1 and β2 (shown as T; the alternate splice donor sites in β1 and β2 are represented by an asterisk. *C:* PCR amplification of the CGβ1 and 2 genes. Oligonucleotide primers used for PCR are shown above and below as arrows. Below the CGβ gene are the amplified regions in the RT-PCR products, denoted by the bars: (i) Expected PCR product if altered splice site in CGβ genes 1/2 was used; (ii) PCR products generated from selection of alternative splice sites in CGβ genes 1/2. Sizes of these products in base pairs is given at right. Stippled regions correspond to the open reading frames of the transcripts.

identical for CGβ3, 5, 7, and 8. While there are differences in the 5′ nontranslated sequences of the first exon (2, 6, 7), Northern blotting with oligonucleotide probes containing a few base pair mismatches would lack the specificity to distinguish and quantitate the unique mRNA species. To address this issue, we employed the *reverse transcription-polymerase chain reaction* (RT-PCR) technique with oligonucleotide primers containing gene-specific sequences. These primers were used to generate unique cDNAs from placental and choriocarcinoma mRNAs by RT-PCR. The data from such experiments confirm the transfection assay of CGβ subclones (5). The level of expression was β5 > β3 = β8 > β7, β1/2.

Our data clarified another issue regarding the organization of the cluster. Based on genomic clones, it was previously reported that the CGβ cluster comprised seven genes: The additional gene, designated CGβ6, had a different 5′ translated sequence in exon I (3, 7). Although several subsequent studies indicated only a six-gene cluster, this difference was not resolved. Clones were obtained during amplification of CGβ gene 7 cDNA that had the CGβ6 sequence. Our data suggest that CGβ6 is an allele of gene 7 with differences in the 5′ nontranslated sequence.

As shown previously, the sequence of CGβ genes 1 and 2 diverges from that of the other CGβ genes 10 base pair (bp) upstream from the initiation of methionine codon in exon I. Both contain an upstream insert and have a single base change at the conserved donor splice site of the first intron. The inserted sequence, which occurred during gene duplication, may affect promoter activity. The lack of expression of these genes using transfection assay implicated them as pseudogenes. However, RT-PCR detected transcripts of 244, 410, and 420 bp from CGβ1/2 in the placental polysomal RNA pool; these RNAs arose by use of alternative splice sites (8) (Fig. 11.1B, Fig. 11.1C). Cloning and sequencing of these amplified cDNAs showed that the 244-bp fragment was spliced 47 bp in the 5′ direction from the nonconsensus donor site (8). The donor sites of the 410- and 420-bp cDNA were 119 and 129 bp in the 3′ direction, respectively, from the modified splice site; the acceptor site is the same as for the other CGβ transcripts. As seen in the case for the β globin gene in thalasemia patients, cryptic splice sites can be utilized, albeit much less efficiently (9).

The sequence of the transcript corresponding to the 410-bp PCR product corresponds to an open reading frame of 30 amino acids, beginning at the CGβ initiation methionine and extending into sequence that is intron I for CGβ5 (4); it therefore would differ substantially from the CGβ protein. Although this peptide has the same signal sequence as β5, a 30-amino acid peptide cannot enter the endoplasmic reticulum and presumably will not be secreted. The transcript corresponding to the 244-bp PCR product could encode a 140-amino acid protein, also different from CGβ. A search of protein data bases revealed no entries with significant homology to the open reading frames of the two CGβ1,2 transcripts. The functional role for these β1,2 transcripts is not known, nor is the extent of their translation, although β1,2 transcripts in human trophoblast are associated with polysomes (7).

Thus at least five, and possibly all, of the genes of the CGβ family are active in vivo, but three of the CGβ genes are expressed preferentially. Although the role for CGβ multigenic expression is not clear, the promoter activities of CGβ genes 3, 5, 7, and 8 are different. Thus, there is preferential transcription of these genes during pregnancy and in tumorigenesis. The variation in promoter structure may account for the presence of CG transcripts in nontrophoblast tissue, such as the pituitary.

## References

1. Boothby M, Ruddon R, Anderson C, McWilliams D, Boime I. A single gonadotropin α-subunit gene in normal tissue and tumor-derived cell lines. J Biol Chem 1981;256:5121–7.
2. Policastro P, Daniels-McQueen S, Carle G, Boime I. A map of the hCG-β-LHβ gene cluster. J Biol Chem 1986;261:5907–16.
3. Boorstein WR, Vamvakopoulos NC, Fiddes JC. Human chorionic gonadotropin β-subunit is encoded by at least eight genes arranged in tandem, inverted pairs. Nature 1982;300:419–22.
4. Talmadge K, Boorstein WR, Vamvakopoulos NC, Gething M-J, Fiddes JC. Only three of the seven human chorionic gonadotropin beta subunit genes can be expressed in the placenta. Nucleic Acids Res 1984;12:8415–36.
5. Otani T, Otani F, Krych M, Chaplin DD, Boime I. Identification of a promoter region in the CG beta gene cluster. J Biol Chem 1988;263:7322–9.
6. Policastro P, Ovitt CE, Hoshina M, Fukuoka H, Boothby MR, Boime I. The β-subunit of human chorionic gonadotropin is encoded by multiple genes. J Biol Chem 1983;258:11492–9.
7. Talmadge K, Vamvakopoulos NC, Fiddes JC. Evolution of the genes for the beta subunits of human chorionic gonadotropin, luteinizing hormone. Nature 1984;307:37–40.
8. Bo M, Boime I. Identification of the transcriptionally active genes of the chorionic gonadotropin beta gene cluster in vivo. J Biol Chem 1992;267: 3179–84.
9. Treisman R, Orkin S, Maniatis T. Nature 1983;302:591–6.

# 12

# Delineation of Subunit and Receptor Contact Sites by Site-Directed Mutagenesis of hCGβ

DAVID PUETT, JIANING HUANG, AND HAIYING XIA

Considerable data have been amassed on the four glycoprotein hormones, *chorionic gonadotropin* (CG), *luteinizing hormone* (LH), *follicle stimulating hormone* (FSH), and *thyroid stimulating hormone* (TSH) (1). It is well established that, within a species, the single α-subunit binds to distinct β-subunits to form bioactive holoproteins, which exhibit specificity for three G protein-coupled *receptors* (r) (2), LH/CGr (3), FSHr (4), and TSHr (5). Since the glycoprotein hormones contain a common α-subunit, and since both subunits are believed to participate in receptor binding (6, 7), interesting questions and models can be developed regarding subunit contact sites and hormone receptor contact sites, particularly those responsible for conferring specificity. The absence of a crystallographic structure for any member of the glycoprotein hormone family has, however, prevented a detailed understanding of these contact regions.

Much of the available information on regions of the β-subunit involved in subunit assembly and receptor binding has been derived from studies on chemical modifications (8, 9), synthetic peptides (10–15), enzymatic cleavages (16, 17), and in a few cases from patients where defective glycoprotein hormones could be attributed to specific mutations (18, 19). Another approach that we and others have used to delineate the role of specific amino acid residues in hormone function is that of site-directed mutagenesis. This chapter reviews some of the recent results from our laboratory on mutagenesis of hCGβ, the amino acid sequence of which is shown in Figure 12.1, and attempts to place these data in the context of findings from other laboratories using complementary approaches.

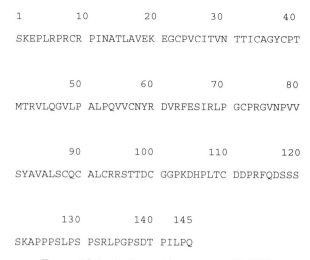

```
1          10          20          30          40

SKEPLRPRCR  PINATLAVEK  EGCPVCITVN  TTICAGYCPT


          50          60          70          80

MTRVLQGVLP  ALPQVVCNYR  DVRFESIRLP  GCPRGVNPVV


          90         100         110         120

SYAVALSCQC  ALCRRSTTDC  GGPKDHPLTC  DDPRFQDSSS


         130         140  145

SKAPPPSLPS  PSRLPGPSDT  PILPQ
```

FIGURE 12.1. Amino acid sequence of hCGβ.

## Experimental Procedures

The cDNA for hCGβ was kindly provided by Dr. John Fiddes (California Biotechnology, Inc., Mountain View, CA). A number of mutagenesis techniques were used, including those approaches inherent in the Bio-Rad Laboratories (Richmond, CA) Muta-gene M13 in vitro mutagenesis kit (20), the CLONTECH Laboratories, Inc. (Palo Alto, CA) Transformer site-directed mutagenesis kit (21), and PCR (22). The mutant and wild-type cDNAs were subcloned into a Prsv expression vector (23), which was transiently transfected (24) into CHO cells containing a stably integrated gene for bovine α. The cells were kindly provided by Dr. John Nilson (Case Western Reserve University, Cleveland, OH) and have been described (25), as have details on cell maintenance and collection of the medium (26). Two radioimmunoassays were used to measure the concentrations of total expressed β and heterodimer in the medium, from which an estimate could be made of holoprotein formation (23, 26, 27). Relative receptor affinities were determined with MA-10 cells, kindly provided by Dr. Mario Ascoli (University of Iowa, Iowa City, IA) and were based on competitive binding with [125I]hCG as described elsewhere (26). The results are reported as mean ± SEM and for comparative purposes are normalized to 100%, that is, total binding. The binding affinities of the mutual hormones to the LH/CGr are relative to bovine α-hCGβ wild type, which was included in all assays.

# Results

## *N-Terminal and C-Terminal Deletion Mutants of hCGβ*

In order to identify a minimum length, bioactive core fragment of hCGβ, a series of N-terminal, C-terminal, and combined N/C-terminal deletion mutants were prepared and characterized (Table 12.1). The N-terminal fragments 1–121 (23, 28), 1–114 (29, 30), and 1–110 (31) associated with α essentially, as well as hCGβ wild type—that is, amino acid residues 1–145—and the resulting heterodimers exhibited comparable binding affinities and steroidogenic potencies to that of the wild-type hormone. The N-terminal fragment 1–100 also bound to α, but the potency of the heterodimer was reduced (26). These results suggest a role, either direct or indirect, for one or more amino acid residues between positions 101 and 110 in receptor binding. A shorter N-terminal fragment, consisting of residues 1–92, was unable to bind to α (26), thus indicating that amino acid residues 93–100 are important in chain folding or subunit binding. Interestingly, this is the region proposed by Ward et al. (32) to be a determinant loop involved in receptor binding and specificity.

The N-terminal region of the CG and LHβ-subunits contains an additional six or seven amino acid residues not present in the β-subunits of FSH and TSH (33). The C-terminal fragment 8–145 was prepared and found, somewhat surprisingly, to exhibit similar properties to hCGβ wild

TABLE 12.1. Amino- and carboxyl-terminal deletion mutants of hCGβ.

| hCGβ fragments | α binding | LH/CGr binding | Reference |
|---|---|---|---|
| 1–121 | Nor | Nor | 23, 28 |
| 1–114 | Nor | Nor | 29, 30 |
| 1–110 | Nor | Nor | 31 |
| 1–100 | Nor | Dec | 26 |
| 1–92 | Dec | —[a] | 26 |
| 8–145 | Nor | Nor | 31 |
| 8–110 | Nor | Nor | 31 |
| 8–100 | Dec | —[b] | 31 |

[a] No heterodimer was detected and no receptor binding was found.

[b] The small amount of heterodimer that formed exhibited some activity.

*Note:* hCGβ deletion mutants were prepared by site-directed mutagenesis to yield the fragments shown in the first column (hCGβ fragments). The second column (α binding) denotes whether binding of the deletion mutant to α is normal (Nor), compared to hCGβ wild type, or decreased (Dec). The third column (LH/CGr binding) indicates whether binding of the heterodimer to the receptor is normal (Nor), compared to bovine α-hCGβ wild type, or decreased (Dec).

type in α binding and LH/CGr binding by the resulting heterodimer (31). Other data, based on synthetic peptides (6, 12) and mutagenesis of Lys-2 (27), have implicated the N-terminal region in receptor binding, but clearly it is not required.

Two combined N/C-terminal deletion mutants, consisting of amino acid residues 8–145 and 8–110, were characterized and found to be similar to hCGβ wild type in α binding, while a fragment containing residues 8–100 bound weakly to α (31). Figure 12.2 shows the results of the competitive LH/CGr binding assay for the holoprotein bovine α associated with the 8–110 fragment, which is the shortest hCGβ core fragment yet defined with activity comparable to that of hCGβ wild type. From these results we can conclude that amino acid residues 1–7 and 111–145 of hCGβ are not required for α binding or for binding of the holoprotein to LH/CGr.

## Replacement of Invariant Amino Acid Residues in β-Subunits

Analysis of the known amino acid sequences of mammalian glycoprotein hormone β-subunits (33) shows that all twelve Cys residues are con-

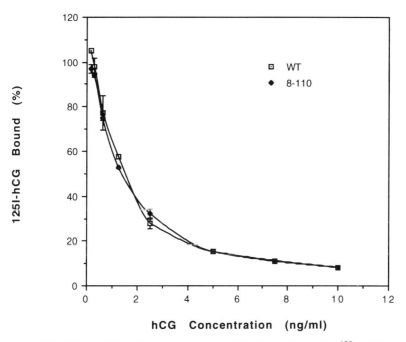

FIGURE 12.2. Competitive binding assay to MA-10 cells with [$^{125}$I]hCG and heterologous heterodimers of bovine α bound to hCGβ wild type (WT) and to a combined N/C-terminal deletion mutant of hCGβ(des[1–7, 111–145]) (8–110). Comparable results were originally published in reference 31.

served, as are eight non-Cys residues: Gly-36, Tyr-37, Gln-54, Pro-70, Gly-71, Val-84, Ala-85, and Asp-99. These invariant amino acid residues may serve as important conformational determinants, binding sites for the α-subunits, or sites involved in binding/activation of the receptor, but they obviously would not be involved in determining receptor specificity. Some data are available on these invariant amino acid residues. For example, Boime et al. (34) mutagenized the hCGβ cDNA to replace each of the Cys residues with Ala and found dramatic effects on subunit assembly. Based on replacements with Arg and Asp, Azuma et al. (35) reported, also from site-directed mutagenesis, that Gly-36 is important in subunit assembly. Earlier studies on a patient with congenital TSH deficiency also indicated an important role for this Gly (18). Tyr-37 has also been implicated in α binding from chemical modification studies (8, 9). An individual presenting with hypogonadism was found to have a point mutation in his LHβ gene at the codon for Gln-54, GAC→GGC, which led to a replacement of Gln with Arg (19). This single change was reported to have no demonstrable effect on α binding, but the heterodimer containing the mutant β-subunit was inactive. Thus, the available data suggest that many of these invariant amino acid residues are important in subunit folding and function.

We replaced Tyr-37 with Phe and with Leu and found that, relative to hCGβ wild type, α binding was not altered with hCGβ(Phe-37), but was reduced with hCGβ(Leu-37) (36). Both mutant heterodimers exhibited about the same relative affinity for LH/CGr as the wild-type hormone, but the steroidogenic potencies were somewhat less, indicating a possible modulatory role for the hydroxyl group at position 37. More recently we found that replacement of Tyr-37 with Asp essentially abolished α binding.

Replacement of Gln-54 with Arg resulted in a mutant hCGβ that formed very little heterodimer; interestingly the small amount that formed was bioactive. In contrast to the results of Weiss et al. (19), our data suggest that Gln-54 is involved in holoprotein formation, not receptor binding.

Several invariant amino acid residues were replaced with rather different groups (for example, Pro-70→Glu, Gly-71→Asp, and Val-84→Asp) with no apparent consequence on subunit assembly or receptor binding by the resulting holoproteins. Whether these particular residues alter the kinetics of chain folding or α-subunit binding remains to be determined. Of considerable interest was the finding that replacement of amino acid residues Arg-68-Leu-69-Pro-70-Gly-71 with Ala-Ala-Ala-Ala greatly reduced the total amount of the expressed mutant hCGβ. However, the small amount of heterodimer that formed was about equipotent with the wild-type hormone. (Positions 68 and 69 contain Arg and Leu in the mammalian LH/CGβ structures; the equivalents in FSHβ are Lys-or-Arg and Val-or-Leu, respectively; and in TSHβ Glu and Ile occupy these positions. Thus, the sequences of the mammalian β-subunits are of the general type: ionizable side chain-hydrophobic side chain-Pro-Gly.)

Replacement of Ala-85 with Ile reduced α binding, and replacement with Asp nearly abolished it. Interestingly, the small amount of each mutant holoprotein that formed was assayed and found to be active. These results indicate a role for Ala-85 in subunit assembly.

Asp-99 was replaced with Asn, Glu, and Arg (37). The three mutant hCGβs bound to α as effectively as hCGβ wild type, and heteroproteins containing hCGβ(Asn-99) and hCGβ(Glu-99) were active, but of reduced potency relative to hCGβ wild type. In contrast, α-hCGβ(Arg-99) was essentially inactive (37, 38). These data suggest a key role for Asp-99 in LH/CGr binding.

The available data on the invariant residues of the β-subunits are given in Table 12.2. Some of these appear to function in α binding, while others are more important in receptor binding.

## Replacement of Conserved Amino Acid Residues in LH/CGβ-Subunits

A number of amino acid residues in the known mammalian LH/CGβ-subunits are invariant or highly conserved (cf 33). Site-directed mutagenesis was used to replace a number of these residues, and the results are summarized in Table 12.3. Of the 13 amino acid residues examined, Thr-40, Tyr-59, and Thr-98 appear to be involved in holoprotein formation, and a number of the residues were found to influence receptor binding, including Lys-2, Arg-43, Arg-94, Ser-96, Thr-98, and Lys-104. Whether these replacements alter α binding or LH/CGr binding directly or indirectly cannot be ascertained by the available data. Interestingly, a number of amino acid residues that are invariant in the mammalian LH/CGβ-subunits (for example, Pro-7, Val-56, and Ser-87), can be replaced with quite different residues with no apparent effect on subunit assembly or receptor binding.

## Discussion

The shortest known fragment of hCGβ with in vitro activity comparable to that of the intact subunit (145 amino acid residues) comprises amino acid residues 8–110. This combined N/C-terminal deletion mutant retains the twelve Cys, which presumably pair appropriately to form the six disulfides of the native subunit (39). It is not necessarily surprising that removal of amino acid residues 1–7 of hCGβ has no effect on holoprotein formation since FSHβ and TSHβ lack the N-terminal extension present in CGβ and LHβ, yet bind to the same complementary α-subunit.

TABLE 12.2. Replacement of invariant amino acid residues in hCGβ.

| β residue | Replacement | α binding | LH/CGr binding |
|-----------|-------------|-----------|----------------|
| Gly-36 | Arg[a] | Dec | —[b] |
| | Asp[a] | Dec | —[b] |
| Tyr-37 | Phe[c] | Nor | Nor |
| | Leu[c] | Dec | Nor |
| | Asp | Dec | —[d] |
| Gln-54[e] | Arg | Nor | Dec |
| Gln-54 | Lys | Dec | —[d] |
| Pro-70 | Glu | Nor | Nor |
| Gly-71 | Asp | Nor | Inc |
| Val-84 | Ile | Nor | Nor |
| | Asp | Nor | Nor |
| Ala-85 | Ile | Dec | Inc |
| | Asp | Dec | —[d] |
| Asp-99 | Asn[f] | Nor | Dec |
| | Glu[f] | Nor | Dec |
| | Arg[f] | Nor | Dec |

[a] Reference 35.
[b] Not determined, but no significant heterodimer was found.
[c] Reference 36.
[d] The very small amount of heterodimer that formed exhibited some activity.
[e] Reference 19 (refers to hLHβ).
[f] Reference 37.

*Note:* The amino acid residues listed in the first column (β residue) are invariant in the known mammalian glycoprotein hormone β-subunits. All data are based on hCGβ with one exception involving a Gln-54→Arg replacement in hLHβ. The second column (Replacement) indicates the amino acid residue substituted for that present in the wild-type β-subunit. The third column (α binding) denotes whether the particular amino acid residue replacement yields a mutant β-subunit exhibiting normal (Nor), compared to hCGβ wild type, binding to α or is decreased (Dec). The fourth column (LH/CGr binding) indicates whether binding of the mutant hCGβ-containing heterodimer to the receptor is normal (Nor), decreased (Dec), or increased (Inc).

On the other hand, results from synthetic peptides (6, 12) and site-directed mutagenesis (27) suggest a role for the N-terminal region in receptor binding. This apparent dilemma may be rationalized on the basis that hormone receptor binding involves multiple contact sites, some of which may be eliminated without a dramatic effect on the overall association free energy. In this context, the removal of a weakly binding region may prove to be of less consequence than replacement with a single amino acid residue in the region that interferes with receptor binding, for example, the Lys-2→Glu mutant that we prepared (27). Also, it must be emphasized that the protocols we and others employ to

TABLE 12.3. Replacement of conserved amino acid residues in hCGβ.

| hCGβ residue | Occurrence | Replacement | α binding | LH/CGr binding |
|---|---|---|---|---|
| Lys-2 | Lys[1], Arg[11] | Glu[a] | Nor | Dec |
| Pro-7 | Pro[12] | Ala | Nor | Nor |
| Thr-32 | Thr[4], Ser[8] | Arg | Nor | Nor |
| Thr-40 | Thr[3], Ser[9] | Arg | Dec | —[b] |
| Arg-43 | Arg[12] | Lys[c] | Nor | Dec |
|  |  | Leu[c] | Nor | Dec |
| Val-56 | Val[12] | Arg | Nor | Nor |
| Tyr-59 | Tyr[12] | Asp | Dec | —[b] |
| Phe-64 | Phe[12] | Asp | Nor | Nor |
| Ser-81 | Ser[12] | Arg | Nor | Nor |
| Tyr-82 | Tyr[1], Phe[10], Val[1] | Arg | Nor | Nor |
| Ser-87 | Ser[12] | Arg | Nor | Nor |
| Arg-94 | Arg[10], Gln[2] | Lys[c] | Nor | Nor |
|  |  | Asp[c] | Nor | Dec |
| Arg-95 | Arg[3], Leu[7], Ile[2] | Ser[d] | Nor | Dec |
| Ser-96 | Ser[10], Lys[2] | Asp[d] | Nor | Dec |
| Thr-97 | Thr[5], Ser[6], Gln[1] | Tyr[d] | Nor | Nor |
| Thr-98 | Thr[5], Ser[7] | Asp[d] | Dec | Dec |
|  |  | Arg[d] | Dec | —[b] |
| Lys-104 | Lys[3], Arg[9] | Glu[a] | Nor | Dec |

[a] Reference 27.

[b] The small amount of heterodimer was active and appeared to be of reduced potency.

[c] Reference 42.

[d] Reference 43.

*Note:* The amino acid residue present in hCGβ wild type is given in the first column (hCGβ residue), and the second column (Occurrence) summarizes the frequencies of the stated residues in the 12 known amino acid sequences of the mammalian CG and LHβ-subunits. The number of times each residue appears in given as a superscript. The third column (Replacement) indicates the amino acid residue substituted for that of the wild type. The fourth column (α binding) denotes if holoprotein formation is normal (Nor)—that is, like that of expressed hCGβ wild type—or decreased (Dec). (No instance was found in which a significant increase occurred.) The fifth column (LH/CGr binding) indicates whether binding of the heterologous heterodimer containing the mutant hCGβ-subunit to the receptor is normal (Nor)—that is, comparable to tht of bovine α-hCGβ wild type—or decreased (Dec).

characterize these mutant hormones depend on radioimmunoassays to estimate concentrations, and potencies thus reflect a ratio of biologic/immunologic activity. Consequently, small differences in measured potencies are not necessarily significant, and conversely, mutant proteins with slightly different potencies will probably not be distinguished from each other or from the wild-type hormone. (The use of at least two types of antibodies in our studies provides some assurance that alteration of an epitope would be noted.)

The results on two of the C-terminal deletion fragments of hCGβ, 1–100 and 1–110, suggest a role for amino acid residues 101–110 in receptor binding. One contributing factor is expected to be Cys-110, which has been proposed to form a disulfide with Cys-26 (39). Of interest was the observation by Ruddon et al. (40) that this is the last disulfide to form in hCGβ, and its closure seems to be enhanced by α binding. Thus, the 26–110 disulfide may constrain the active conformer of hCGβ. We have also shown that replacement of Lys-104 with Glu has no demonstrable effect on holoprotein formation, but receptor binding by the mutant gonadotropin was diminished (27). In addition to Lys-104 and Cys-110, several other amino acid residues in the 101–110 sequence are conserved in the LH/CGβ-subunits, for example, Gly-101, Pro-107, and Leu-108. Thus, several amino acid residues may contribute to the stabilizing/binding contribution of this region. Of interest was the observation by Moyle et al. (41) that this region of the β-subunit is important in conferring receptor (LH/CG vs. FSH) specificity.

Another interesting comparison of the N/C-terminal deletion mutants of hCGβ involves fragments 8–100 and 1–100. As discussed above, the latter exhibits apparently normal α binding and reduced receptor binding. The former, under the conditions used, binds α only weakly, although the limited amount of heterodimer that forms retains some activity (31). These results could be interpreted by attributing a role of amino acid residues 1–7 in hCGβ to α binding. Yet, as discussed above, the 8–110 fragment behaves similarly to hCGβ wild type in the in vitro systems used. One could thus argue that either amino acid residues 1–7 or 101–110 are necessary to achieve hCGβ wild-type-like α binding and receptor binding (by the holoprotein). In this context, we recently described a pseudo-internal repeat in hCGβ (27, 31) consisting of a basic amino acid residue (Lys) adjacent to an acidic amino acid residue (Glu or Asp) (e.g., Lys-2-Glu-3 and Lys-104-Asp-105), which is contiguous with Pro-4-Leu-5 or separated from Pro-107-Leu-108 by His-106 (cf. Fig. 12.1). Perhaps the presence of one of these regions is necessary for proper β folding and α binding.

Boime et al. (34) have carefully screened the role of the various Cys residues in hCGβ with Ala replacements, and we and others have shown that at least half of the eight invariant non-Cys amino acid residues appear to function in α binding (e.g., Gly-36 (35), Tyr-37 (36), and Ala-85) (Table 12.2) or receptor binding (e.g., Asp-99) (37). The role of Gln-54 is controversial. Weiss et al. (19) found it to be involved in receptor binding with hLH, whereas we found it to influence α binding with hCGβ (Table 12.2).

Of considerable surprise was the observation that Pro-70, Gly-71, and Val-84 could be replaced with acidic amino acid residues with no apparent effect on α binding or receptor binding by the resultant heterodimer. Changes in kinetic processes would be overlooked in the present experimental approach, although it is noteworthy that replacement of Arg-68-Leu-69-Pro-70-Gly-71 with Ala-Ala-Ala-Ala resulted in a mutant hCGβ that was apparently secreted to a much lesser degree than hCGβ wild type or, perhaps, was degraded more rapidly.

Of the numerous conserved amino acid residues in the CG/LHβ-subunits that we replaced with site-directed mutagenesis (cf Table 12.3), several appear to participate in α binding (e.g., Thr-40, Tyr-59, and Thr-98), while others seem to be involved in receptor binding by the holo-protein (e.g., Lys-2, Arg-43, Arg-94, Ser-96, Thr-98, and Lys-104). These are obviously not exclusive, since Thr-98 appears in both categories. Combining the various single-site replacement data, one would infer the following amino acid residues to be involved with α binding and with receptor binding: α: Gly-36, Tyr-37, Thr-40, Gln-54, Tyr-59, Ala-85, and Thr-98; and r: Lys-2, Arg-43, Gln-54, Arg-94, Arg-95, Ser-96, Thr-98, Asp-99, and Lys-104.

These results suggest that α and receptor binding sites may be in close proximity to each other in certain regions of hCGβ. This suggestion may have important implications on the nature and constraints of the hormone-receptor complex. The data summarized in Tables 12.1–12.3 and in several references (18, 19, 35, 36, 42, 43) are consistent with the participation, based on the amino acid sequence and numbering of hCGβ, of the CAGY sequence (residues 34–37), the Keutmann loop (residues 38–57), and the Ward loop (residues 93–100) in subunit assembly and receptor binding. Additional peptide and antibody mapping techniques are also being used to identify other potential receptor contact regions (6, 7), but our results indicate that, in addition, such domains may also represent sites for subunit assembly.

*Acknowledgments.* We wish to dedicate this paper to the three scientists honored at the 1993 Serono Symposium on Glycoprotein Hormones, Drs. Harold Papkoff, Robert J. Ryan, and Darrell N. Ward. These individuals, along with Dr. John G. Pierce, who was honored at his retirement several years ago, have contributed enormously to the structural aspects of the glycoprotein hormone field and advanced our knowledge to the point where studies such as those presented at this Serono symposium could be undertaken. We also thank others from Dr. Puett's laboratory who have contributed to the data presented herein; these include Dr. Fang Chen, Ms. Tsuey-Ming Chen, Dr. Samer El-Deiry, Ms. Lizette M. Fernandez, Dr. Makoto Ujihara, Mr. Yi Wang, and Dr. Hiroaki Yoshida. Our appreciation is also extended to Drs. Mario Ascoli, Steven Birken, Robert E. Canfield, John Fiddes, and John Nilson for providing reagents and cells necessary for these studies. This research was supported by NIH Grant DK-33973.

# References

1. Ryan RJ, Charlesworth MC, McCormick DJ, Milius RP, Keutmann HT. The glycoprotein hormones: recent studies of structure-function relationships. FASEB J 1988;2:2661–9.

2. Merz WE. Properties of glycoprotein hormone receptors and post-receptor mechanisms. Exp Clin Endocrinol 1992;100:4–8.

3. Ascoli M, Segaloff DL. On the structure of the luteinizing hormone/chorionic gonadotropin receptor. Endocr Rev 1989;10:27–44.

4. Sprengel R, Braun T, Nikolics K, Segaloff DL, Seeburg PH. The testicular receptor for follicle stimulating hormone: structure and functional expression of cloned cDNA. Mol Endocrinol 1990;4:525–30.

5. Vassart G, Dumont JE. The thyrotropin receptor and the regulation of thyrocyte function and growth. Endocr Rev 1992;13:596–611.

6. Keutmann HT. Receptor-binding regions in human glycoprotein hormones. Mol Cell Endocrinol 1992;86:C1–6.

7. Dias J. Progress and approaches in mapping the surfaces of human follicle-stimulating hormone: comparison with other human pituitary glycoprotein hormones. Trends Endocrinol Metab 1992;3:24–9.

8. Pierce JG, Parsons TF. Glycoprotein hormones: structure and function. Annu Rev Biochem 1981;50:465–95.

9. Gordon WL, Ward DN. Structural aspects of luteinizing hormone actions. In: Ascoli M, ed. Luteinizing hormone action and receptors. Boca Raton, FL: CRC Press, 1985:173–97.

10. Keutmann HT, Charlesworth MC, Mason KA, Ostrea T, Johnson L, Ryan RJ. A receptor-binding region in human choriogonadotropin/lutropin β subunit. Proc Natl Acad Sci USA 1987;84:2038–42.

11. Keutmann HT, Mason KA, Kitzmann K, Ryan RJ. Role of the β93-100 determinant loop sequence in receptor binding and biological activity of human luteinizing hormone and chorionic gonadotropin. Mol Endocrinol 1989:526–31.

12. Salesse R, Bidart JM, Troalen F, Bellet D, Garnier J. Peptide mapping of intersubunit and receptor interactions of human choriogonadotropin. Mol Cell Endocrinol 1990;68:113–9.

13. Santa-Coloma TA, Reichert LE Jr. Identification of a follicle-stimulating hormone receptor-binding region in hFSH-β-(81–95) using synthetic peptides. J Biol Chem 1990;265:5037–42.

14. Santa-Coloma TA, Dattatreyamurty B, Reichert LE Jr. A synthetic peptide corresponding to human FSH β-subunit 33–53 binds to FSH receptor, stimulates basal estradiol biosynthesis, and is a partial antagonist of FSH. Biochemistry 1990;29:1194–200.

15. Santa-Coloma TA, Reichert LE Jr. Determination of α-subunit contact regions of human follicle-stimulating hormone β-subunit using synthetic peptides. J Biol Chem 1991;266:2759–62.

16. Birken S, Kolks MAG, Amr S, Nisula B, Puett D. Structural and functional studies of the tryptic core of the human chorionic gonadotropin β-subunit. Endocrinology 1987;121:657–66.

17. Bousfield GR, Ward DN. Selective proteolysis of ovine lutropin or its β subunit by endoproteinase arg-C. J Biol Chem 1988;263:12602–7.

18. Hayashizaki Y, Hiraoka Y, Endo Y, Miyai K, Matsubara K. Thyroid-stimulating hormone (TSH) deficiency caused by a single base substitution in the CAGYC region of the β-subunit. EMBO J 1989;8:2291–6.

19. Weiss J, Axelrod L, Whitcomb RW, Harris PE, Crowley WF, Jameson JL. Hypogonadism caused by a single amino acid substitutuion in the β subunit of luteinizing hormone. N Engl J Med 1992;326:179–83.

20. Kunkel TA. Rapid and efficient site-specific mutagenesis without phenotypic selection. Proc Natl Acad Sci USA 1985;82:488–92.

21. Deng WP, Nickoloff JA. Site-directed mutagensis of virtually any plasmid by eliminating a unique site. Anal Biochem 1992;200:81–8.

22. Jones DH, Howard BH. A rapid method for site-specific mutagenesis and directional subcloning by using the polymerase chain reaction to generate recombinant circles. BioTechniques 1990;8:178–83.

23. El-Deiry S, Kaetzel D, Kennedy G, Nilson J, Puett D. Site-directed mutagenesis of the human chorionic gonadotropin β-subunit: bioactivity of a heterologous hormone, bovine α-human des-(122–145) β. Mol Endocrinol 1989;3:1523–8.

24. Chaney WG, Howard DR, Pollard JW, Sallustio S, Stanley P. High-frequency transfection of CHO cells using polybrene. Somatic Cell Mol Genet 1986; 3:237–44.

25. Kaetzel DM, Browne JK, Wondisford F, Nett TM, Thomason AR, Nilson JH. Expression of biologically active bovine luteinizing hormone in Chinese hamster ovary cells. Proc Natl Acad Sci USA 1985;82:7280–3.

26. Chen F, Puett D. Delineation via site-directed mutagenesis of the carboxyl-terminal region of human choriogonadotropin β required for subunit assembly and biological activity. J Biol Chem 1991;266:6904–8.

27. Xia H, Huang J, Chen T-M, Puett D. Lysines 2 and 104 of the human chorionic gonadotrophin β subunit influence receptor binding. J Mol Endocrinol 1993.

28. El-Deiry S, Chen T-M, Puett D. Comparison of steroidogenic potencies of homologous and heterologous gonadotropins in rat and mouse Leydig cells. Mol Cell Endocrinol 1991;76:105–13.

29. Matzuk MM, Hsueh JW, Lapolt P, Tsafriri A, Keene JL, Boime I. The biological role of the carboxyl-terminal extension of human chorionic gonadotropin β-subunit. Endocrinology 1990;126:376–83.

30. Chen W, Bahl OP. Recombinant carbohydrate variant of human choriogonadotropin β-subunit (hCGβ) descarboxyl terminus (115–145). J Biol Chem 1991;266:6246–51.

31. Huang J, Chen F, Puett D. Amino/carboxyl-terminal deletion mutants of choriogonadotropin β. J Biol Chem 1993.

32. Moore WT Jr, Burleigh BD, Ward DN. Chorionic gonadotropins: comparative studies and comments on relationships to other glycoprotein hormones. In: Segal SJ, ed. Chorionic gonadotropin. New York: Plenum Press, 1980:89–126.

33. Ward DN, Bousfield GR, Moore KII. Gonadotropins. In: Cupps PT, ed. Reproduction in domestic animals. 4th ed. New York: Academic Press, 1991:25–80.

34. Suganuma N, Matzuk MM, Boime I. Elimination of disulfide bonds affects assembly and secretion of the human chorionic gonadotropin β subunit. J Biol Chem 1989;264:19302–7.

35. Azuma C, Miyai K, Saji F, et al. Site-specific mutagenesis of human chorionic gonadotropin (hCG)-β subunit: influence of mutation on hCG production. J Mol Endocrinol 1990;5:97–102.

36. Xia H, Fernandez LM, Puett D. Replacement of the invariant tyrosine in the CAGY region of the human chorionic gonadotropin β subunit. Mol Cell Endocrinol 1993;92:R1–5.

37. Chen F, Wang Y, Puett D. Role of the invariant aspartic acid 99 of human choriogonadotropin β in receptor binding and biological activity. J Biol Chem 1991;266:19357–61.
38. Chen F, Puett D. A single amino acid residue replacement in the β subunit of human chorionic gonadotrophin results in the loss of biological activity. J Mol Endocrinol 1992;8:87–9.
39. Mise T, Bahl OP. Assignment of disulfide bonds in the β subunit of human chorionic gonadotrophin. J Biol Chem 1981;256:6587–92.
40. Huth JR, Mountjoy K, Perini F, Ruddon RW. Intracellular folding pathway of human chorionic gonadotropin β subunit. J Biol Chem 1992;267:8870–9.
41. Campbell RK, Dean-Emig DM, Moyle WR. Conversion of human choriogonadotropin into a follitropin by protein engineering. Proc Natl Acad Sci USA 1991;88:760–4.
42. Chen F, Puett D. Contributions of arginines-43 and -94 of human choriogonadotropin β to receptor binding and activation as determined by oligonucleotide-based mutagenesis. Biochemistry 1991;30:10171–5.
43. Huang J, Ujihara M, Xia H, Chen F, Yoshida H, Puett D. Mutagenesis of the "determinant loop" region of human choriogonadotropin β. Mol Cell Endocrinol 1993;90:211–8.

# Part IV

## Protein Processing and Posttranslational Events

# 13

# Folding of the β-Subunit of hCG and Its Role in Assembly of the α-β Heterodimer

RAYMOND W. RUDDON, JEFFREY R. HUTH, ELLIOTT BEDOWS, KIMBERLY MOUNTJOY, AND FULVIO PERINI

It has been known for some time that the assembly of the α- and β-subunits of *human chorionic gonadotropin* (hCG) within cells that produce it is not totally efficient. As a result, uncombined α- and β-subunits, as well as α-β heterodimer, are secreted (1–3). Efficiency of α-β assembly in vitro, using purified urinary forms of the subunits, has also been reported to be low, with a kd of 0.6 μM (4).

The ratio of secreted free β compared to intact α-β heterodimer is greater for choriocarcinoma cells and in urine from choriocarcinoma patients than the ratio secreted by normal placental explants or found in normal pregnancy urine (3), suggesting a defect in α-β assembly in choriocarcinoma cells. This has led to the use of free β detected by immunologic methods as a tumor marker for gynecologic and other cancers (5, 6).

Some posttranslational modifications of the subunits, particularly the α-subunit, that could potentially alter the assembly competence of the subunits have been observed. An additional O-linked oligosaccharide has been found on the free α- (compared to dimer α) subunit from LH- (7) and hCG- (8, 9) producing cells. The free α-subunit from hCG-producing cells can also be phosphorylated (10), and the efficiency of combination with β varies with the amount of processing of the N-linked oligosaccharides of α (11). The α molecules with larger N-linked, complex oligosaccharides combine less efficiently with β in vitro (11). The rate of intracellular oligosaccharide processing of the N-linked glycans of the free β-subunit is somewhat slower than that of the β-subunit combined with α, yet free α, free β, and α-β dimer are all secreted from hCG-producing *choriocarcinoma* (JAR) cells at a similar rate (2). Finally, the free β-subunits purified from JAR cell lysates or from culture medium combine equally well with urinary α in vitro compared to dimer β (prepared from combined α-β-subunits) purified from the same sources (12).

None of the posttranslational modifications noted above, however, can explain the regulation of α-β assembly inside cells for the following reasons. First, α-β dimer assembly occurs rapidly, within minutes, while the α- and β-subunits are still within the endoplasmic reticulum and contain high mannose (man 8–9) oligosaccharides (2). Second, all of the above-noted posttranslational modifications (i.e., processing of high-mannose N-linked glycans, addition of an extra O-linked glycan, and phosphorylation) occur later in time and in the cellular secretory pathway (i.e., post-ER compartments) than the completion of assembly (2, 9–11). Third, free β isolated from cells or culture medium combines with an excess of α in vitro under equilibrium conditions just as well as β isolated from the intact α-β dimer (12).

This apparent conundrum, coupled with a classic example of serendipity, led us to explore events in the folding of the β-subunit as a potential mechanism regulating α-β heterodimer assembly. In this chapter, we describe how we detected incompletely folded intracellular precursor forms of the β-subunit and showed that disulfide bond formation correlates with the conformational changes in the β-subunit that lead to its becoming assembly competent. We have been able to define the *precursor* (p)-product relationship between various intracellular forms and to show that the formation of specific disulfide bonds occurs at discrete steps in the β folding pathway. We have now also been able to show that the intra-cellular folding pathway can be reproduced in vitro, given the appropriate redox conditions. Furthermore, we have shown that the rate of β folding and assembly in vitro can be increased by addition of *protein disulfide isomerase* (PDI) and may be assisted by molecular chaperones.

## Materials and Methods

The materials and methods used in the experiments described here have, for the most part, been previously reported and will only be described briefly here.

### Cell Culture

JAR human choriocarcinoma cells were grown at 37 °C in Dulbecco's modified Eagle's medium (GIBCO) with 10% fetal bovine serum as described previously (2). *Chinese hamster ovary* (CHO) cells transfected with wild-type or mutated hCGβ genes alone or cotransfected with the glycoprotein hormone α gene were grown in F-12 medium supplemented with 10% fetal bovine serum, the neomycin analog G-418 (GIBCO), and antibiotics as previously described (13). All mutated hCGβ genes used to transfect the CHO cells described in this chapter have had both of the Cys residues of each designated *disulfide* (S–S) bond changed to Ala.

## Metabolic Labeling of Cells with Radioactive Substrates

JAR or CHO cells grown to 80% to 95% confluency in 100 mm plastic dishes were pulse labeled for 40 sec to 5 min as indicated below with L-[$^{35}$S]cysteine (~1100 Ci/mmol; Du Pont–New England Nuclear) at a concentration of 200–500 µCi/ml in serum-free medium lacking cysteine (2). All pulse incubations were carried out as described previously (2), and the cells were incubated for the chase times indicated in the text. Cells were harvested by rinsing with cold *phosphate buffered saline* (PBS) and immediately lysed in 5 ml of PBS containing detergents (1.0% Triton X-100, 0.5% sodium deoxycholate, and 0.1% *sodium dodecyl sulfate* [SDS]), protease inhibitors (20 mM EDTA and 2 mM phenylmethane-sulfonyl fluoride), and 50 mM iodoacetic acid (pH 8.0) to alkylate the free sulfhydryl groups of the β folding intermediates. Cell lysates were incubated 20–30 min at 22 °C in the dark, followed by disruption through a 22 ga needle (3 times), centrifuged for 1 h at 100,000 × g, and frozen at −70 °C for further use.

## Immunoprecipitation of Cell Lysates and Culture Media

The immunoreactive forms of hCGβ were immunoprecipitated with a polyclonal antibody (the G-10 antibody) that recognizes all forms of the β-subunit. In some experiments designed to indicate whether pβ1 had been converted to pβ2, radiolabeled cell lysates were initially incubated in the presence of a monoclonal antibody, 91-M (Chemicon), specific for pβ2-free. The epitope for the 91-M antibody appears to reside on a site covered when the α-subunit binds to hCGβ. This antibody does not recognize the partially folded pβ1 molecule. After a second immuno-precipitation with the 91-M antibody to immunoprecipitate residual free hCGβ, lysates were reacted with the G-10 antibody. All immuno-precipitations were carried out for 16 h at 4 °C with rotation in the dark. Immune complexes were precipitated with Protein A-Sepharose (Sigma) and prepared for *sodium dodecyl sulfate–polyacrylamide gel electro-phoresis* (SDS-PAGE) or reversed-phase HPLC as previously described (2, 14, 15).

## SDS-PAGE

Radiolabeled hCG forms that bound to washed Protein A-Sepharose beads were eluted with 2× concentrated SDS gel sample buffer (125 mM Tris.HCl [pH 6.8] containing 2% SDS, 20% glycerol, and 40 µg/ml bromophenol blue). Samples that were run under nonreducing conditions were boiled for 4 min in sample buffer, applied to polyacrylamide gradient slab gels (5%–20%), and run and analyzed as described previously (2).

## Purification of hCGβ Folding Intermediates and HPLC Analysis

The hCGβ folding intermediates pβ1, pβ2-free, and pβ2-combined were purified using the two-step process (immunoprecipation followed by $C_4$ reversed-phase HPLC) described by Huth et al. (14). Briefly, pβ1 and pβ2 were immunoprecipitated from cell lysates with the G-10 antibody, and all immunocomplexes were precipitated with Protein A-Sepharose beads as described above. To dissociate precipitated immunocomplexes, pellets were treated with 6 M guanidine HCl, pH 3, for 16 h at room temperature with 100 μg of myoglobin as carrier. Following low-speed centrifugation to remove Protein A-Sepharose beads, the guanidine eluates were injected onto a Vydac 300 Å $C_4$ reversed-phase column equilibrated with 0.1% *trifluoroacetic acid* (TFA). The column was eluted isocratically with 18% acetonitrile in 0.1% TFA, followed by a 0.48%/min gradient of 80% acetonitrile in 0.1% TFA for 90 min (14). Fractions containing hCGβ forms were concentrated by Speed-Vac centrifugation and pooled for tryptic peptide analysis.

## Strategy for the Identification of Peptides Following Tryptic Digestion

Human CGβ contains 6 disulfide bonds and 13 Arg and Lys residues that are arranged such that all of the Cys-containing tryptic peptides remain attached to each other as a result of their covalent disulfide bridges. If, however, particular S–S bonds were not formed in a given hCGβ folding intermediate, specific Cys-containing peptides would be released from the disulfide-linked hCGβ core. For example, if the 26–110 bond was unformed, then hCGβ peptide 105–114 (containing Cys-110) would be released. The pattern of trypsin-released peptides as distinguished from the disulfide-linked peptides by HPLC reveals incomplete bond formation (14).

Cells were lysed in the presence of 50 mM iodoacetate to trap the cysteine-free thiols of unformed hCGβ disulfide bonds. These *carboxy-methylated cysteine* (CM-Cys)-containing peptides were separated from the disulfide linked β-subunit core by $C_8$ reversed-phase HPLC (see below). Identity of the trypsin-releasable, CM-Cys-containing peptides was confirmed by comparing the elution times of these peaks with the elution times of Cys-containing peptides from the tryptic peptide maps of hCGβ obtained from [³H]leucine/[³⁵S]cysteine-labeled JAR cells and by amino acid sequencing (14).

## Tryptic Digestions

Nonreduced hCGβ forms were digested for 16 h in silanized polypropylene tubes containing 100–200 µg myoglobin, 0.03% DPCC-treated trypsin (Sigma), 5 mM $CaCl_2$, and 50–150 mM Tris-HCl, pH 8. The digestion was continued for 4 h with the addition of two aliquots each of 50 µg of trypsin (0.06% final concentration) (14). Digestions were stopped by freezing at −70 °C. Samples were stored at −70 °C until HPLC analysis of the peptides was carried out.

## HPLC Purification of Tryptic Peptides

Human CGβ tryptic digests were injected onto a Brownlee $C_8$ 300 Å reversed-phase column equilibrated with 0.1% TFA. The column was eluted isocratically for 3 min with 0.1% TFA, followed by a 0.32%/min acetonitrile gradient in 0.1% TFA for 100 min (14). One-minute fractions were collected in silanized polypropylene tubes. Tubes into which disulfide-linked peptides eluted contained 5 µg myoglobin as carrier. Samples were concentrated by Speed-Vac centrifugation and stored at −70 °C.

## Ion Exchange HPLC

A reversed-phase HPLC peak that eluted at 37 min and contained two peptides was resolved by ion exchange HPLC using a Waters DEAE anion exchange column (14). The peptides were eluted during a 30 min 0–300 mM NaCl gradient at a constant pH of 7 in 20 mM phosphate buffer. One-minute fractions were collected and analyzed by scintillation counting to mark the elution positions of [$^{35}$S]Cys-containing peptides.

## Gas Phase Sequence and Amino Acid Analysis

Amino acid analysis and sequencing of nondisulfide-linked tryptic peptides were used to prove that the cysteines in tryptic peptides were carboxy-methylated (14). Confirmation of carboxymethylation demonstrated that a cysteine was not disulfide paired in the intact intermediate in the cell lysate and did not simply result from S–S bond rearrangement during the tryptic digestion at alkaline pH.

## Preparation of Non-Alkylated [$^{35}$S]cysteine-pβ1 and [$^{35}$S]cysteine-pβ2

Nonalkylated [$^{35}$S]cysteine-pβ1 was prepared by labeling ten 100 mm Petri dishes of 90%–100% confluent JAR cells with 400 µCi/ml [$^{35}$S]cysteine for 4.5 min and immediately lysing them in the detergent-PBS buffer that lacked iodoacetate. [$^{35}$S]cysteine-pβ1 was purified by immunoprecipitation and HPLC purification as described above. Based on radioimmunoassay measurements of the amount of hCGβ in JAR cells (16) and estimating that 10% of hCGβ is pβ1 since the $t_{1/2}$ of pβ1 in JAR cells is 4 min (i.e., the rate of folding to pβ2) and the $t_{1/2}$ of pβ2 (i.e., the secretion rate) is 35 min (2), the specific activity of [$^{35}$S]cysteine-pβ1 was estimated to be 6–9 × $10^4$ CPM/ng. Nonalkylated [$^{35}$S]cysteine-pβ2 was prepared by labeling five 100 mm Petri dishes of confluent JAR cells with 200 μCi/ml [$^{35}$S]cysteine for 30 min and immediately lysing them in detergent-PBS that lacked iodoacetate. Typically, 1 × $10^6$ CPM of pβ2 were purified from 2.5 × $10^7$ JAR cells, yielding an estimated specific activity of 3 × $10^4$ CPM/ng of pβ2.

## In Vitro Refolding of hCGβ

2000 CPM of [$^{35}$S]cysteine-pβ1 were incubated in a 40 μl reaction that consisted of 50 mM triethanolamine, pH 7.4, 1.7 mM cysteamine (the reductant), and 1.34 mM cystamine (the oxidant). pβ1 was folded at 37 °C for 0–8 h, after which folding was stopped by the addition of 5 μl of 900 mM iodoacetate in 450 mM Tris-HCl, pH 8.7, yielding a final concentration of 100 mM iodoacetate, 50 mM Tris-HCl, and final pH of 8.7. After 5 min of alkylation in the dark at room temperature, 50 μl of nonreducing gel electrophoresis buffer was added and samples were stored at −20 °C until analysis by nonreducing SDS-PAGE.

## In Vitro Assembly of pβ2 and Urinary hCGα

2500 CPM of [$^{35}$S]cysteine-pβ2 were incubated in a 40 μl reaction that consisted of 50 mM triethanolamine, pH 7.4, and 0, 0.1, 1.0, or 10 μM hCGα for 1 h at 37 °C. Where indicated in Figure 13.7, 1.7 mM cysteamine (the reductant) and 1.34 mM cystamine (the oxidant) were also included in the reaction. The assembly reactions were stopped by the addition of 5 μl of 500 mM iodoacetate in 50 mM Tris, pH 8.7, for 10 min and then the addition of 50 μl of ice-cold gel electrophoresis sample buffer. Samples were analyzed on nonreducing SDS-PAGE gels at 4 °C to prevent dissociation of the hCG dimer (17).

## Results and Discussion

The α- and β-subunits of hCG are secreted both as a combined, noncovalently linked dimer form as well as uncombined, free forms by human trophoblastic cells. We have utilized the cultured choriocarcinoma cell line JAR to determine what regulates the combination of the two subunits.

It should be noted that hCG produced by JAR cells has full biologic activity as determined by MA-10 Leydig cell radioreceptor and biologic activity assays (18). In fact, the biologic activity/immunologic reactivity ratio was 1.48 for hCG secreted by JAR cells compared to a B/I ratio of 1.0 for a urinary hCG standard (Whitcomb, Ruddon, unpublished observation).

The hCG subunits produced by JAR cells were biosynthetically labeled with [$^{35}$S]cysteine by a pulse-chase protocol, purified by immunoprecipitation with specific antisera that recognize free or combined subunits, and separated by SDS-PAGE under nonreducing conditions.

Figure 13.1 shows the pulse-chase kinetics of conversion of an early intracellular precursor form (pβ1) into a conformationally different and slower migrating form (pβ2) that combines with the precursor α-subunit (pα) to form the α-β dimer (19). Based on the rate of incorporation and disappearance of radioactivity from the gel bands (Fig. 13.2), the following precursor product relationship was observed:

$$\text{p}\beta1 \rightarrow \text{p}\beta2 \text{ (free)} \nearrow \text{p}\alpha\text{-p}\beta2 \text{ (combined)} \rightarrow \text{mature } \alpha\text{-}\beta$$
$$\text{p}\alpha \underline{\hspace{2cm}}\big/$$

The $t_{1/2}$ for the conversion of pβ1 to pβ2 was 4 min in JAR cells (20) and 5 min in CHO cells (15).

Based on the observation that the differences between pβ1 and pβ2 disappear when the immunopurified pβ-subunits are analyzed by SDS-

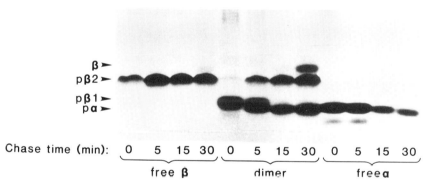

FIGURE 13.1. Pulse-chase kinetics of hCG dimer and free α- and β-subunits in JAR cells. JAR cells were labeled with [$^{35}$S]cysteine for 2 min and chased for specified times, and the cell lysates were sequentially immunoprecipitated with antiserum specific for the free β-subunit, α-β dimer, and the free α-subunit. The immunoprecipitated forms were separated by SDS-PAGE under nonreducing conditions. β = mature β-subunit; pβ2 = precursor form of β, part of which combines with α; pβ1 = early precursor form of β; and pα = precursor form of α-subunit. Reprinted with permission from Ruddon, Krzesicki, Saccuzzo Beebe, Loesel, Perini, and Peters (19), © The Endocrine Society, 1989.

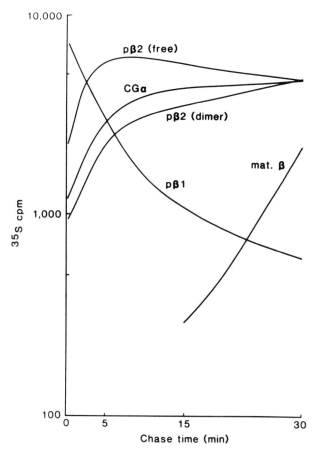

FIGURE 13.2. Quantitative evaluation of pulse-chase kinetics. Bands of radioactivity were cut from the kinetics gel shown in Figure 13.1, depolymerized in $H_2O_2$, and analyzed for radioactivity. Mat. β = mature β-subunit. Reprinted with permission from Ruddon, Krzesicki, Saccuzzo Beebe, Loesel, Perini, and Peters (19), © The Endocrine Society, 1989.

PAGE under reducing conditions (20), we concluded that the difference between the pβ forms was due in large part to differences in intramolecular S–S bond formation. Thus, we examined the content of free cysteine sulfhydryl groups in the various β forms (Fig. 13.3). This led to the conclusion that the pβ forms did have different amounts of free SH groups, indicating that they varied in the content of their intramolecular S–S bonds, of which native hCGβ has 6 (21). It was also observed that the intracellular α-subunit had no detectable free cysteine sulfhydryls, indicating that all five of the S–S bonds of α had formed before those of β were completed.

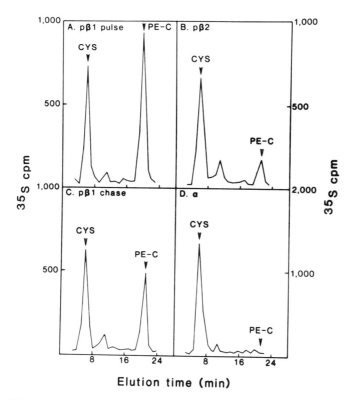

FIGURE 13.3. Determination of free thiols and cystine disulfides in the intracellular precursor forms of hCG-subunits. JAR cells were pulsed with [$^{35}$S]cysteine for 2 min and chased for 5 min. Cell lysates were prepared in the presence of 50 mM vinylpyridine and immunoprecipitated with the anti-α-β dimer, and the subunits were separated by SDS-PAGE under nonreducing conditions. The gel bands were eluted and hydrolyzed in 6 N HCl in the presence of 50 mM β-mercaptoethanol, and pyridylethyl [$^{35}$S]cysteine was separated from [$^{35}$S]cysteine by HPLC as previously described (17). PE-C = pyridylethylcysteine. Reprinted with permission from Ruddon, Krzesicki, Norton, Saccuzzo Beebe, Peters, and Perini (20).

Because of the well-documented role of S–S bond formation in protein folding (22, 23), we speculated that the conformational changes in β-subunit leading to formation of an assembly-competent form were accompanied by (and perhaps even directed by) formation of intramolecular S–S bonds. We also concluded, based on our experiments (14, 15, 19, 20), that the folding of the β-subunit and not of the α-subunit was rate limiting for α-β assembly. Neither the amount of S–S bonds nor the conformation as detected by SDS-PAGE changed for the pα-subunit prior to or during α-β assembly, strongly suggesting that pα folds and completes its S–S bonds before pβ.

Folding of the β-subunit and assembly of the α-β heterodimer in JAR choriocarcinoma cells and in β gene-transfected CHO cells has proven to be a unique model for studying protein folding and assembly inside of cells. Although protein folding has been studied extensively in vitro, very few models for intracellular protein folding have been examined (24).

The malignant trophoblastic cell line JAR was initially used as a model system to study β-subunit folding, since we had used this model to identify conformational intermediates in the production of an assembly-competent form of the hCGβ-subunit. The first goal of these studies was to identify the folding intermediates in the pathway leading to the formation of an assembly-competent state of β. As noted above, the earliest biosynthetic precursor of the hCGβ-subunit detectable in JAR cells pulse labeled for 2 min is pβ1 (Fig. 13.1). That form lacked at least half of the six intrachain disulfide bonds observed in the fully processed dimer form of β (25). pβ1 was rapidly ($t_{1/2} \sim 4$ min) converted into pβ2, which had a greater complement of intrachain S−S bonds and combined with the α-subunit. We have identified the late-forming S−S bonds involved in the transition of pβ1 into the assembly-competent form, pβ2. The late-forming S−S bonds were identified in JAR cell lysates that had been pulse labeled with [$^{35}$S]cysteine for 2 or 5 min, followed by trapping of the cysteine thiols with iodoacetate before immunopurification of the β-subunit forms. Immunopurified pβ1 was eluted from the immunoconjugated Protein A-Sepharose beads with guanidine HCl and treated with trypsin under nonreducing conditions to liberate [$^{35}$S]cysteine-containing peptides from the disulfide-linked β core polypeptide. These tryptic peptides were then separated by HPLC and sequenced to determine the location of the carboxymethyl-[$^{35}$S]cysteine residues. We concluded that the late-forming S−S bonds are the ones involved in stabilizing the conformation of the β-subunit that is required for combination with α to form the biologically functional α-β heterodimer (25).

Identification of the S−S bonds formed in each of the intracellular β folding intermediates was more definitively shown by short pulse-labeling experiments in which the various β forms were first purified by HPLC and the CM-Cys-containing peptides identified by tryptic maps as described above. Identification of tryptic peptides released from the β precursors under nonreducing conditions was used to identify which S−S bonds were formed in each pβ intermediate and to determine in what order they were formed (Fig. 13.4). The amount of a peptide that was released indicated the extent of S−S bond formation involving the cysteine(s) in that peptide. These experiments showed that, of the six S−S bonds in hCGβ, bonds 34−88 and 38−57 form first. The rate-limiting event of folding involves the formation of the S−S bonds between cysteines 9−90 and cysteines 23−72, resulting in an assembly-competent conformation. Disulfide bond 93−100, the formation of which appears to be necessary for stabilization of the hCG heterodimer, forms next. Finally, S−S bond 26−110 is com-

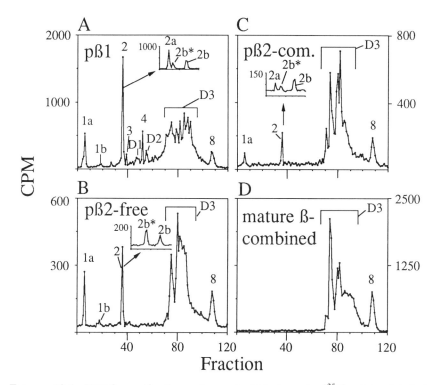

FIGURE 13.4. HPLC purification of nondisulfide-bound, [$^{35}$S]cysteine-labeled peptides from tryptic digests of nonreduced β-subunit folding intermediates and mature β-combined. *A:* Results of the HPLC purification of the peptides from a tryptic digest of nonreduced pβ1, an early β-subunit folding intermediate that can be purified from JAR cells (Fig. 13.1). The counts/min of [$^{35}$S]cysteine (y-axis) in a 5% aliquot of each HPLC fraction (x-axis) are plotted to determine the elution positions of nondisulfide-linked, [$^{35}$S]cysteine-containing tryptic peptides. The identities of the peptides in these peaks were confirmed by comparing their elution positions to the positions of the peptides in the β-subunit tryptic peptide map as previously described (14). The identification of these peptides reveals the cysteines that were not disulfide linked in pβ1. Inset: ion-exchange HPLC purification of the peptides that coeluted in peak 2 (y-axis, counts/min of [$^{35}$S]cysteine; x-axis, HPLC fraction number, 1–30). The peptide peaks in the pβ1 analysis indicate seven cysteines that were not disulfide linked completely: peaks 1a and 1b, peptide β-(96–104), Cys$^{100}$; peak 2a, peptide β-(69–74), Cys$^{72}$; peaks 2b* and 2b, peptide β-(105–114), Cys$^{110}$; peak 3, peptide β-(87–94), Cys$^{88}$, Cys$^{90}$, and Cys$^{93}$; and peak 4, peptide β-(9–20), Cys$^9$. Peaks D1–D3 contained disulfide-linked peptides. *B:* Nondisulfide-linked peptides indicate two cysteines that were not disulfide-bound in pβ2-free, the second intermediate in the β-subunit folding pathway: peaks 1a and 1b, peptide β-(96–104), Cys$^{100}$; and peaks 2b* and 2b, peptide β-(105–114), Cys$^{110}$. *C:* Three cysteines are shown to be nondisulfide-bound in pβ2-combined, the third intermediate in the β-subunit folding pathway: peak 1a, peptide β-(96–104), Cys$^{100}$; peak 2a, peptide β-(69–74), Cys$^{72}$; and peaks 2b* and 2b, peptide β-(105–114), Cys$^{110}$. *D:* Absence of peaks other than D3 and 8 shows that all cysteines were disulfide-linked in mature β-combined. Reprinted with permission from Huth, Mountjoy, Perini, and Ruddon (14).

pleted after assembly with the α-subunit, suggesting that completion of folding of the COOH terminus in the β-subunit occurs after assembly with the α-subunit. We concluded from these experiments that the extent of S–S bond formation correlates with the conformational changes that occur in the folding of the β-subunit leading to an assembly-competent state and that the formation of specific S–S bonds are involved in each step of the folding pathway (14).

Next, we wanted to measure the intracellular rates of formation of the six S–S bonds in the hCGβ-subunit to determine whether the folding pathway of this molecule can be described by a simple sequential model (26). If such a model is correct, the formation of S–S bonds, which is indicative of tertiary structural changes during protein folding, should occur in a discrete order. The individual rates of disulfide bridging were determined by identifying the extent of S–S bond formation in hCGβ intermediates purified from JAR choriocarcinoma cells that had been metabolically labeled for 40 or 120 sec and chased for 0 to 25 min (Fig. 13.5). With these short labeling times, an additional intracellular precursor form, pβ1-early, was identified. pβ1-early is defined as the earliest precursor form detected in cells labeled with [$^{35}$S]cysteine for very short times (e.g., 40 sec). pβ1-early is converted into pβ1-late, the next intermediate in the pathway. The results of these kinetic studies define a folding pathway in which the S–S bonds between cysteines 34–88, 38–57, 9–90, and 23–72 stabilize, in a discrete order, the putative domain(s) involving amino acids 1–90 of hCGβ. However, the S–S bonds 93–100 and 26–110 begin to form before the complete formation of the S–S bonds that stabilize the amino terminal 1–90 domain(s) and continue to form after complete formation of these S–S bonds. This suggests that hCGβ does not fold by a simple sequential pathway. The order of completion of each of the six S–S bonds of hCGβ is: 34–88 ($t_{1/2} = 1$–$2$ min), 38–57 ($t_{1/2} = 2$–$3$ min), 9–90 and 23–72 ($t_{1/2} = 4$–$5$ min), 93–100, and 26–110. These experiments also showed that 60%–100% of each of the six S–S bonds form posttranslationally. Furthermore, nonnative S–S bonds were not detected during intracellular folding of hCGβ (26).

In order to demonstrate that the above-described folding pathway was not peculiar to choriocarcinoma cells, we employed CHO cell lines transfected with either the wild-type hCGβ gene alone (CHO β cells) or in conjunction with the gene expressing the α-subunit (CHO α-β cells) to study the folding pathway of the hCGβ-subunit (15). In both CHO β and CHO α-β cells, the earliest detectable hCGβ precursor, pβ1, which had two of the six potential S–S bonds (34–88 and 38–57) formed, was converted to pβ2, a form that, following the formation of S–S bonds between cysteines 9–90 and 23–72, migrated more slowly than pβ1 by SDS-PAGE under nonreducing conditions. The $t_{1/2}$ for the conversion of pβ1 to pβ2 in both CHO α-β and CHO β cells was 5 min, demonstrating that the α-subunit had no effect on the rate of this conversion. Further-

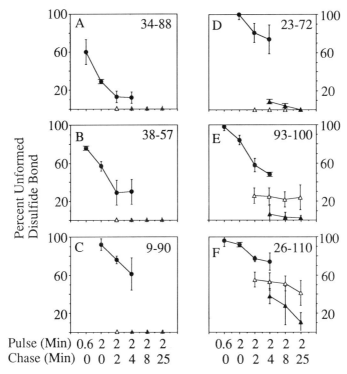

Pulse (Min) 0.6 2 2 2 2 2        0.6 2 2 2 2 2
Chase (Min)  0 0 2 4 8 25         0 0 2 4 8 25

FIGURE 13.5. Kinetics of formation of each of the six disulfide bonds in hCGβ. The methods used to quantitate disulfide bond formation have been previously described (23). The results are expressed as the percent that a disulfide bond was unformed, since nonpaired cysteines and not intact disulfide bonds were measured in these experiments. The disulfide bond is indicated in the upper right of each panel, and data are shown for pβ1 (solid circle), pβ2-free (open triangle), and pβ2-combined (solid triangle) in each graph. The graphs are arranged in the order in which the disulfide bonds formed. Except for the 0.6 min pulse and 0 chase time point, which was done twice, all of the results are shown as an average of three independent experiments ± SD. Carboxymethylation of cysteines in trypsin-released peptides was confirmed by amino acid analysis. Reprinted with permission from Huth, Mountjoy, Perini, Bedows, and Ruddon (26).

more, the trypsin-releasable peptides generated from nonreduced pβ1 or pβ2 were the same in both CHO α-β and CHO β cells. Thus, both the rate and order of S–S bond formation during the conversion of the folding intermediate pβ1 into pβ2 are the same as in JAR cells and are the same whether or not the α-subunit is present. In addition, it was observed in different clones of CHO cells, with different ratios of α-to-β expression, that the higher the glycoprotein hormone α-subunit to β-subunit ratio, the greater the rate and extent of hCG heterodimer assembly (15).

In order to confirm the conclusions from the kinetic experiments, which strongly suggested that the formation of individual S–S bonds of β are involved at each step of the β-subunit folding pathway, we examined folding of β in CHO cells transfected with hCGβ genes in which the cysteine residues involved in the formation of each of the six intramolecular S–S bonds of β were replaced by alanine (27). These mutants were prepared by site-directed mutagenesis in the laboratory of Dr. Irving Boime (Washington University, St. Louis, MO) (13). When the cysteines required for any of the first three hCGβ S–S linkages to form (bonds 34–88, 38–57, or 9–90) were substituted by alanines, folding did not proceed beyond the earliest detectable folding intermediate (pβ1-early). In the absence of the next S–S bond to form (bond 23–72), pβ1-early was converted into a second folding intermediate (pβ1-late), but conversion to the following intermediate (pβ2-free) was inhibited. When either of the final two S–S bonds (bonds 93–100 or 26–110) were removed, conversion of pβ1-late to pβ2-free was detected, but conversion of pβ2-free to the next folding intermediate, pβ2-combined, was not seen. These data support the hypothesis that individual S–S bonds are involved in discrete steps in the hCGβ folding pathway (27).

Taken together, the data from JAR cells and β gene-transfected CHO cells support the following model for folding of the β-subunit (the S–S bonds formed at each step are indicated):

$$
\begin{array}{ccc}
34\text{--}88 & 38\text{--}57 & \\
p\beta0 \longrightarrow p\beta1\text{-early} \longrightarrow p\beta1\text{-late} & & 9\text{--}90 \\
& & 23\text{--}72 \\
26\text{--}100 & 93\text{--}100 & \\
\text{mature } \alpha\text{-}\beta \longleftarrow p\alpha\text{-}p\beta2 \longleftarrow p\beta2\text{-free} & & \\
& p\alpha &
\end{array}
$$

In this model, pβ0 is the hypothetical unfolded, completely reduced molecule that may exist very briefly in the cell, since most of the S–S bond formation occurs posttranslationally (26). pβ1-early results from formation of the 34–88 S–S bond and is detected in cell lysates by immunopurification and SDS-PAGE or HPLC separation from other pβ forms. A shift to a slightly faster migrating form on SDS-PAGE indicates the formation of pβ1-late, which contains S–S bonds 34–88 and 38–57 (26). Completion of the 9–90 and 23–72 S–S bonds forms pβ2-free, the rate-limiting step in pβ folding (14, 15, 19, 20). The conversion of pβ1-late to pβ2-free results in a large conformational shift, as detected by a change in antibody reactivity and migration on SDS-PAGE (Fig. 13.1). Formation of the 93–100 and 26–110 bonds appears to be required for stabilization of the α-β dimer, and complete formation of these bonds does not occur until assembly of the pα-pβ heterodimer occurs (Fig. 13.5). Assembly of pα-pβ is completed in the ER while both subunits still bear high mannose-type oligosaccharides (2, 20). Processing of the

N-linked oligosaccharides, formation of complex, sialylated N-linked glycans, and addition of O-linked glycans occur in the Golgi, leading to the formation of mature α-β that is secreted from the cell.

The next question asked was whether β-subunit folding occurred via the same pathway in vitro as in intact cells. We have now been able to reproduce the rate-limiting event in the hCGβ folding pathway in vitro using the oxidants cystamine (Fig. 13.6) and glutathione (Huth et al., submitted). Although oxidized glutathione has been detected in the endoplasmic reticulum and is proposed to establish the redox state of this organelle (28), the observation that cystamine also promotes folding of hCGβ supports the hypothesis proposed by Ziegler and Poulsen (29) that cystamine is the oxidizing equivalent in the ER. In addition to the rate-limiting event, the other folding and assembly steps were reconstituted in vitro. Starting with reduced and denatured hCGβ, the order of S–S bond formation in the in vitro folding pathway was found to be the same as in the intracellular folding pathway in JAR choriocarcinoma cells and in CHO cells transfected with the hCGβ gene. Furthermore, PDI was found to catalyze the in vitro folding of hCGβ without changing the order of S–S bond formation (Huth et al., submitted).

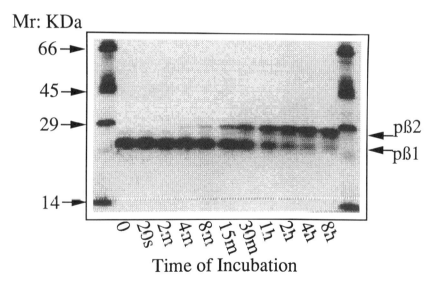

FIGURE 13.6. In vitro folding of pβ1 to pβ2-free in a cysteamine/cystamine redox buffer: fluorogram of a nonreducing SDS-PAGE separation of the hCGβ intermediates. The first and last lanes, show [14C]-labeled molecular weight markers (Sigma): bovine serum albumin (Mr = 66,000), chicken egg albumin (Mr = 45,000), carbonic anhydrase (Mr = 29,000), and α-lactalbumin (Mr = 14,000). The remaining lanes show the conversion of pβ1 to pβ2-free following incubations for the times indicated (see "Materials and Methods" for the buffer conditions).

The assembly of pβ2 with urinary α was also facilitated in a redox buffer composed of cysteamine and cystamine (Fig. 13.7) or of glutathione and PDI (Huth et al., submitted). In intact cells, as noted above, assembly of the α-β heterodimer occurs before all the intramolecular S–S bonds are formed. We found that assembly in vitro was increased by PDI after reduction of two of the carboxyterminal S–S bonds of hCGβ. These results strongly suggest that, both in intact cells and in vitro, partially unfolded hCGβ is more assembly competent than is fully folded hCGβ. In the PDI-catalyzed reaction, a kd of <1 nM was calculated for α-β assembly (Huth et al., submitted). This contrasts with the kd of 0.6 μM calculated by Strickland and Puett (4) for the in vitro assembly of urinary α- and β-subunits and suggests that assembly under conditions more closely resembling the intracellular environment may more accurately reflect the true efficiency of α-β dimerization.

Future experiments will be focused on determining the role of molecular chaperones in fostering the folding of β and the assembly of the α-β dimer. We are also interested in preparing sufficient amounts of the intermediates in the β folding pathway to carry out structural studies by

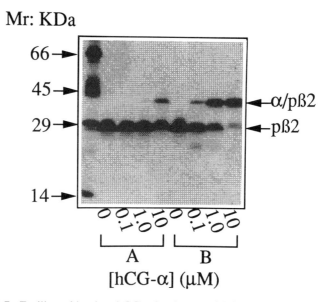

FIGURE 13.7. Facilitated in vitro hCG subunit assembly by a cysteamine/cystamine redox buffer: fluorogram of a 4 °C nonreducing SDS-PAGE separation of the [35S]cysteine-pβ2 and the pα-pβ2 dimer. Lane 1 shows [14C]-labeled molecular weight markers; A lanes show the extent of assembly with hCGα after a 1 h incubation in the absence of the redox buffer; and B lanes show the extent of assembly with hCGα in the presence of the redox buffer (see "Materials and Methods" for the exact buffer conditions).

NMR. To this end, we are developing expression systems that produce high amounts of β-subunit.

*Acknowledgments.* The authors thank Dr. Irving Boime (Departments of Pharmacology and Obstetrics and Gynecology, Washington University, St. Louis) for the hCG gene-transfected CHO cells and Dr. Randall Whitcomb (Reproductive Endocrine Unit, Massachusetts General Hospital) for assay of the biologic activity of hCG produced by JAR cells. We also thank Joan Amato for her careful preparation of the manuscript.

This research is supported by National Cancer Institute Grant CA32949 to Raymond W. Ruddon.

## References

1. Ruddon RW, Hartle JR, Peters BP, Anderson C, Huot RI, Stromberg K. Biosynthesis and secretion of chorionic gonadotropin subunits by organ cultures of first trimester human placenta. J Biol Chem 1981;256:11389–92.
2. Peters BP, Krzesicki RF, Hartle RJ, Perini F, Ruddon RW. A kinetic comparison of the processing and secretion of the αβ dimer and the uncombined α and β subunits of chorionic gonadotropin synthesized by human choriocarcinoma cells. J Biol Chem 1984;259:15123–30.
3. Cole LA, Hartle RJ, Laferla JJ, Ruddon RW. Detection of the free beta subunit of human chorionic gonadotropin (HCG) in cultures of normal and malignant trophoblast cells, pregnancy sera, and sera of patients with choriocarcinoma. Endocrinology 1983;113:1176–8.
4. Strickland TW, Puett D. The kinetic and equilibrium parameters of subunit association and gonadotropin dissociation. J Biol Chem 1982;257:2954–60.
5. Marcillac I, Troalen F, Bidart J-M, et al. Free human chorionic gonadotropin β subunit in gonadal and nongonadal neoplasms. Cancer Res 1992;52:3901–7.
6. Alfthan H, Haglund C, Roberts P, Stenman U-H. Elevation of free β subunit of human choriogonadotropin and core β fragment of human choriogonadotropin in the serum and urine of patients with malignant pancreatic and biliary disease. Cancer Res 1992;52:4628–33.
7. Parsons TF, Bloomfield GA, Pierce JG. Purification of an alternate form of the α subunit of the glycoprotein hormones from bovine pituitaries and identification of its O-linked oligosaccharide. J Biol Chem 1983;258.240–4.
8. Cole LA, Perini F, Birken S, Ruddon RW. An oligosaccharide of the O-linked type distinguishes the free from the combined form of hCG α subunit. Biochem Biophy Res Comm 1984;122:1260–7.
9. Peters BP, Krzesicki RF, Perini F, Ruddon RW. O-glycosylation of the α-subunit does not limit the assembly of chorionic gonadotropin αβ dimer in human malignant and nonmalignant trophoblast cells. Endocrinology 1989; 124:1602–12.
10. Saccuzzo JE, Krzesicki RF, Perini F, Ruddon RW. Phosphorylation of the secreted, free α subunit of human chorionic gonadotropin. Proc Natl Acad Sci USA 1986;83:9493–6.

11. Saccuzzo Beebe J, Krzesicki RF, Norton SE, Perini F, Peters BP, Ruddon RW. Identification and characterization of subpopulations of the free α-subunit that vary in their ability to combine with chorionic gonadotropin-β. Endocrinology 1989;124:1613–24.

12. Saccuzzo Beebe J, Huth JR, Ruddon RW. Combination of the chorionic gonadotropin free β-subunit with α. Endocrinology 1990;126:384–91.

13. Suganuma N, Matzuk MM, Boime I. Elimination of disulfide bonds affects assembly and secretion of the human chorionic gonadotropin β subunit. J Biol Chem 1989;264:19302–7.

14. Huth JR, Mountjoy K, Perini F, Ruddon RW. Intracellular folding pathway of human chorionic gonadotropin β subunit. J Biol Chem 1992;267:8870–9.

15. Bedows E, Huth JR, Ruddon RW. Kinetics of folding and assembly of the human chorionic gonadotropin β subunit in transfected Chinese hamster ovary cells. J Biol Chem 1992;267:8880–6.

16. Ruddon RW, Hanson CA, Addison NJ. Synthesis and processing of human chorionic gonadotropin subunits in cultured choriocarcinoma cells. Proc Natl Acad Sci USA 1979;76:5143–7.

17. Schlaff S. An acrylamide gel recombination assay for glycoprotein hormone subunits. Endocrinology 1976;98:527–33.

18. Whitcomb RW, Schneyer AL. Development and validation of a radioligand receptor assay for measurement of luteinizing hormone in human serum. J Clin Endocrinol Metab 1990;71:591–5.

19. Ruddon RW, Krzesicki RF, Saccuzzo Beebe J, Loesel L, Perini F, Peters BP. Conformational intermediates in the production of the combinable form of the β-subunit of chorionic gonadotropin. Endocrinology 1989;124:862–9.

20. Ruddon RW, Krzesicki RF, Norton SE, Saccuzzo Beebe J, Peters BP, Perini F. Detection of a glycosylated, incompletely folded form of chorionic gonadotropin β subunit that is a precursor of hormone assembly in trophoblastic cells. J Biol Chem 1987;262:12533–40.

21. Mise T, Bahl OP. Assignment of disulfide bonds in the β subunit of human chorionic gonadotropin. J Biol Chem 1981;256:6587–92.

22. Creighton TE. Disulfide bonds as probes of protein folding pathways. Methods Enzymol 1986;131:83–106.

23. Weissman JS, Kim PS. Reexamination of the folding of BPTI: predominance of native intermediates. Science 1991;253:1386–93.

24. Weissman JS, Kim PS. The pro region of BPTI facilitates folding. Cell 1992:841–51.

25. Saccuzzo Beebe J, Mountjoy K, Krzesicki RF, Perini F, Ruddon RW. Role of disulfide bond formation in the folding of human chorionic gonadotropin β subunit into an αβ dimer assembly-competent form. J Biol Chem 1990;265:312–7.

26. Huth JR, Mountjoy K, Perini F, Bedows E, Ruddon RW. Domain-dependent protein folding is indicated by the intracellular kinetics of disulfide bond formation of human chorionic gonadotropin β subunits. J Biol Chem 1992;267:21396–403.

27. Bedows E, Huth JR, Suganuma N, Bartels CF, Boime I, Ruddon RW. Disulfide bond mutations affect the folding of the human chorionic gonadotropin (hCG)-β subunit in transfected CHO cells. J Biol Chem (in press).

28. Hwang C, Sinskey AJ, Lodish HF. Oxidized redox state of glutathione in the endoplasmic reticulum. Science 1992;257:1496–502.
29. Ziegler DM, Poulsen LL. Protein disulfide bond synthesis: a possible intracellular mechanism. Trends Biochem Sci 1977;2:79–81.

# 14

# Structure and Function of the Gonadotropin Free α Molecule of Pregnancy

Diana L. Blithe

There appears to be only a single gene for the gonadotropin α-subunit in humans (1, 2), but the α polypeptide may combine noncovalently with any of several β-subunits to form one of the heterodimeric hormones, LH, FSH, TSH or hCG. Less commonly appreciated is the notion that substantial quantities of α are secreted in uncombined or "free α" forms.

Pregnancy free α molecule is a major placental product. In maternal serum, the levels of free α are relatively low early in the first trimester, rising steadily to reach average concentrations of approximately 350 ng/ml (24 nmol/L) at around 24 weeks of gestation, and they are maintained at this level until near term (3, 4). It is noteworthy that the serum concentrations of hCG reach a peak at 8–12 weeks of gestation and then decline to about 1200 ng/ml (32 nmol/L) for the remainder of the pregnancy (3, 4). The metabolic clearance rate of hCGα-subunit is 49.7 ml/min/m$^2$ (5), whereas the MCR of hCG is 1.9 ml/min/m$^2$ (6); thus, since free α has a clearance rate similar to that of hCGα-subunit (7), the placental production of free α in the third trimester is calculated to be greater than 19-fold that of hCG on a molar basis.

Interestingly, the amount of free α molecule in amniotic fluid follows a different pattern from that in maternal serum. In amniotic fluid, where the concentrations of free α are 10-fold greater than maternal serum levels, free α rises to a maximum by week 15 of pregnancy and begins to decline after 20 weeks (8). Since the source of free α for both amniotic fluid and maternal serum is predominantly trophoblast cells, it is not clear why its distribution into the two compartments should follow a different time course.

Intact hCG can be dissociated into its α- and β-subunits, and the two subunits, hCGα and hCGβ, can reassociate to form the active heterodimeric hormone. A distinguishing characteristic of the pregnancy free α molecule is that it is unable to combine with the hCGβ-subunit (9).

## Glycosylation of α

Posttranslational modifications to the peptide have resulted in a family of α glycoforms. The carbohydrate moieties on free α represent approximately 30%–40% of the molecule. The human α polypeptide consists of 92 amino acids and contains two consensus sequences for *asparagine-linked* (N-linked) glycosylation. Both consensus sequence sites (asn 52 and asn 78) are glycosylated (10).

Studies investigating the sizes of free α and hCGα using gel chromatography before and after glycosidase treatment suggested that differences might exist between the glycosylation of free α and combined hCGα-subunit (11, 12). In order to compare the oligosaccharides on free α with the oligosaccharides on the α-subunit of hCG (designated hCGα), quantities of free α and intact hCG were purified from pregnancy urine pooled throughout the second and third trimesters of individual pregnancies (9). The α-subunit of hCG was dissociated from the β-subunit and the dissociated hCGα-subunit and free α molecules were purified by immunoaffinity chromatography. The oligosaccharides from both free α and hCGα were characterized by size (by molecular sieve chromatography), charge content (by DEAE binding properties), structure (determined by lectin affinity chromatography), and carbohydrate composition (13).

The study of glycosylated hormones is complicated by the existence of multiple glycoforms of any naturally occurring glycoprotein hormone. The term microheterogeneity has been introduced to describe the phenomenon of numerous glycoforms originating from a single polypeptide glycosylation site. After attachment of an oligosaccharide unit to a consensus sequence on the polypeptide, the oligosaccharide is processed by a large number of enzymes located in the endoplasmic reticulum and the Golgi apparatus (reviewed in 14, 15). Processing involves removal of outer sugars and sequential addition of new sugars; however, there is no template to ensure fidelity of structure. The final product depends on which enzymes interact with the oligosaccharide, and a single glycosylation site generally yields a diversity of structures.

Examples of the general classes of oligosaccharide structures associated with N-linked glycosylation are shown in Figure 14.1. The basic common structure for all of the oligosaccharides is a trimannosyl core linked through two N-acetylglucosamine residues to asparagine. Complex structures are defined as those containing one or more antennae (defined as a branch starting with N-acetylglucosamine) on each mannose arm, and they represent the most highly processed forms. High mannose structures

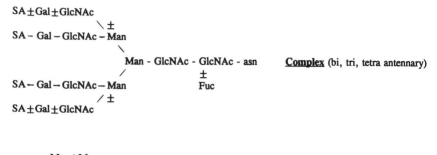

Complex (bi, tri, tetra antennary)

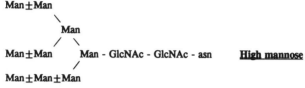

High mannose

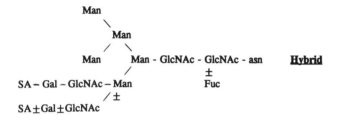

Hybrid

SA  -  sialic acid
Gal  -  galactose
GlcNAc  -  N-acetylglucosamine
Man  -  mannose
Fuc  -  fucose
±  -  denotes possible additional branches
asn - asparagine

FIGURE 14.1. Common potential N-linked structures resulting from oligosaccharide processing.

lack antennae and represent the most unprocessed forms. The combination of a complex antenna on one arm of the trimannosyl core and terminal mannose residues on the other arm is called a hybrid structure to distinguish it from either a high-mannose or a complex structure.

The predominant oligosaccharide structures found on hCGα were hybrids consisting of a single complex antenna linked to mannose on one arm of the trimannosyl core (Fig. 14.2) and 0, 1, or 2 additional mannose residues on the other arm (13). The truncated hybrid (lacking additional mannose residues on the trimannosyl core) is an unusual structure, yet it

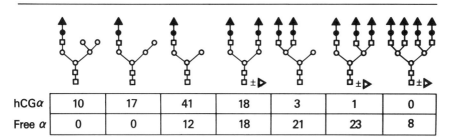

| | | | | | | |
|---|---|---|---|---|---|---|
| hCGα | 10 | 17 | 41 | 18 | 3 | 1 | 0 |
| Free α | 0 | 0 | 12 | 18 | 21 | 23 | 8 |

FIGURE 14.2. Estimated amounts (%) of oligosaccharide structures proposed for hCGα-subunit and for free α. The structures were proposed on the basis of lectin binding properties, DEAE binding properties, glycosidase sensitivity or resistance, and compositional analysis (13). Open square = N-acetylglucosamine; open circle = mannose; solid circle = galactose; open triangle = fucose; and solid triangle = sialic acid. The numbers in the table indicate the percentage of each oligosaccharide in the total population. Additional components that were present in minor amounts exhibited lectin and DEAE binding properties that were not consistent with the structures shown; thus, those components were not assigned specific structures.

represents the most abundant oligosaccharide moiety found on the hCGα-subunit (13). The structures proposed in Figure 14.2 for hCGα have been confirmed by NMR analysis (16).

In contrast to the unbranched structures that were found on hCGα, the oligosaccharides on free α had two or more antennae (9). Additionally, a substantial number of the oligosaccharides on free α contained fucose (13). One explanation for the differential processing of the two α peptides is that when the α-subunit combines with the β-subunit prior to encountering processing enzymes in the Golgi apparatus, the interaction of the β-subunit results in restricted access of the enzymes to the oligosaccharides on hCGα. On the basis of the observed structures, the enzymes that appear to be restricted are mannosidase II, which removes the last two mannose branches from the trimannosyl core (17), N-acetylglucosaminyl transferase II, which adds the starting residue for a second antenna (18), and fucosyltransferase, which adds fucose to the asparagine-linked N-acetylglucosamine (19). It is thought that all of these enzymes require a similar catalytic site on the oligosaccharide (15) and, thus, it is likely that access to that site is restricted when the α-subunit is combined with the β-subunit. However, if combination with β fails to occur, then the processing of free α is unrestricted and the heterogeneity of the glycoforms that are generated will reflect the activities of the processing enzymes in the Golgi at the time of synthesis.

Thus, a markedly different distribution of oligosaccharides was found on free α compared with the population of oligosaccharides present on the combined α-subunit of hCG (13). The differences between the free α oligosaccharides and those of the combined α-subunit occur despite synthesis of both free α and intact hCG in the same cell (20). In view of these dramatic differences, could the presence of branches on the N-linked oligosaccharide structures on free α be responsible for its inability to combine with β-subunit? Although free α of bovine pituitary origin has been shown to contain an O-linked oligosaccharide that can interfere with combination with β-subunit (21), free α from human pregnancy urine does not contain O-linked carbohydrate (7, 13). Yet, the free α molecules in pregnancy are unable to combine with available β-subunits (9).

In order to address the question of whether N-linked glycosylation could interfere with combinability, in vitro cell culture conditions were employed. JEG choriocarcinoma cells produce both hCG and free α (22). The free α molecules that were secreted were unable to combine with hCGβ-subunit. To demonstrate that the N-linked oligosaccharide structures were responsible for the failure to combine, the glycosylation of free α was modified. JEG cells were incubated with Swainsonine, a drug that inhibits α-mannosidase II (23), preventing removal of two peripheral mannose residues and resulting in the formation of hybrid oligosaccharide structures on all of the glycosylation sites that would normally have contained complex oligosaccharide structures (24). The result was a twofold increase in the amount of secreted hCG (25). In addition, although the Swainsonine-treated cells also secreted free α molecules, a substantial portion of these free α molecules could combine with hCGβ-subunit (25). Thus, by restricting the processing of the oligosaccharides on free α to hybrid structures, free α could be maintained in a combinable form.

## Changes in Glycosylation as a Function of Gestational Age

The glycosylation of free α changes as gestation advances (26). Free α obtained from early gestation (weeks 7–12) was compared with free α obtained from late gestation (weeks 28–32) in five individual pregnancies. The oligosaccharide structures on the free α molecules were examined using lectin affinity chromatography on *Concanavalin A* (ConA) and *Lens culinaris* (Lch). Chromatography of free α on ConA resulted in three populations: (i) unbound material, in which neither oligosaccharide on free α could bind to ConA due to the presence of two antennae on a single mannose residue (27); (ii) weakly bound material, in which one oligosaccharide could not bind to ConA while the other oligosaccharide bound with weak affinity; and (iii) tightly bound material, in which either one oligosaccharide binds to ConA with high affinity or both oligosaccharides bind with weak affinity, but together result in tight binding under the elution conditions used. In all five pregnancies there was a decrease in tightly bound forms of free α and a concomitant increase in the unbound forms of free α as gestation advanced (Fig. 14.3). The mean difference in the percentage of ConA tightly bound forms between early and late pregnancy was $17.0 \pm 2.4\%$ ($P < 0.01$). Thus, the free α molecules produced in late pregnancy contain more highly branched oligosaccharides than those produced in early pregnancy.

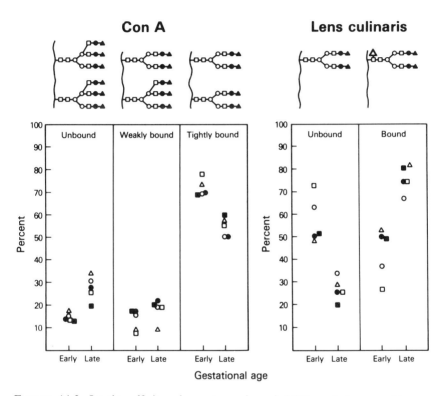

FIGURE 14.3. Lectin affinity chromatography of hCG and free α. Free α molecules were subjected to affinity chromatography on ConA and Lch. The cartoon figures above the boxes represent potential oligosaccharide structures that would account for the binding properties of the material and the symbols in the oligosaccharide cartoons represent the same sugar residues as those for Figure 14.2. With respect to the data points, each symbol represents one pregnant individual, and the same symbol is used to represent material from both early and late gestation. The elution buffer for unbound material was 0.2 M ammonium acetate, pH 7.4, 0.1 mM $CaCl_2$, 0.1 mM $MgCl_2$, 0.1 mM $MnCl_2$, 0.1% BSA, 0.1% $NaN_3$. Weakly bound material was eluted from ConA with elution buffer containing 10 mM α-methyl-D-glucoside. Tightly bound material was eluted from ConA and from Lch with elution buffer containing 500 mM α-methyl-D-mannoside. Free α levels were assayed by RIA (26).

Lch has a binding specificity similar to that of ConA, except for an additional requirement of a fucose residue attached to the asparagine-linked N-acetylglucosamine (28). Chromatography of free α on Lch resulted in two populations: (i) unbound material, in which the oligosaccharides either did not contain fucose or failed to bind because of interference at the mannose binding site, and (ii) bound material, in which at least one oligosaccharide on free α contained fucose and access to the mannose binding site was not impaired. In all five pregnancies, there was a dramatic increase in the amount of fucosylation of free α as gestation advanced (Fig. 14.3). The mean difference in the percentage of Lch-bound forms between early and late pregnancy was $30.1 \pm 4.8\%$ ($P < 0.01$). The most likely explanation for increased fucosylation in late pregnancy is that the amount of fucosyl transferase is elevated in the α producing cells compared to the amount of this enzyme in early pregnancy. Whether this increase is associated with a specific point in gestational development or is the result of a gradual increase between weeks 12 and 28 of gestation remains to be determined.

In both lectin analyses, there was no overlap between the early samples and the late samples in the five individuals that were examined. It is likely that the glycosylation pattern of pregnancy free α will reflect the time of gestation, rather than relate to characteristics of an individual pregnancy. It is possible that such a "fingerprint" may have diagnostic potential if aberrations in glycosylation are identified with particular diseases associated with pregnancy, such as diabetes mellitus.

## Biologic Role for Free α in Pregnancy

Although free α has been demonstrated to be a major placental product, it has no activity at the hCG receptor (29) and was thought to be a by-product of pregnancy. However, recent data indicated that free α purified from pregnancy urine stimulated secretion of prolactin from primary cultures of human decidual cells (30). The amount of prolactin secreted during a 24 h incubation was concentration dependent over a range of increasing doses of α from 0.2 to 20 ng/ml with an $ED_{50}$ of 2 ng/ml (Fig. 14.4). These concentrations are well within the physiologic levels of free α in maternal serum, which average 350 ng/ml during the third trimester. Purified hCGα-subunit reference preparations (CR119 and CR123) also stimulated secretion of prolactin with a similar dose-response curve to that of free α. Incubation of decidual cells with a reference preparation of intact hCG (CR123) at a concentration of 260 ng/ml resulted in some stimulation of prolactin secretion; however, the observed stimulation was completely attributable to contamination of the preparation with free α or dissociated hCGα-subunit. Purified hCGβ-subunit did not stimulate activity in the decidual cell cultures. The effect of α on the stimulated

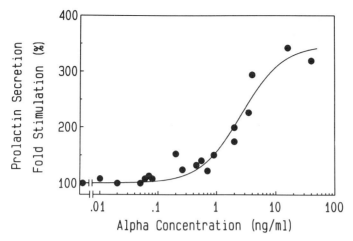

FIGURE 14.4. Dose response curve for α stimulation of hPRL secretion from human decidual cells. Normal term placenta and associated membranes were obtained by cesarean section or vaginal delivery. Decidual tissue was dissected from chorion membranes and subjected to enzymatic digestion. Isolation and culture of decidual cells was performed as described by Handwerger et al. (36). After a 48 h preincubation in RPMI-1640 containing 5% fetal calf serum, the medium was replaced with serum-free RPMI 1640 containing the indicated concentrations of α. After 24 h of incubation, the medium was removed and assayed for hPRL by specific homologous RIA (37) using material provided by the National Hormone and Pituitary Program, NIDDK. Data points represent stimulation obtained by purified free α or a reference preparation of hCGα-subunit (CR119). The data point to the left of the break in the x-axis represents control medium in which no α was added.

release of prolactin was not due to a generalized stimulation of protein synthesis and secretion, since no increase was observed in the release of $^{35}$S-labeled proteins compared to controls. In addition, the observed increase in prolactin secretion was not due to a toxic effect of α, since there was no visible effect on cell viability, and the cellular enzymes, LDH and alkaline phosphatase, were not detected in the culture medium (30).

Several lines of evidence are consistent with free α molecules having a physiologic role in the regulation of prolactin secretion in vivo. In both maternal serum and amniotic fluid, the gestational profiles of prolactin and free α are very similar. In maternal serum, the concentrations of prolactin and free α increase steadily throughout pregnancy, with maximal levels reached near term (3, 4, 31). In amniotic fluid, prolactin and free α both rise to maximal levels by week 15 of pregnancy and begin to decline after 20 weeks (8, 32). Thus, in two separate compartments, free α and prolactin have concordant gestational profiles. Although the synthesis and secretion of decidual and pituitary prolactin have previously been considered to be regulated by quite different mechanisms, it may be that free α plays a role in both systems. In the rat pituitary, LH and dissociated LHα-subunit have been shown to stimulate differentiation of lactotrophs (33). In addition, GnRH, which is also known to stimulate differentiation of pituitary lactotrophs (34), may cause pulsatile secretion of pituitary free α molecules (35), suggesting that free α may be the factor that actually stimulates differentiation of the pituitary lactotrophs.

Thus, placental free α molecules appear to have a functional role in pregnancy of stimulating decidual prolactin secretion. Free α molecules, previously believed to be an inert placental by-product, may prove to be a "new" glycoprotein hormone.

## *References*

1. Fiddes JC, Goodman HM. Isolation, cloning and sequence analysis of the cDNA for the α-subunit of human chorionic gonadotropin. Nature 1980; 281:351–6.
2. Boothby M, Ruddon RW, Anderson C, McWilliams D, Boime I. A single gonadotropin α-subunit gene in normal tissue and tumor-derived cell lines. J Biol Chem 1981;256:5121–7.
3. Ashitaka Y, Nishimura R, Futamura K, Ohashi M, Tojo S. Serum and chorionic tissue concentrations of human chorionic gonadotropin and its subunits during pregnancy. Endocrinol Jpn 1974;21:547–50.
4. Reuter AM, Gaspard UJ, Deville JL, Vrindts-Gevaert Y, Franchimont P. Serum concentrations of human chorionic gonadotrophin and its alpha and beta subunits. Clin Endocrinol 1980;13:305–17.
5. Wehmann RE, Nisula BC. Metabolic clearance rates of the subunits of human chorionic gonadotropin in man. J Clin Endocrinol Metab 1979; 48:753–9.
6. Wehmann RE, Nisula BC. Metabolic and renal clearance rates of purified human chorionic gonadotropin. J Clin Invest 1981;68:184–94.
7. Blithe DL, Nisula BC. Similarity of the clearance rates of free α-subunit and α-subunit dissociated from intact human chorionic gonadotropin, despite differences in sialic acid contents. Endocrinology 1987;121:1215–20.
8. Ozturk M, Brown N, Milunsky A, Wands J. Physiological studies of human chorionic gonadotropin and free subunits in the amniotic fluid compartment compared to those in maternal serum. J Clin Endocrinol Metab 1988; 67:1117–21.
9. Blithe DL, Nisula BC. Variations in the oligosaccharides on free and combined alpha subunits of human choriogonadotropin in pregnancy. Endocrinology 1985;117:2218–28.
10. Kessler MJ, Reddy MS, Shah RH, Bahl OP. Structures of N-glycosidic carbohydrate units of human chorionic gonadotropin. J Biol Chem 1979; 254:7901–8.
11. Fein HG, Rosen SW, Weintraub BD. Increased glycosylation of serum hCG and subunits from eutopic and ectopic sources: comparison with placental and urinary forms. J Clin Endocrinol Metab 1980;50:1111–20.

12. Posillico EG, Handwerger S, Tyrey L. Demonstration of intracellular and secreted forms of large human chorionic gonadotrophin alpha subunit in cultures of normal placental tissue. Placenta 1983;4:439–48.
13. Blithe DL. Carbohydrate composition of the alpha subunit of human choriogonadotropin and the free alpha molecules produced in pregnancy: most free alpha and some combined hCG-alpha molecules are fucosylated. Endocrinology 1990;126:2788–99.
14. Kornfeld R, Kornfeld S. Assembly of asparagine-linked oligosaccharides. Annu Rev Biochem 1985;54:631–64.
15. Schachter H. Biosynthetic controls that determine the branching and microheterogeneity of protein-bound oligosaccharides. Biochem Cell Biol 1986; 64:163–81.
16. Weishaar G, Hiyama J, Renwick AGC. Site-specific N-glycosylation of human chorionic gonadotropin: structural analysis of glycopeptides by one- and two-dimensional $^1$H NMR spectroscopy. Glycobiology 1991;1:393–404.
17. Tabas I, Kornfeld S. The synthesis of complex-type oligosaccharides: identification of an α-D-mannosidase activity in a late stage of processing of complex-type oligosaccharides. J Biol Chem 1978;253:7779–86.
18. Harpaz H, Schachter H. Control of glycoproteiin synthesis: bovine colostrum UDP-N-acetylglucosamine: α-D-mannoside β2-N-acetylglucosaminyltransferase I. Separation from UDP-N-acetylglucosamine: α-D-mannoside β2-N-acetylglucosaminyltransferase II, partial purification and substrate specificity. J Biol Chem 1980;255:4885–93.
19. Longmore GD, Schachter H. Product-identification and substrate-specificity studies of the GDP-L-fucose: 2-acetamido-2-deoxy-β-D-glucoside (Fuc goes to Asn-linked GlcNAc) 6-α-L-fucosyltransferase in a Golgi-rich fraction from porcine liver. Carbohydr Res 1982;100:365–92.
20. Corless CL, Bielinska MB, Ramabhadran TV, et al. Gonadotropin α subunit: differential processing of free and combined forms in human trophoblast and transfected mouse cells. J Biol Chem 1987;262:14197–203.
21. Parsons TF, Pierce JG. Free α-like material from bovine pituitaries: removal of its O-linked oligosaccharide permits combination with lutropin-β. J Biol Chem 1984;259:2662–6.
22. Benveniste R, Conway MC, Puett D, Rabinowitz D. Heterogeneity of the human chorionic gonadotropin α-subunit secreted by cultured choriocarcinoma (JEG) cells. J Clin Endocrinol Metab 1979;48:85–91
23. Tulsiani DRP, Harris TM, Touster O. Swainsonine inhibits the biosynthesis of complex glycoproteins by inhibition of Golgi Mannosidase II. J Biol Chem 1982;257:7936–9.
24. Kang MS, Elbein AD. Alterations in the structure of the oligosaccharides of vesicular stomatitis virus G protein by Swainsonine. J Virol 1983;46:60–9.
25. Blithe DL. N-linked oligosaccharides on free alpha interfere with its ability to combine with human chorionic gonadotropin-beta subunit. J Biol Chem 1990;265:21951–6.
26. Skarulis MC, Wehmann RE, Nisula BC, Blithe DL. Glycosylation changes in human chorionic gonadotropin and free alpha subunit as gestation progresses. J Clin Endocrinol Metab 1992;75:91–6.
27. Kornfeld R, Ferris C. Interaction of immunoglobulin glycopeptides with Concanavalin A. J Biol Chem 1975;250:2614–9.

28. Kornfeld K, Reitman ML, Kornfeld R. The carbohydrate binding specificity of pea and lentil lectins: fucose is an important determinant. J Biol Chem 1981;256:6633–40.
29. Canfield RE, Morgan FJ, Kammerman S, Bell JJ, Agnosto GM. Studies of human chorionic gonadotropin. Recent Prog Horm Res 1971;27:121–64.
30. Blithe DL, Richards RG, Skarulis MC. Free alpha molecules from pregnancy stimulate secretion of prolactin from human decidual cells: a novel function for free alpha in pregnancy. Endocrinology 1991;129:2257–9.
31. Tyson JE, Hwang P, Guyda H, Friesen HG. Studies of prolactin secretion in human pregnancy. Am J Obstet Gynecol 1972;113:14–20.
32. Clements JA, Reyes FI, Winter JSD, Faiman C. Studies on human sexual development, IV. Fetal pituitary and serum, and amniotic fluid concentrations of prolactin. J Clin Endocrinol Metab 1977;44:408–13.
33. Begeot M, Hemming FJ, Dubois PM, Combarnous Y, Dubois PM, Aubert ML. Induction of pituitary lactotrope differentiation by luteinizing hormone α subunit. Science 1984;226:566–8.
34. Begeot M, Hemming FJ, Martinat N, Dubois MP, Dubois PM. Gonadotropin releasing hormone (GnRH) stimulates immunoreactive differentiation of pituitary lactotropes. Endocrinology 1983;112:2224–6.
35. Spratt DI, Chin WW, Ridgway EC, Crowley WF. Administration of low dose pulsatile gonadotropin-releasing hormone to GnRH-deficient men regulates free alpha-subunit secretion. J Clin Endocrinol Metab 1986;62:102–8.
36. Handwerger S, Harmon I, Costello A, Markoff E. Cyclic AMP inhibits the synthesis and release of prolactin from human decidual cells. Mol Cell Endocrinol 1987;50:99–106.
37. Sinha YN, Selby FW, Lewis UJ, Vanderlaan WP. A homologous radioimmunoassay for human prolactin. J Clin Endocrinol Metab 1973;36:509–16.

# 15

# Glycosylation and Glycoprotein Hormone Function

Jacques U. Baenziger

*Lutropin* (LH), *follitropin* (FSH), *thyrotropin* (TSH), and *chorionic gonadotropin* (CG) share a common α-subunit and have highly homologous β-subunits. Even though these glycoproteins have closely related structures at the primary, secondary, and tertiary levels, the structures of their Asn-linked oligosaccharides differ. LH and TSH bear oligosaccharides that terminate with the unique sequence SO$_4$-4GalNAcβ1, 4GlcNAcβ1,2Manα (S4GGnM), whereas FSH and CG bear oligosaccharides that terminate with the sequence sialic acid α2,3/6Galβ1, 4GlcNAcβ1,2Manα (1–6). The presence of unique oligosaccharide structures on specific members of the glycoprotein hormone family raises many questions about the regulation of their synthesis and their biologic significance. We have determined that the synthesis of these sulfated oligosaccharides reflects the activity of two highly specific enzymes, a GalNAc-transferase and a sulfotransferase. The expression of these transferases is hormonally regulated, and the sulfated oligosaccharides produced play a key role in the expression of LH bioactivity in vivo by controlling its circulatory half-life following release into the blood.

## Synthesis of Oligosaccharides Terminating with SO$_4$-4GalNAcβ1,4GlcNAcβ1,2Manα

The synthetic pathway for the sulfated and sialylated oligosaccharides found on the glycoprotein hormones is illustrated in Figure 15.1. The structures enclosed within the box are intermediates in the synthetic pathway that are common for sulfated and sialylated oligosaccharides. The identical oligosaccharide intermediates act as acceptors for the addition of GalNAc and Gal by their respective glycosyltransferases. As shown in Figure 15.2, a GalNAc-transferase is present in pituitary mem-

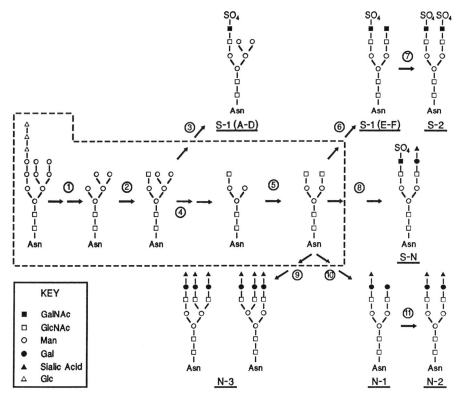

FIGURE 15.1. Proposed pathway for the synthesis of sulfated and sialylated oligosaccharides on the pituitary glycoprotein hormones.

branes that discriminates between identical oligosaccharide acceptors on LH, CG, and other glycoproteins, including FSH. Galactosyltransferase does not display protein specificity using the glycoproteins in Figure 15.3 as acceptor (7).

We have examined the specificity of the GalNAc-transferase (7–9) and determined that it recognizes a tripeptide motif found in the α-subunit, as well as the β-subunits of LH and hCG. This tripeptide motif consists of a ProXaaArg/Lys located 6–9 residues N-terminal to a glycosylated Asn (Fig. 15.3) and is absent from the β-subunit of FSH due to an amino-terminal deletion within the FSHβ gene as compared to LHβ and CGβ. Combination of FSHβ with the α-subunit prevents recognition of the α-subunit tripeptide motif by the GalNAc-transferase (Fig. 15.3). The apparent Km for GalNAc addition to the synthetic intermediate $GlcNAc_2Man_3GlcNAc_2Asn$ on proteins containing the tripeptide motif is 4–13 μM, as compared to 1–2 mM for addition to the same oligosaccharide acceptor on proteins that do not contain the tripeptide motif (7).

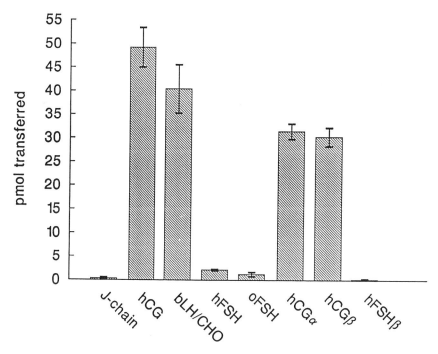

FIGURE 15.2. Comparison of glycoproteins as substrates for the GalNAc-transferase. Each glycoprotein was digested with neuraminidase and β-galactosidase to produce oligosaccharides with the structure of the product of reaction 5 in Figure 15.1. Assays contained 4 μM substrate, 720 μM UDP-GalNAc, and 385 μU/ml of GalNAc-transferase activity. J-chain contains one Asn-linked oligosaccharide, whereas hCG, hFSH, and oFSH contain four, and bLH/CHO contains three. The isolated subunits hCGα, hCGβ, and hFSHβ each contain two Asn-linked oligosaccharides.

hCGβ        S-K-E-P-L-R-P-R  C  R-P-I-N-A-T-L-A-V-E-K
bLHβ        S-R-G-P-L-R-P-L-C-Q-P-I-N-A-T-L-A-A-E-K
hFSHβ                N-S-C-E-L-T-N-I-T-I-A-I-E-K
hCGα  R-A-Y-P-T-P L R-S-K-K-T-M-L-V-Q-K-N-V-T-S-E

FIGURE 15.3. Alignment of peptides known to be recognized by the glycoprotein hormone-specific GalNAc-transferase and comparison with the amino-terminal sequence of hFSHβ. The amino acid sequences represented in the single letter code are aligned with respect to glycosylated Asn residues (dashed box). The ProXaaArg/Lys motif required for recognition by the GalNAc-transferase is underlined. The amino acid sequences correspond to amino acids 1–20 for hCGβ, 1–20 for bLHβ, 1–14 for hFSHβ, and 35–56 for human α-subunit.

In contrast to the GalNAc-transferase, the sulfotransferase requires only the presence of the trisaccharide sequence GalNAcβ1,4GlcNAcβ1,2Manα for transfer of sulfate to the 4-hydroxyl of the GalNAc (10, 11). The GalNAc-transferase and the sulfotransferase are both expressed in pituitary, whereas neither is detected in human placental tissue, accounting for the absence of sulfated oligosaccharides on hCG despite the presence of the tripeptide recognition motif (8, 11).

LH synthesis and secretion by gonadotropes increase and decrease in response to estrogen levels (12). Using sensitive and highly specific assays for the PXR/K-specific GalNAc-transferase (13) and GGnM-4-sulfotransferase (8), we examined the effect of ovariectomy and ovariectomy with estrogen replacement on the levels of transferase expression in pituitary. Like LH, the levels of PXR/K-specific GalNAc-transferase and GGnM-4-sulfotransferase in rat pituitaries increase in response to ovariectomy. Administration of exogenous 17β-estradiol to ovariectomized rats results in a return of LH to basal levels in the pituitary. Similarly, the levels of PXR/K-specific GalNAc-transferase and GGnM-4-sulfotransferase also return to basal levels in response to exogenous estradiol. Thus, the PXR/K-specific GalNAc-transferase and GGnM-4-sulfotransferase respond to estrogen levels in the same manner and to a similar extent as LH. As a result, the Asn-linked oligosaccharides on LH terminate with GalNAc-4-SO$_4$, regardless of the level of LH synthesis by the gonadotrope.

## Functional Significance of Sulfated Oligosaccharides on LH

The presence of unique sulfated oligosaccharides on LH from a number of different animal species raises the question of its functional significance. The importance of this question is amplified by the knowledge that expression of the GalNAc- and sulfotransferases responsible for the synthesis of these sulfated oligosaccharides is modulated by estrogen levels, assuring that LH oligosaccharides terminate with the S4GGnM sequence at all levels of LH synthesis. We have examined three potential functions that could be attributed to these unique oligosaccharides: (i) directing the sorting of LH to the appropriate storage granule, (ii) modulating the binding to or activation of the LH/CG receptor, or (iii) controlling the circulatory half-life of LH following release into the blood.

Terminal glycosylation of LH can be altered in cultured cells by the use of selective inhibitors of oligosaccharide processing and by inhibition of sulfate addition. Using such agents, we have not been able to demonstrate that either the presence of sulfate or the structure of the oligosaccharides on LH play an essential role in directing the hormone to granules in primary cultures of bovine gonadotropes. Even though LH and FSH are synthesized within the same cells, they are segregated to separate granules within the gonadotrope. Terminal glycosylation does not appear to play an essential role in directing LH into the regulated pathway of secretion, but may have an impact on the segregation of LH and FSH.

We have also examined the impact of terminal glycosylation on binding to and activation of the LH/CG receptor using cultured MA-10 cells (14, 15). Removal of the terminal sulfate from native bovine LH does not have any impact on binding to the LH/CG receptor, activation of adenylcyclase, or production of progesterone. Recombinant bovine LH expressed in CHO cells bears oligosaccharides terminating with sialic acid α2,3Gal, rather than GalNAc-4-SO$_4$. Recombinant LH is less potent than native LH; however, following enzymatic removal of the terminal sialic acid, recombinant LH has the same potency as native LH. Thus, the presence of terminal sialic acid reduces the potency of LH, whereas the presence of terminal sulfate has no effect at the level of the LH/CG receptor in vitro. Since sulfate and sialic acid both are negatively charged, the reduced potency seen with sialic acid is more likely to be related to steric effects than charge. Furthermore, the detailed structural features of the oligosaccharides on LH do not appear to have a marked impact on binding to or activation of the LH/CG receptor.

Terminal glycosylation of the Asn-linked oligosaccharides on LH has a major impact on its circulatory half-life and site of clearance from the blood (15, 16). Native LH is cleared from the blood 5-fold more rapidly

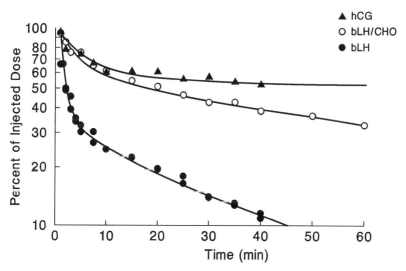

FIGURE 15.4. Plasma clearance rates for bovine LH (bLH), recombinant bovine LH (bLH/CHO), and hCG. Hormones labeled with $^{125}$I (0.5–2.0 × 10$^6$ cpm, 50–200 ng) were injected into heparinized anesthetized rats. Samples (200 μl) were drawn from the carotid artery at the times indicated. Radioactivity is expressed as the percent of the injected dose per ml of plasma.

than recombinant LH bearing oligosaccharides terminating with sialic acid-Gal (Fig. 15.4). The fact that the only difference between native LH and recombinant LH resides in the identity of their terminal sugars indicates that the GalNAc-4-SO$_4$ is responsible for the rapid clearance of native LH. Based on this observation, we were able to demonstrate that LH is rapidly removed from the circulation by a receptor that is present in hepatic endothelial/Kupffer cells. This receptor specifically recognizes oligosaccharides with the terminal sequence SO$_4$-4GalNAcβ1, 4GlcNAcβ1,2Manα. LH is bound by the S4GGnM receptor with an apparent Km of $1.6 \times 10^{-7}$ M and is rapidly internalized and degraded. Hepatic endothelial cells express >500, 000 S4GGnM binding sites per cell at their surface, indicating that this receptor system has the capacity to remove from the blood large amounts of LH and other glycoproteins bearing oligosaccharides terminating with S4GGnM.

What is the functional significance of the S4GGnM receptor and the sulfated oligosaccharides on LH? We have compared the ability of different forms of LH to stimulate ovulation in mice (15). The potency of LH to stimulate ovulation correlates with both its affinity for the LH/CG receptor and its circulatory half-life. As a result, the level of circulating LH that must be attained to stimulate ovulation is directly related to its rate of clearance from the blood, as well as its affinity for the LH/CG receptor. A characteristic of LH secretion is its pulsatile pattern (17–19), which requires rapid clearance of released LH from the blood. Since the LH/CG receptor is down-regulated following ligand binding (20, 21), this pulsatile pattern of appearance in the blood may reflect a requirement to maintain the maximum number of LH/CG receptors in the ground state at the time of exposure to a new pulse of LH. Thus, reduced potency due to rapid clearance from the blood would be compensated for by maximal activation of cAMP production at the level of the LH/CG receptor.

Additional evidence for the importance of terminal glycosylation for expression of LH bioactivity in vivo comes from a comparison of equine LH and CG, which have identical α- and β-subunit peptides, but are synthesized in pituitary and placenta, respectively. Equine LH bears oligosaccharides terminating with SO$_4$-4GalNAcβ1,4GlcNAcβ1,2Manα, whereas equine CG oligosaccharides terminate with sialic acid α-Galβ1, 4GlcNAcβ1,2Manα (22). Equine LH, but not equine CG, is recognized by the S4GGnM receptor and is rapidly cleared from the circulation. Thus, the major difference between equine CG and LH lies in their glycosylation and its affect on circulatory half-life, suggesting rapid clearance is essential for ovulation, but not for maintenance of pregnancy.

## Conclusions

The sulfated oligosaccharides found on LH are critical for the expression of its in vivo bioactivity. The sulfated oligosaccharides are recognized by an abundant receptor in hepatic endothelial/Kupffer cells, which mediates the rapid clearance of LH from the circulation. The GalNAc-transferase responsible for synthesis of these structures recognizes a tripeptide motif, ProXaaArg/Lys, which is found on both the α- and β-subunits of LH. In contrast, the sulfotransferase requires only the trisaccharide sequence GalNAcβ1,4GlcNAcβ1,2Manα for transfer. Both transferases are regulated by estrogen levels in a manner similar to that of LH, assuring that LH bears oligosaccharides terminating with $SO_4$-4GalNAcβ1, 4GlcNAcβ1,2Manα and will be cleared by the S4GGnM receptor.

*Acknowledgments.* This work was supported by NIH Grants R37-CA21923 and R01-DK41738.

## References

1. Baenziger JU, Green ED. Structure, synthesis, and function of the asparagine-linked oligosaccharides on pituitary glycoprotein hormones. In: Ginsberg V, Robbins PW, eds. Biology of carbohydrates; vol 3. London: JAI Press, 1991:1–46.
2. Green ED, Baenziger JU. Asparagine-linked oligosaccharides on lutropin, follitropin, and thyrotropin, I. Structural elucidation of the sulfated and sialylated oligosaccharides on bovine, ovine, and human pituitary glycoprotein hormones. J Biol Chem 1988;263:25–35.
3. Green ED, Baenziger JU. Asparagine-linked oligosaccharides on lutropin, follitropin, and thyrotropin, II. Distributions of sulfated and sialylated oligosaccharides on bovine, ovine, and human glycoprotein hormones. J Biol Chem 1988;263:36–44.
4. Green ED, Van Halbeek H, Boime I, Baenziger JU. Structural elucidation of the disulfated oligosaccharide from bovine lutropin. J Biol Chem 1985;260:15623–30.
5. Endo Y, Yamashita K, Tachibana Y, Tojo S, Kobata A. Structures of the asparagine-linked sugar chains of human chorionic gonadotropin. J Biochem (Tokyo) 1979;85:669–79.
6. Mizuochi T, Kobata A. Different asparagine-linked sugar chains on the two polypeptide chains of human chorionic gonadotropin. Biochem Biophys Res Commun 1980;97:772–8.
7. Smith PL, Baenziger JU. Recognition by the glycoprotein hormone-specific N-acetylgalactosaminetransferase is independent of hormone native conformation. Proc Natl Acad Sci USA 1990;87:7275–9.
8. Smith PL, Baenziger JU. A pituitary N-acetylgalactosamine transferase that specifically recognizes glycoprotein hormones. Science 1988;242:930–3.
9. Smith PL, Baenziger JU. Molecular basis of recognition by the glycoprotein hormone-specific N-acetylgalactosamine-transferase. Proc Natl Acad Sci USA 1992;89:329–33.
10. Skelton TP, Hooper LV, Srivastava V, Hindsgaul O, Baenziger JU. Characterization of a sulfotransferase responsible for the 4-O-sulfation of terminal β-N-acetyl-D-galactosamine on asparagine-linked oligosaccharides of glycoprotein hormones. J Biol Chem 1991;266:17142–50.

11. Green ED, Morishima C, Boime I, Baenziger JU. Structural requirements for sulfation of asparagine-linked oligosaccharides of lutropin. Proc Natl Acad Sci USA 1985;82:7850–4.

12. Dharmesh SM, Baenziger JU. The levels of lutropin, ProXaaArg/Lys-specific GalNAc-transferase, GalNAcβ1,4GlcNAcβ1,2Manα-4-sulfotransferase are coordinately regulated in the pituitary (in preparation).

13. Mengeling BJ, Smith PL, Stults NL, Smith DF, Baenziger JU. A microplate assay for analysis of solution phase glycosyltransferase reactions: determination of kinetic constants. Anal Biochem 1991;199:286–92.

14. Smith PL, Kaetzel D, Nilson J, Baenziger JU. The sialylated oligosaccharides of recombinant bovine lutropin modulate hormone bioactivity. J Biol Chem 1990;265:874–81.

15. Baenziger JU, Kumar S, Brodbeck RM, Smith PL, Beranek MC. Circulatory half-life but not interaction with the lutropin/chorionic gonadotropin receptor is modulated by sulfation of bovine lutropin oligosaccharides. Proc Natl Acad Sci USA 1992;89:334–8.

16. Fiete D, Srivastava V, Hindsgaul O, Baenziger JU. A hepatic reticuloendothelial cell receptor specific for $SO_4$-4-GalNAcβ1,4GlcNAcβ1,2Manα that mediates rapid clearance of lutropin. Cell 1991;67:1103–10.

17. Veldhuis JD, Carlson ML, Johnson ML. The pituitary gland secretes in bursts: appraising the nature of glandular secretory impulses by simultaneous multiple-parameter deconvolution of plasma hormone concentrations. Proc Natl Acad Sci USA 1987;84:7686–90.

18. Veldhuis JD, Evans WS, Johnson ML, Wills MR, Rogol AD. Physiological properties of the luteinizing hormone pulse signal: impact of intensive and extended venous sampling paradigms on its characterization in healthy men and women. J Clin Endocrinol Metab 1986;62:881–91.

19. Veldhuis JD, Johnson ML. A novel general biophysical model for simulating episodic endocrine gland signaling. Am J Physiol 1988; 255:E749–59.

20. Wang H, Segaloff DL, Ascoli M. Lutropin/choriogonadotropin downregulates its receptor by both receptor-mediated endocytosis and a cAMP-dependent reduction in receptor mRNA. J Biol Chem 1991;266:780–5.

21. Rodriguez MC, Xie Y-B, Wang H, Collison K, Segaloff DL. Effects of truncations of the cytoplasmic tail of the luteinizing hormone/chorionic gonadotropin receptor on receptor-mediated hormone internalization. Mol Endocrinol 1992;6:327–36.

22. Smith PL, Bousfield GS, Kumar S, Fiete D, Baenziger JU. Equine lutropin and chorionic gonadotropin bear oligosaccharides terminating with $SO_4$-4-GalNAc and sialic acidα2,3Gal respectively. J Biol Chem 1993;268:795–802.

# Part V

## Receptor Structure/Function

# 16

# Structure and Regulated Expression of the TSH Receptor Gene: Differences and Similarities to Gonadotropin Receptors

Motoyasu Saji, Shoichiro Ikuyama, Hiroki Shimura,
Toshiaki Ban, Shinji Kosugi, Akinari Hidaka,
Fumikazu Okajima, Yoshie Shimura, Cesidio Giuliani,
Giorgio Napolitano, Kazuo Tahara, Takashi Akamizu,
and Leonard D. Kohn

Glycoprotein hormones—*lutropin/chorionic gonadotropin* (LH/CG), *follicle stimulating hormone* (FSH), and *thyrotropin* (TSH)—have a high degree of primary and tertiary structure similarity, yet are target tissue specific (1–5). The structural similarities of the hormones are complemented by functional similarities (1–5). Thus, all are well recognized to activate the cAMP signal transducing system, and all modulate both growth and differentiation of their respective target tissues. It is increasingly evident, in addition, that all depend on the action of other hormones, such as insulin and *insulin-like growth factor I* (IGF-I), to regulate growth and differentiated function (6–8).

As early as 1975, studies in our laboratory (9) suggested that the *TSH receptor* (TSHR) was structurally related to gonadotropin receptors; this idea was reinforced (10) when the cloning of the receptors in 1989–1990 allowed the full extent of the similarities to be appreciated. The similarities in structure are certainly paralleled by differences; for example, the TSHR structure is prominently associated with organ-specific autoimmune disease, whereas the gonadotropin receptor structure is not (11). Nevertheless, the differences have become assets in defining the structure and function of all glycoprotein hormone receptors. For example, the ready pool of TSHR autoantibodies has complemented site-directed mutagenesis studies that have defined (i) ligand binding sites on the extracellular domain of the receptors and (ii) the interaction of these sites with G protein coupling domains.

This chapter summarizes studies in three areas: (i) the ligand binding sites on the extracellular domain of the receptor and their coupling to G protein signals; (ii) regulation of receptor expression; and (iii) elements and factors regulating the TSHR promoter that are relevant to understanding tissue-specific expression, autoregulation of receptor expression by hormone-induced signals, and the multihormonal control of growth and function. Data reviewed focus primarily on studies of the rat TSHR; the applicability of rat TSHR studies to human TSHR will become apparent. The bases for the use of the rat TSHR should, however, be noted. First, the rat FRTL-5 thyroid cell line in continuous culture exists as a well-described and widely used model of TSH-dependent growth and function, TSH linkage to multiple signals, and the impact of other hormones on growth, function, TSH, and *TSHR autoantibody* (TSHRAb) action (7, 8, 12–15). Second, FRTL-5 cells are also the most widely used bioassay system to detect human TSHRAb (7, 8, 14, 15).

## Ligand Binding Sites on the Extracellular Domain of the TSH Receptor and Their Coupling to G Protein Signals

Like the LH/CG and FSH receptors, the TSHR has a long hydrophilic extracellular domain followed by a region with 7 hydrophobic, membrane spanning domains and a short cytoplasmic tail (Fig. 16.1) (16–31). The major difference is that the TSHR has approximately 64 more amino acid residues than the LH/CG receptor, most of which are located in two inserts of the extracellular domain. A long insert containing 50–60 of these amino acids is located in the region of residues 300–400,[1] and a short insert at residues 38–45 (Fig. 16.1, dark residues).

The TSHR cDNA contains a long open reading frame encoding a protein comprised of 764 amino acids, $M_r$ approximately 86,500 based on the amino acid content (Fig. 16.1). The first in-frame ATG is followed by a 21-residue hydrophobic sequence (Fig. 16.1), which is a signal peptide important for processing. The hydrophilic region contains five potential N-linked glycosylation sites in the rat receptor (Fig. 16.1), but six in the human receptor. Trypsinization of thyroid cells releases a 15–30 kd receptor fragment into the medium that can bind TSH and results in a cell that loses TSH or TSHRAb-increased cAMP levels (9, 32). Thus, consistent with evidence from site-directed mutagenesis (see below) and direct measurements of binding in gonadotropin receptors (21, 33), the extracellular domain is the primary site with which the hormone and autoantibodies interact. The cytoplasmic surface of the transmembrane region contains the G protein coupling sites and is homologous with that

---

[1] All residue numbers are determined by counting from the methionine start site.

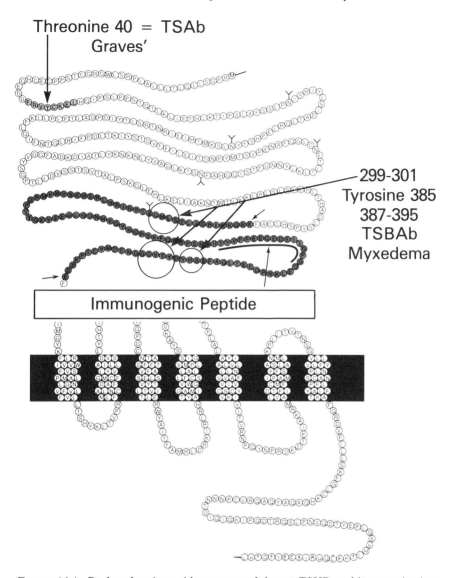

FIGURE 16.1. Deduced amino acid sequence of the rat TSHR and its organization into a long, hydrophilic external domain, a 7-transmembrane domain, and a short cytoplasmic domain. Five potential glycosylation sites (extracellular domain, Y) are noted. The dark circles define the residues in the TSHR where there is no or little homology to gonadotropin receptors and where the extra amino acids of the TSHR reside. The location of three critical domains for autoantibodies to the receptor are noted: a critical immunogenic peptide, residues 352–366; a residue critical for stimulating TSHRAb (TSAb) activity, Graves' disease, and hyperthyroidism; and residues (circled) critical for blocking TSHRAb (TSBAb) activity, idiopathic myxedema, and hypothyroidism.

for adrenergic and cholinergic receptors (10, 34–36). The fundamental difference between adrenergic or cholinergic receptors (34, 35) and receptors for glycoprotein hormones (16–31) is the large extracellular domain (Fig. 16.1).

There is a single TSHR gene (27, 37, 38). The human TSHR gene was mapped to chromosome 14, and the mouse to chromosome 12 (27, 37). Yet, transfected into nonthyroid cells, TSHR cDNA results in cells wherein TSH stimulates both the cAMP and $PIP_2$ signal cascades (39, 40); adrenergic receptors—that is, $\beta$- and $\alpha_1$-AR—couple to only one of these signals, respectively (34–36).

The major approach used to define the TSH binding sites on the extracellular domain, their coupling to the transmembrane domain, and the activation of G proteins involved (i) site-directed mutagenesis and (ii) the definition of the epitopes for TSHRAbs (40–47). We deleted or mutated a portion of the extracellular domain of the TSHR and, as appropriate, substituted residues from the gonadotropin receptors (40–49). The mutant receptor was then transfected into a Cos-7, TSHR-negative cell line, and the ability of the transfected cell to bind or respond to TSH, stimulating TSHRAbs, or blocking TSHRAbs was measured.

Stimulating TSHRAbs increase cAMP levels, are present in patients with Graves' disease, and are associated with goiter (growth) and hyperthyroidism (hyperfunction) (13–15). Blocking TSHRAbs are found in patients with idiopathic myxedema; they inhibit TSH binding, block the action of TSH, as well as stimulating TSHRAbs in cAMP assays, and are associated with atrophic thyroid glands, as well as hypothyroidism (13–15). Studies with monoclonal antibodies to the TSHR predicted that stimulating TSHRAbs and receptor autoantibodies that inhibited TSH binding were different, had different receptor epitopes, and were competitive agonists or antagonists of TSH (50–54), respectively. Use of the antibodies thus allowed the definition of TSH agonist and antagonist sites on the extracellular domain; more important, the use of multiple antibodies, as well as TSH, resulted in full or nearly full retention of one activity by the mutant receptor. This insured that the receptor was synthesized and incorporated into the bilayer, along with Western blotting by a specific antibody to the TSHR (45).

The importance of this point is emphasized in an early study (55) of the short insert in the TSHR, relative to gonadotropin receptors, residues 38–45. Deletion of residues 38–45 or substitutions with serine and alanine, which tried to preserve the generally hydrophilic characteristics of this region, resulted in the loss of all activities in transfected cells (48). This was interpreted as the loss of an important high-affinity TSH binding site (55). Subsequent studies (46, 56) showed that the loss of TSH binding and activity, as well as stimulating TSHRAb activity, was caused by the loss of cysteine 41—that is, the loss might well be explained by a change in receptor tertiary structure and conformation. As noted below, subsequent studies (43–49, 57–59) showed this region was important as a component of the stimulating TSHRAb epitope—that is, it was better linked to a low-affinity rather than high-affinity TSH binding site; it appears to be an agonist rather than an antagonist site.

Residues 303–382 of the TSHR, which are in the region with little homology to gonadotropin receptors, can be deleted with no loss of TSH binding or function, yet they include the most hydrophilic, immunodominant portion of the extracellular domain (Fig. 16.1), residues 352–372 (41, 42, 60–62, 140). Thus, better than 80% of antibody activity in rabbits immunized with the extracellular domain of the receptor reacted with peptide 352–366 or 357–372 (62, 140). In addition, peptide 352–366 reacted with IgG preparations from >80% of hyperthyroid Graves patients (41, 42). The peptides could also induce the formation of receptor autoantibodies related to blocking, but not *TSH binding inhibiting* (TBIAb) activity (140) and to sites that are important both for TBIAb and blocking activity, which flank residues 303–382 (42, 46). Finally, antibodies to peptide 352–366 can specifically identify three major forms of the TSHR on Western blots of detergent-solubilized membrane preparations from Cos-7 cells transfected with full-length rat and human TSHR cDNA (45): 230, 180, and 95–100 kd (Fig. 16.2). The 95–100 kd protein appears to be the processed, glycosylated, functional receptor on the cell surface; the 230 kd protein is a nonprocessed early synthetic form of the functional TSHR; the 180 kd protein appears to be a processed intermediate between the 230 kd early synthetic form and the 95–100 kd functional receptor, rather than a dimer of the latter (45). In sum, this region of the TSHR, which has little or no homology with gonadotropin receptors (Fig. 16.1), is not important for function, but is critical for TSHR immunoreactivity in autoimmune thyroid disease.

Residues important both for high-affinity TSH binding and blocking TSHRAb activity are tyrosine 385 and residues 295–306 and 387–395, particularly cysteines 301 and 390 (Fig. 16.1) (42, 43, 46, 47, 49). Thus, deletion or mutation of these residues eliminates measurable high-affinity TSH binding and causes a 50-fold decrease in the ability of TSH to increase cAMP levels (42, 43, 46). Stimulating TSHRAb activity is, however, almost fully retained relative to the wild-type transfectants. Preservation of stimulating TSHRAb activity allows the conclusion that the receptor is incorporated within the membrane; Western blot data are consistent with this conclusion. The identification of these sites as blocking TSHRAb sites was established when IgG preparations from the sera of hypothyroid patients with idiopathic myxedema were shown to lose the ability to inhibit the stimulating TSHRAb activity that was preserved (42, 43, 46).

Whereas the high-affinity TSH binding site is on the C-terminal portion of the external domain of the TSHR, stimulating TSHRAb epitopes that define an agonist TSH site appear to be on the N-terminal region of the

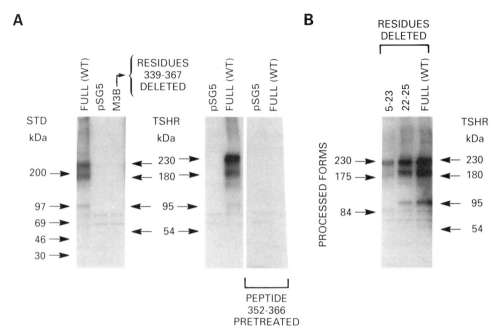

FIGURE 16.2. Multiple TSHR forms exist in detergent-solubilized crude mem-brane preparations from Cos-7 cells transfected with TSHR and Western blotted with an antibody to the immunogenic peptide identified in Figure 16.1. *A:* Cos-7 cells were transfected with full length (Full) or wild-type (WT) rat TSHR cDNA, with pSG5 vector, or with a mutant TSHR (M3B) cDNA where the antibody-reactive peptide is deleted, that is, residues 339–367 (45). Membranes were solu-bilized in Laemmli sample buffer containing 5% β-mercaptoethanol for 30 min at room temperature. Ten g membrane protein was applied to each lane and SDS-PAGE was performed using a 4%–12% gel. After blotting, nitrocellulose mem-branes were sequentially incubated with a rabbit antibody against a synthetic TSHR peptide (residues 352–366) and [$^{125}$I]-labeled donkey anti-rabbit Ig F(ab')$_2$. In the right panel, the antiserum was preincubated with 700 ng/ml peptide for 60 min at 37°, centrifuged at 3000 × g for 5 min, then diluted and used as in the left panel; the first two lanes in the right panel depict reactions with antisera sham treated with peptide, centrifuged, and diluted identically. Substitution of bovine serum albumin, 1 mg/ml, did not duplicate the action of free peptide. Molecular weight assignments for TSHR proteins are derived from [$^{14}$C]methylated protein standards whose locations are noted on the left of the figure. Values are the mean determined from multiple (>30) experiments. *B:* the deletion of residues 5–23 prevents in vitro processing and glycosylation (45), whereas mutants with residues 22–25 deleted exhibit normal in vitro processing and glycosylation. The 95–100 kd receptor form disappears, consistent with its being the glycosylated, processed form of the receptor in the bilayer (45).

extracellular domain of the TSHR (46, 47, 49): threonine 40 and residues 30–33, 34–37, 42–45, 52–56, 58–61 (Fig. 16.1). Mutation of each resulted in the loss of stimulating TSHRAb activity by, respectively, 12, 2, 12, 11, 11, and 2 of 12 Graves IgGs known to be able to increase cAMP levels in FRTL-5 rat thyroid cells. In contrast, mutation of each results in a receptor that exhibits a nearly normal ability of TSH to increase cAMP levels, albeit curvilinear or lower affinity binding isotherms characteristic of agonist sites. The identification of the stimulating TSHRAb epitope on the N-terminal portion of the TSHR is consistent with results from TSHR-LH/CG receptor chimeras, which indicate that an important stimulating TSHRAb site is between residues 8–89 (44, 141). It is also consistent with studies using peptides from this region that have shown immunization in chickens and rabbits can induce the formation of antibodies with stimulating TSHRAb activity (57–59).

In no case are these residues epitopes for blocking TSHRAbs in patients with idiopathic myxedema and hypothyroidism. Thus, studies of the TSHR-LH/CG receptor chimeras (44) showed that the activity of blocking TSHRAbs isolated from the sera from patients with idiopathic myxedema is not lost coincident with the loss of the stimulating TSHRAb determinant. In addition, as noted above, separate studies (42, 43, 46) established that the blocking TSHRAb epitope was on the C-terminal portion of the extracellular domain and involved the immunodominant peptide, residues 352–374, 295–306, 387–395, and tyrosine 385 (Fig. 16.1). Conclusive evidence that the determinants on the N-terminal region of the TSHR were stimulating, not blocking, TSHRAb epitopes emerged when blocking TSHRAb-positive IgG preparations from patients with idiopathic myxedema were shown to preserve their ability to inhibit TSH-increased cAMP levels in mutants that lost stimulating TSHRAb activity (46, 47, 49). Again, studies with peptides (63, 64, 140) are consistent with the identification of blocking TSHRAb activity on the C-terminal portion of the TSHR, near the immunodominant peptide and the flanking residues defining the high-affinity TSH binding sites (Fig. 16.1).

Residues that make up the sites for the stimulating and blocking TSHRAbs are identical or homologous in all TSHRs (Fig. 16.3); most are, in contrast, not homologous with sequences in the gonadotropin receptors (Fig. 16.3) and are in the regions where the TSHR has inserts that create the nonhomology. Since the major TSHR autoantibody determinants are TSHR specific, this accounts for their specific association with autoimmune thyroid disease and their inability to perturb gonadal function. Since there are different epitopes for stimulating and blocking TSHRAbs, this accounts for the different types of thyroid autoimmune disease associated with receptor autoantibodies: hyperthyroidism (Graves) or hypothyroidism (idiopathic myxedema), respectively.

The model in Figure 16.4 approximates the stimulating and blocking TSHRAb determinants on the N- and C-termini of the external domain

TSAb SITE

| Receptor | | Residue Number |
|---|---|---|

RATTSHR  M - R P G S L L Q L T L L L A L - - - - - - - - - - P R S L W G R G C T S P P C E C H Q **E D D F R V T**  40
HUMTSHR  M - R P A D L L Q L V L L L D L - - - - - - - - - - P R D L G G M G C S S P P C E C H Q **E E D F R V T**  40
DOGTSHR  M - R P P P L L H L A L L L A L - - - - - - - - - P R S L G G K G C P S P P C E C H Q **E D D F R V T**  40
RATLHR   M G R R V P A L R Q L L V L A V L L L K P S Q L Q S R E L S G S R C P E P - C D C A P D G A L R - -  47
PORLHR   M R R R S L A L R L L L A L - - L L L P P P L - - P Q T L L G A P C P E P - C S C R P D G A L R - -  43
HUMANLHR M K Q R F S A L Q L L K L L - - L L L Q P P L - - P R A L R E A L C P E P - C N C V P D G A L R - -  43

RATTSHR  **C K E L H Q I P S L P P S T Q T** L K L I E T H L K T I P S L A F S S L P N I S R I Y L S I D A T L Q  90
HUMTSHR  **C K D I Q R I P S L P P S T Q T** L K L I E T H L R T I P S H A F S N L P N I S R I Y V S I D L T L Q  90
DOGTSHR  **C K D I H R I P T L P P S T Q T** L K F I E T Q L K T I P S R A F S N L P N I S R I Y L S I D A T L Q  90
RATLHR   C - - - - - - P G P R A G L A R L S L T Y L P V K V L P S Q A F R G L N E V V K I E I S Q S D S L E  91
PORLHR   C - - - - - - P G P R A G L S R L S L T Y L P I K V L P S Q A F R G L N E V V K I E I S Q S D S L E  87
HUMANLHR C - - - - - - P G P T A G L T R L S L A Y L P V K V I P S Q A F R G L N E V I K I E I S Q I D S L E  87

TSBAb AND IMMUNOGENIC PEPTIDE SITE

RATTSHR  L P S K G L E H L K E L I A K N T W T L K K L P L S L S F L H L T R A D L S Y P S H C C A F K N Q K  290
HUMTSHR  L P S K G L E H L K E L I A R N T W T L K K L P L S L S F L H L T R A D L S Y P S H C C A F K N Q K  290
DOGTSHR  L P S K G L E H L K E L I A R N T W T L K K L P L S L S F L H L T R A D L S Y P S H C C A F K N Q K  290
RATLHR   L P S H G L E S I Q T L I A L S S Y S L K T L P S K E K F T S L L V A T L T Y P S H C C A F - - - -  286
PORLHR   L P S Y G L E S I Q T L I A T S S Y S L K K L P S R E K F T N L L D A T L T Y P S H C C A F - - - -  282
HUMANLHR L P S Y G L E S I Q R L I A T S S Y S L K K L P S R E T F V N L L E A T L T Y P S H C C A F - - - -  282

RATTSHR  K I R G I L E S **L M C** N E S S I R N L R E R K S V N V M R G P V Y Q E Y E E G L G D N H V G Y K Q N  340
HUMTSHR  K I R G I L E S **L M C** N E S S M Q S L R Q R K S V N A L N S P L H Q E Y E E N L G D S I V G Y K E K  340
DOGTSHR  K I R G I L E S **L M C** N E S S I R S L R Q R K S V N T L N G P F D Q E Y E E Y L G D S H A G Y K D N  340
RATLHR   - - - - - - - - - - - - - - - - R N L P K K E - - Q N F S F S I F E N F S K - - - - - - - - - Q C E  309
PORLHR   - - - - - - - - - - - - - - - - R N L P T K E - - Q N F S F S I F K N F S K - - - - - - - - - Q C E  305
HUMANLHR - - - - - - - - - - - - - - - - R N L P T K E - - Q N F S H S I S E N F S K - - - - - - - - - Q C E  305

RATTSHR  S K F Q E G P S N S H **Y Y V F F E E Q E D E I I G F G Q E L K N** P Q E E T L Q A F D S H **Y D Y T V C**  390
HUMTSHR  S K F Q D T H N N A H **Y Y V F F E E Q E D E I I G F G Q E L K N** P Q E E T L Q A F D S H **Y D Y T I C**  390
DOGTSHR  S Q F Q D T D S N S H **Y Y V F F E E Q E D E I I L G F G Q E L K N** P Q E E T L Q A F D S H **Y D Y T V C**  390
RATLHR   S T V R K A D N E T L Y S A I F E E N E - - - - - - - - - - - - - - - - L S G W D - - Y D Y G F C  340
PORLHR   S T A R R P N N E T L Y S A I F A E S E - - - - - - - - - - - - - - - - L S D W D - - Y D Y G F C  336
HUMANLHR S T V R K V S N K T L Y S S M L A E S E - - - - - - - - - - - - - - - - L S G W D - - Y E Y G F C  336

RATTSHR  **G D N E D** M V C T P K S D E F N P C E D I M G Y K F L R I V V W F V S P M A L L G N V F V L F V L L  440
HUMTSHR  **G D S E D** M V C T P K S D E F N P C E D I M G Y K F L R I V V W F V S L L A L L G N V F V L L I L L  440
DOGTSHR  **G G N E D** M V C T P K S D E F N P C E D I M G Y K F L R I V V W F V S L L A L L G N V F V L I V L L  440
RATLHR   S P K T - L Q C A P E P D A F N P C E D I M G Y A F L R V L I W L I N I L A I F G N L T V L F V L L  389
PORLHR   S P K T - L Q C A P E P D A F N P C E D I M G Y D F L R V L I W L I N I L A I M G N M T V L F V L L  385
HUMANLHR L P K T - P R C A P E P D A F N P C E D I M G Y D F L R V L I W L I N I L A I M G N M T V L F V L L  385

FIGURE 16.3. Sequence comparison of TSH and gonadotropin receptors in regions on the extracellular domain where the stimulating TSHRAb (TSAb), blocking TSHRAb (TSBAb), and immunogenic peptide (all shaded) are identified. The TSAb site includes not only threonine 40, but also residues 30–33, 34–37, 42–45, 52–56, and 58–61 (Fig. 16.5) (46, 47, 49). Dots denote homologous residues identified in computer analyses. Dashes denote residues that are presumed to not be present, according to best-fit alignments.

of the TSHR. This seems reasonable since the studies with monoclonal antibodies to the TSHR showed that these sites were competitive agonists and antagonists of TSH, respectively (50–54). Although the region between residues 303 and 382 can be deleted with no loss of TSHR activity, its association with both an important immunogenic and a blocking antibody determinant of the receptor suggests it is looped and is a separate peptide domain near cysteines 301 and 390 (Fig. 16.4). The loop places the immunodominant peptide adjacent to the high-affinity TSH binding site involving tyrosine 385 and residues 295–306 and 387–395, since all are involved in blocking TSHRAb activity.

This model defines a high-affinity, ligand binding antagonist site on the extracellular domain of the TSHR and a different, distinct, but spatially approximated low-affinity agonist site for TSH. The TSH interaction with the high-affinity binding determinants can be presumed to induce the conformational change in the receptor evidenced in previous studies (9, 65, 66). The conformational change brings TSH into contact with receptor determinants in the agonist site of the extracellular domain and transmits this information to the transmembrane domain to initiate signal transduction.

This model (Fig. 16.4) allows tests and predictions to define sites important for the coupling of the extracellular domain to contact points on the 7-transmembrane region of the TSHR that are important for G protein interactions and signal transduction. For example, in the $\beta_2$ *adrenergic receptor* (AR) the two cysteines in the exoplasmic loops of the transmembrane domain are necessary for agonist binding and action (35, 67, 68). These cysteines—that is, cysteines 494 and 569 in the TSHR (Fig. 16.4)—are conserved in the glycoprotein hormone receptor family (16–31). Transfected cells containing a TSHR whose cysteines 494 or 569 in the exoplasmic loop are mutated to serine exhibit no TSH binding and no response to TSH or a Graves IgG in cAMP assays, by comparison to cells transfected with wild-type receptor (48). The mutations do not alter the composition or amount of TSHR forms detected on Western blots of membranes from cells; the lost activity cannot, therefore, be related to abnormal receptor synthesis, processing, or incorporation into the bilayer. Thus, these two conserved cysteines in the exoplasmic loops of the transmembrane domain are important for TSH and stimulating TSHRAb signal transduction, despite the fact the primary binding sites for all ligands are on the extracellular domain. The two cysteines of the exoplasmic loop in the $\beta_2$-AR are postulated to form a disulfide bond and be important in receptor tertiary structure (35, 67, 68).

When TSHR cDNA is transfected into cells that normally contain no TSHR, the cell develops a TSH-induced cAMP signal (22–27). Glycoprotein hormone receptors are, however, members of a family of receptors for physiologically important hormones that also couple G proteins associated with stimulation of the $PIP_2$ signal cascade (69). Thus, for example, transfected TSHR cDNA confers both cAMP and $PIP_2$ responses in CHO and Cos-7 cells (39, 40), and evidence has been presented for dual coupling of the murine LH/CG receptor to phosphoinositide breakdown and $Ca^{++}$ mobilization, as well as adenylyl cyclase activity (70). Mutation of alanine 623 in the third cytoplasmic loop of the TSHR to glutamic acid or lysine has been shown to result in a loss of TSH- or Graves IgG-induced $PIP_2$, but not cAMP signaling (40). Alanine 623 is

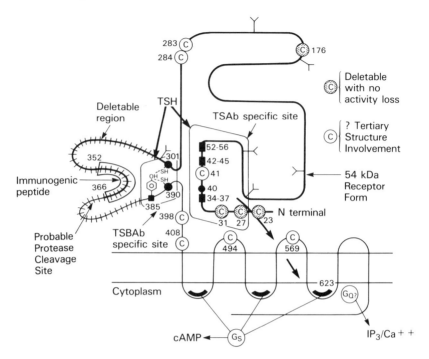

FIGURE 16.4. Model of TSHR with the TSH site and the TSHR autoantibody epitopes defined in a three-dimensional array. Determinants for blocking TSHRAbs (TSBAb) and stimulating TSHRAbs (TSAb) are approximated to comprise the TSHR site. The former are implicated in the expression of disease by patients with idiopathic myxedema and hypothyroidism; the latter are implicated in patients with Graves' disease and hyperthyroidism. They are presumed to comprise the high-affinity TSH binding site and agonist site for TSH, respectively, based on the competitive antagonism or agonism of TSH with, respectively, monoclonal inhibiting and stimulating receptor antibodies. The loop between residues 303 and 382 is x-marked and separated from the remainder of the external domain, since residues within it can be deleted with no loss in receptor function. This loop includes residues 352–366, which comprise the immunogenic peptide used to produce the receptor antibody in Figure 16.2. The immunogenic peptide is approximated near the blocking TSHRAb site because immunization with it or an adjacent peptide can produce blocking TSHRAbs that do not inhibit TSH binding and because they can, with time, produce antibodies reactive with peptides containing residues identified as blocking TSHRAb and high-affinity TSH binding. The model does not preclude the possibility that other sites contribute to the TSHR site—that is, residues that react with peptides common to TSH and chorionic gonadotropin, which have been identified on antiidiotypic receptor antibodies or by inhibition of TSH binding using TSHR peptides (see text). These are presumed to lie within residues 200–285 in particular. We identify cysteines that are (open circles) or are not (cross-hatched circles) likely to affect tertiary structure. The two cysteines on the exoplasmic loops of the transmembrane domain are assumed to carry the conformational signal of the receptor ligand in-

conserved in most adrenergic receptors, as well as receptors for glycoprotein hormones (10, 35). To characterize further the importance of
other TSHR residues in the third loop for $PIP_2$ signaling, we mutated
(142) TSHR residues to sequences in N- and C-termini of the third loop
of the $\alpha_1$- and $\beta_2$-AR, which computer analysis had identified as homologous to those in the TSHR (10, 35).

When transfected cells were evaluated for the effects of TSH on $PIP_2$
and cAMP signals, we could conclude the following (142). The C- and N-
terminal 5 residues of the third cytoplasmic loop of the TSHR, not only
alanine 623, are important for TSH- and Graves IgG-induction of the
$PIP_2$, but not their induction of the cAMP signal. In addition, they are
important for the regulation of basal cAMP levels, but have only minimal
effects on basal inositol phosphate levels. Mutation of residues 610–613
enhances agonist-increased $PIP_2$ signaling only. Residues 617–620 are
also important for regulation of constitutive cAMP levels but, more
importantly, their mutation to either $\alpha_1$- or $\beta_2$-AR residues results in a
receptor that can be compared to the alanine 293 mutation in the third
cytoplasmic loop of the $\alpha_{1B}$-AR. This $\alpha_{1B}$-AR mutation also increases
constitutive inositol phosphate levels and improves the agonist $EC_{50}$ (71).
Thus, nearby mutations in the third cytoplasmic loop of both receptors
result in an activated conformation, one that will bind and activate a G
protein in the absence of ligand. An interesting implication of these data
is that the native structure of the third cytoplasmic loop in both TSHR
and $\alpha_{1B}$-AR may be important to prevent wild-type receptor from interacting with G proteins without ligand stimulation. All of these residues
are conserved in the TSHR species sequenced to date; the 7 N- and C-
terminal residues of the TSHR third cytoplasmic loop are conserved in
gonadotropin receptors. These results, therefore, may be applicable to
the gonadotropin receptor field. We speculate that the third cytoplasmic
loop should be a region to examine in cases of tumors with constitutively
elevated signal levels, but no G protein abnormalities.

Mutation of alanine 623 and 293 of the TSHR and $\alpha_{1B}$-AR, respectively, results in the unusual phenomenon of increased receptor affinity
for ligand, but decreased $B_{max}$. In the case of the $\alpha_{1B}$-AR (71), the lower
kd and associated decrease in $B_{max}$ is believed to result from an altered

◄─────────────────────────────────────────────────

teraction with the extracellular domain of the receptor; it is alternatively possible
that these cysteines are involved in ligand binding, as proposed in adrenergic
receptors. The third intracytoplasmic loop, particularly alanine 623, but also its N-
and C-terminal 4 residues, is identified as the critical link for hormone and
TSHRAb coupling to a G protein important for signal transducing activity involved in $PIP_2$ cascade. It also interacts with $G_s$, but only regulates constitutive or
basal cAMP levels. Intracytoplasmic loops 1 and 2 are presumed to be coupled to
$G_s$ (139). Figure 16.5 further delineates possible coupling to G proteins.

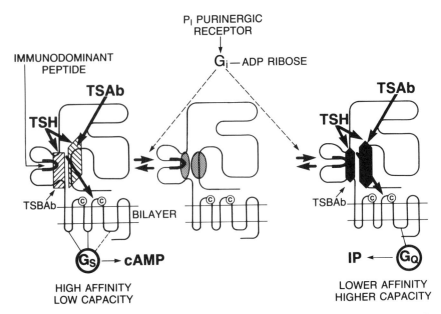

FIGURE 16.5. Speculative model of a single TSHR that binds to different G proteins. The two species are in equilibrium; this equilibrium is assumed to include a receptor molecule with no G protein bound. One species is proposed to have a higher affinity than the other, that coupled to $G_s$; this species is linked to hormone-increased cAMP levels and is presumed to be lower in number than the TSHR molecules coupled to $G_q$. The higher capacity of the $G_q$-coupled receptor is assumed to signify a larger number of these forms. The $G_q$-coupled receptor has a lower affinity for ligand and is coupled to the inositol phosphate (IP) signal cascade. The equilibrium between the two forms is suggested to be set by the $P_1$ purinergic receptor via $G_i$; the regulation of $G_i$ is further suggested to involve ADP-ribosylation (see text). The model suggests that the precoupling of the different G proteins induces different conformations in the extracellular domain, as represented by the hatched and dark boxes vs. the open circles in TSHR with no G protein bound. The blocking TSHR autoantibody (TSBAb) site is a high-affinity antagonist TSH binding site that involves tyrosine 385 and residues 295–306 and 387–395, particularly cysteines 301 and 390. The thyroid stimulating autoantibody (TSAb), which increases cAMP levels, interacts with an agonist site on the extracellular domain; this site involves threonine 40 and residues 30–33, 34–37, 42–45, 52–56, and 58–61. These sites and their properties have been separately defined in Figure 16.4. After interacting with these sites, TSH or the thyroid stimulating autoantibody is presumed to transmit its signal to the transmembrane domain; the two exoplasmic cysteines critical for this process are noted. The stimulating autoantibody that increases cAMP levels might not be identical to the stimulating autoantibody activating the $PIP_2$ signal (50). The mechanism by which $G_i$ regulates the equilibrium is unknown.

conformation associated with coupling to the G protein. In the case of the TSHR, altered ligand binding is also likely to result from a conformation change, given the full preservation of at least one ligand-induced activity in each mutant and the normal Western blots indicating normal receptor synthesis, processing, and incorporation into the bilayer. One model that must be considered to account for (i) these binding data and (ii) dual signaling by a single receptor molecule is that some receptor forms are precoupled to G proteins, such as $G_s$ and $G_q$, and that there is an equilibrium between these receptor forms that have different conformations in the extracellular domain (Fig. 16.5). The TSHR molecule might not, therefore, interact with both $G_s$ and $G_q$ simultaneously, but with only one G protein at a time, which is the normal case for adrenergic receptors. Multiple receptor forms with different conformations and affinities for ligands might account for the curvilinear binding isotherms in normal thyroid cells and membranes (27, 73)—that is, the existence of high-affinity and low-affinity TSH binding sites. They might account, in addition, for the observation that different concentrations of TSH are necessary to induce the cAMP vs. $PIP_2$ signals in vitro (8, 72, 143)—that is, the cAMP signal is linked to the high-affinity, $G_s$-coupled TSHR form, whereas the $PIP_2$ signal is linked to the low-affinity, $G_q$-coupled TSHR form (Fig. 16.5). It may be presumed, in this model, that there are fewer of the $G_s$- than $G_q$-coupled receptor forms, thereby accounting for TSHR binding models with high-affinity, low-capacity and low-affinity, high-capacity sites (27, 73).

This model (Fig. 16.5) has a potentially important implication for regulation of glycoprotein hormone receptor-G protein interactions by other receptors in the cell. The $P_1$ purinergic receptor regulates TSH and Graves IgG induction of the cAMP and $PIP_2$ signals in FRTL-5 cells (72). Thus, adenosine and $P_1$ purinergic agonists (i.e., *phenylisopropyladenosine* [PIA]) inhibit TSH- or Graves IgG-induced cAMP production, but enhance TSH- or Graves IgG-induced inositol phosphate formation, despite the fact that they have little direct effect on either signal (72, 143). The effect of PIA is mediated by a pertussis toxin-sensitive G protein ($G_i$) and is common to $\alpha_1$-adrenergic receptors (72, 74, 75, 143). It is reasonable, therefore, to speculate that the distribution of the different TSHR forms is regulated by the action of the $P_1$ adrenergic receptor and that this involves regulation of G protein interactions with TSHR. Interestingly, the ability of TSH to modulate the cAMP signal system in FRTL-5 thyroid cells has been related to pertussis toxin-sensitive ADP-ribosylation of a $G_i$ family member (76, 77). ADP-ribosylation may, therefore, be involved in regulating this equilibrium (Fig. 16.5).

In sum, Figures 16.4 and 16.5 present a putative model of ligand binding with antagonist and agonist sites on the extracellular domain of the TSHR. Signal coupling involves conformational changes and tertiary structure such that the extracellular domain and/or ligand bound to that domain perturbs the exoplasmic and cytoplasmic portions of the 7-transmembrane domain of the TSHR. This perturbation then alters G protein interactions and generates both signals. We suggest that the data concerning the coupling of the ligand binding sites on the extracellular domain of the TSHR to its transmembrane domain are relevant to all TSH and gonadotropin receptors, since the mutated sequences in the transmembrane domain are nearly identical in all glycoprotein hormone receptors. We would suggest that the data concerning the extracellular domain ligand binding sites are equally relevant to understanding or predicting sites on the extracellular domain of gonadotropin receptors that are important for binding and function.

Thus, one can imagine the extracellular domain of the TSHR in Figure 16.4 in the image of a hand with finger, which is attached to the trans-membrane domain, our body, via an arm (Fig. 16.6). The palm and surrounding fingers create a pocket or groove for the ligand to bind. The fingers can be envisioned to have sites important to position the ligand in the groove; the fingers have determinants that are common to all ligands, as well as determinants specific for one or the other ligand. One finger can be envisioned be TSHR specific, residues 300–395, and to include the immunodominant peptide of the TSHR (Figs. 16.1 and 16.4); it creates the high-affinity, TSH-specific binding site reactive with blocking TSHRAbs. A second finger, the N-terminus of the receptor, contains important determinants reactive with stimulating TSHRAbs (Figs. 16.1 and 16.4); it is an agonist site on the extracellular domain that is important for transducing the signal to the transmembrane domain. This finger may have residues important not only for the agonist action of glycopro-

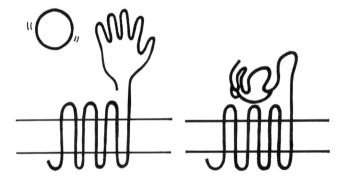

FIGURE 16.6. The TSHR expressed in the configuration of a hand with fingers waiting for ligand and attached to the transmembrane domain (left). After interacting with ligand, the TSHR undergoes a conformational change that now pushes the ligand onto the exoplasmic loops of the transmembrane domain. This induces the G proteins on the cytoplasmic surface to activate the transducing signal pathways.

tein hormone receptors but also for ligand specificity. Thus, the N-terminus LH/CG and FSH receptors have been identified as a critical region for ligand-specific signal expression of that receptor (78). The other fingers have sites common to TSH and gonadotropins, as well as the gonadotropin-specific, high-affinity antagonist binding sites. Several predictions flow from this hypothetical model (Figs. 16.4 and 16.6).

First, if one imagines catching a ball, one can intuitively see that the arrival of the ball into the pocket created by the fingers of the extra-cellular domain will create a conformational shift as the arm bends backward from the arrival of the ball, then forward. Ligand-induced conformational changes in the receptor should, therefore, be important in signal generation. This has been shown in studies using solubilized preparations of the glycoprotein component of the TSHR embedded in liposomes (65, 66). These studies have shown that hCG can be a very good inhibitor of TSH binding; however, hCG will not release carboxy-fluorescein from dye-loaded liposomes, whereas TSH does. Thus, ligand binding sites are related, but the conformational changes induced by the antagonist-agonist site interaction on the extracellular domain are not the same. Second, the glycoprotein hormones are made of related β-subunits and identical α-subunits. The β-subunit was shown to determine tissue specificity (1–5). Studies with peptides (79, 80) have shown that regions of both the α- and β-subunit can inhibit hormone binding. If one envisions the β-subunit interacting with the hand and fingers, one might envision the α-subunit, shifted by conformational perturbations, now to interact with the transmembrane domain. In this hypothesis, therefore, the higher degree of α-subunit homology on the ligands matches the higher homology of the glycoprotein hormone receptors in the trans-membrane domain.

Third, we have previously identified an α-subunit peptide that is identified by computer searches as homologous among glycoprotein hormones, cholera toxin, nonapeptide hormones, cytotoxins, and some neurotoxins (9). This peptide is ADP-ribosylated when incubated with membranes, can increase adenylate cyclase activity in membrane preparations, and is markedly enhanced in its activity when coupled with fluoride (81). The possibility is being examined that it interacts with the transmembrane domain and G protein interaction sites.

Last, sequences of the hypervariable regions of two monoclonal antibodies directed at the TSHR, and selected by the antiidiotypic approach after immunization with TSH, have recently been evaluated (80). The presumption of this approach is that the antiidiotype will mimic the ligand not only in activity, but in sequence as well. Sequences were identified that were similar to α- and β-glycoprotein hormone peptides that had previously been identified as important for gonadotropin as well as TSH interactions with their respective receptors (79). The receptor site with which these antibodies react cannot, therefore, be identical to determinants with which the TSHRAb autoantibodies interact, since the latter are not present on gonadotropin receptors (see above). Rather, these antiidiotypic antibodies must recognize common ligand interaction sites on both receptors, perhaps those in the region between residues 85 and 285. This may involve residues 201–211 and 222–230, which have been implicated in TSHR-LH/CG receptor chimera studies (82). It is notable in this respect that recent studies with TSHR peptides have identified residues 247–266 as second only to residues 307–325 as inhibitors of TSH binding (83). The former is in a region having a high degree of homology in both TSHR and LH/CG receptors; the latter is in the region of one of the high-affinity TSH binding site determinants that are reactive with thyroid-specific blocking TSHRAbs and that are identified in site-directed mutagenesis studies (Fig. 16.4), that is, residues 295–302.

# Receptor Gene Expression: Autoregulation and Relation to Hormones Controlling Growth and Function of the Cell

In FRTL-5 rat thyroid cells, growth and function are dependent on TSH (7, 8, 12). Evidence has accumulated, however, that TSH action involves multiple signals: not only cAMP, but also $PIP_2$, multiple hormones, particularly insulin and IGF-I, and an ordered and sequential action of each (7, 8, 12, 84, 85). Thus, for example, the TSH-induced cAMP signal is necessary, but insufficient for FRTL-5 thyroid cell growth; the cAMP signal requires insulin and IGF-I to be effective (7, 8, 12, 84). One explanation is that insulin/IGF-I is necessary, in addition to the cAMP signal, to regulate genes such as c-*myc* and c-*fos* (86). Insulin/IGF-I are necessary, in addition to the cAMP signal, to induce the cyclooxygenase that uses products of the $PIP_2$ cascade, $Ca^{++}$ and arachidonic acid, to increase thymidine incorporation into DNA (84). The increase in cyclo-oxygenase activity and gene expression immediately precedes the TSH-induced increase in thymidine incorporation into the cell and the related increase in cell number that follows (84). The role of cAMP, insulin/IGF-I, their sequential action as a function of time, and the involvement of such autocrine factors as *basic fibroblast growth factor* (bFGF) have been similarly and elegantly defined for TSHR and gonadotropin receptors (7, 85–91).

TSHR mRNA levels in rat FRTL-5 thyroid cells are regulated at a transcriptional level by the multiplicity of hormones, autocrine factors, and signals required for the growth and function of the cells: TSH, insulin/IGF-I, bFGF, cAMP, and the $PIP_2/Ca^{++}$ cascade (27, 92–94). Autoregulation by cAMP has been best studied to date and involves the interplay of two cAMP-modulated regulatory factors, whose activities are expressed sequentially after TSH addition to FRTL-5 thyroid cells (92). The first increases (92) and the second decreases (27, 92–94) TSHR gene expression. The first is activated by low levels of a cAMP analog (0.2 mM), the second by high (>1 mM) (92); both act transcriptionally (92). The action of the cAMP-modulated negative regulatory factor involves a rapidly synthesized protein as an intermediate, as evidenced by cycloheximide sensitivity within 1 h (92). In contrast, the positive cAMP regulatory factor is cycloheximide insensitive; thus, its activity is magnified within as little as 1 h after the addition of TSH or a cAMP analog plus cycloheximide (92). One positive regulator under cAMP control has been identified as Ku70 (95). Ku70 is a DNA and TSH binding protein and thyroid autoantigen in Graves' disease; it is also an autoantigen in systemic lupus erythematosus (9, 96, 97). The Ku70 mechanism of action is discussed below.

LH/CG and FSH can similarly induce positive and negative modulation of receptor RNA levels via their cAMP signal (98–103); this has been linked to cAMP-signaled regulation of follicle development, luteinization, and ovulation (98–103). Negative regulation of LH/CG receptor by the cAMP signal, like the case for TSHR (38), has also been identified as transcriptional at a promoter level (104); however, posttranscriptional negative regulation has been observed (105). The fibronectin gene in granulosa cells probably has a related regulatory system (106). In this system (106), a negative regulatory factor is induced by FSH/cAMP, acts transcriptionally, and is cycloheximide sensitive. It suppresses CRE-dependent transcriptional activation and requires upstream sequences other than CRE, which suggests that other regulatory elements are involved. It is tempting to speculate that one of the missing factor and promoter elements will be modulated by insulin/IGF-I; this is demonstrated below for TSHR. In sum, positive and negative autoregulation by cAMP is a common feature of glycoprotein hormone receptors; both seem critical for cell cycle progression, growth, and differentiation.

The sensitivity of the negative regulator to alterations in protein synthesis is important to keep in mind. Thus, studies with primary cultures of human thyroid cells have not seen negative regulation (107, 108). However, these systems place cells in 5%–10% fetal calf serum and no TSH for several days prior to experiments. This procedure is well known to result in the irreversible loss of thyroid-specific functions, particularly TSH-regulated iodide transport. At this time, therefore, the loss of the cAMP-induced, transcriptionally active, negative regulatory factor may be associated with loss in functions, given its short functional half-life and high sensitivity to disruption of the protein-synthetic machinery of the cell.

In addition to the autoregulatory action of cAMP, steady state TSHR mRNA levels and gene transcription are increased by insulin/IGF-I (92, 93). The action of insulin/IGF-I cannot be duplicated by hydrocortisone;

the insulin/IGF-I–positive factor involves a rapidly synthesized protein as an intermediate, as evidenced by cycloheximide sensitivity within 4 h (92). The action of insulin/IGF-I affects both the negative and positive cAMP autoregulatory factors. Thus, insulin/IGF-I is required for the negative transcriptional regulation of the TSHR by TSH/cAMP (92, 93), as is the case for TSH/cAMP-induced increases in thyroglobulin (109). The action of the TSH/cAMP-induced positive regulator actually acts in conjunction with insulin/IGF-I. Thus, when TSH is withdrawn from cells and cAMP levels fall toward basal values by 12 h, the addition of low levels of a cAMP analog accelerate the insulin/IGF-I–dependent return of TSHR mRNA levels from several days to less than 24 h (92). The action of this TSH/cAMP-induced positive regulator thus counterbalances the action of the negative regulator and works in concert with insulin/IGF-I. The complex interplay of these two cAMP-activated autoregulatory factors is associated with coincident increases or decreases in cell surface TSHR measured by $^{125}$I-TSH binding (92, 93) and appears, therefore, to be an autoregulatory, feedback mechanism to maintain TSHR activity on the cell surface (Fig. 16.7).

Negative regulation of gene expression by TSH and its cAMP signal is not a usual feature of thyroid-specific genes in a normally functioning thyroid cell. The expression of thyroglobulin and thyroid peroxidase genes, whose products are thyroid autoantigens like TSHR, is increased by TSH and its cAMP signal (Table 16.1). The expression of such nonthyroid-specific genes as malic enzyme or HMGCoA reductase is also increased by TSH and its cAMP signal (Table 16.1). It was, therefore, of interest to find that *major histocompatibility complex* (MHC) class I genes are negatively regulated by TSH/cAMP and that the negative regulation is, like TSHR, dependent on insulin/IGF-I (Table 16.1) (110, 111). It was of further interest that insulin/IGF-I decreases MHC class I mRNA levels, rather than increasing them, as in the case of thyroglobulin or TSHR (Table 16.1). Thus, negative regulation of TSHR and class I mRNA levels by TSH/cAMP appears to be specific. Preliminary experiments indicate that the transcriptional factors that are associated with negative regulation of TSHR are related to those negatively regulating class I and class II MHC genes (Fig. 16.7). These observations may be of interest regarding gonadotropin receptors, whose action has also been linked to MHC class I (10). It is particularly relevant to understanding the basis for thyroid autoimmunity (110, 111) that may evolve when exogenous factors eliminate or reduce negative regulation. The synthesis and turnover of products of all the thyroid-specific genes increases, MHC Class I and II genes and proteins increase, and peptides derived from the thyroid-specific gene products are presented to the immune system, becoming autoantigens as they overwhelm normal self-recognition processes.

The TSHR exhibits tissue specificity. Thus, for example, Northern analyses identified receptor mRNA in rat thyroid, but not rat testis,

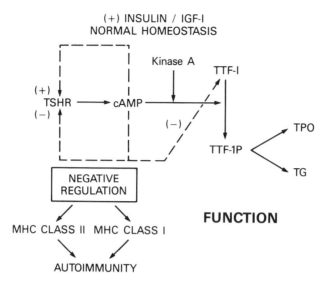

FIGURE 16.7. Proposed autoregulatory model where TSH regulates TSHR mRNA and receptor levels by two separate cAMP-mediated transcriptional factors. The cAMP in the presence of the kinase A phosphorylates TTF-I and phosphorylated TTF-I is the agent that increases thyroglobulin (TG) and thyroid peroxidase (TPO) mRNA and activity levels. In this model, the increase in cAMP will decrease TSHR levels and cause an autoregulated return to a basal homeostatic level of activity. Conversely, a lowering of cAMP levels reverses this and increases TG and TPO activities as needed. TTF-I becomes a critical modulator of function, as well as an important determinant for developmental expression of all three thyroid-specific genes: TSHR, Tg, and TPO. The negative regulation is proposed to be key not only to normal homeostasis, but also to self-recognition of the three thyroid-specific genes. Its loss, associated with increased turnover of the three thyroid specific genes, increased MHC class I expression, and aberrant class II expression, is suggested to initiate the autoimmune state. TTF-I gene expression is also negatively regulated by TSHR-generated cAMP, further establishing the interrelationships of this autoregulatory loop.

brain, liver, lung, or spleen cells; low levels were detected in rat ovary (27). Tissue specificity is common to the expression of gonadotropin receptors. In the thyroid, it appears that tissue-specific expression of the TSHR at a developmental level, as well as tissue-specific expression of thyroglobulin and thyroid peroxidase genes, requires the action of *thyroid transcription factor I* (TTF-I), which is thyroid-specific (112–115). TTF-I has also been identified (116–120) as the cAMP-regulated transcriptional element that controls thyroglobulin biosynthesis and thyroid peroxidase gene expression (Fig. 16.7).

In functioning FRTL-5 cells, there is a dynamic and interactive relationship between TTF-I and TSHR (Fig. 16.7). Thus, TSH increases

TABLE 16.1. Summary of the effect of TSH/cAMP or insulin/IGF-I on RNA levels of the noted genes in rat FRTL-5 cells as measured by Northern analysis.

| Gene | TSH | Insulin/IGF-I | Dependence of cAMP action on insulin/IGF-I |
|------|-----|---------------|--------------------------------------------|
| Class I | ↓ | ↓ | Yes |
| TSH receptor | ↓ | ↑ | Yes |
| Thyroglobulin | ↑ | ↑ | Yes |
| Thyroid peroxidase | ↑ | NC | No |
| Malic enzyme | ↑ | NC | No |

NC = No change detected.
↑ = Increase in RNA levels.
↓ = Decrease in RNA levels.
*Source:* Data are summarized from Saji, Moriarty, Ban, Singer, and Kohn (111).

cAMP levels that, in the presence of kinase A, phosphorylate TTF-I. Phosphorylated TTF-I in the nucleus increases thyroglobulin and thyroid peroxidase activity and, as is discussed below, regulates TSHR promoter activity; a *cAMP response element* (CRE) is not directly involved. An autoregulatory loop emerges as follows. TSH-increased cAMP causes a decrease in TSHR levels. The cAMP levels fall, phosphorylated TTF-I decreases, and phosphorylated TTF-I–mediated regulation of thyroglobulin, thyroid peroxidase, and TSHR gene expression abate. The independent regulation of TSHR gene expression by insulin/IGF-I preserves a constitutive level of TSHR activity. As cAMP levels fall, lower levels of cAMP induce positive regulatory factors that interact with insulin/IGF-I–modulated transcriptional factors to reverse the process (Fig. 16.7). The independent regulation of thyroglobulin by insulin/IGF-I establishes continued preservation of thyroglobulin synthesis for iodide trapping, even if iodide uptake, iodination, and TSHR gene expression are decreased by the autoregulatory action of TSH.

TSHR gene expression is also regulated by a $Ca^{++}$ signal (93). Thus, TSHR mRNA levels in rat FRTL-5 thyroid cells are decreased by treatment with the calcium ionophores A23187 or ionomycin. Down-regulation by $Ca^{++}$ is independent of and additive with cAMP. The $Ca^{++}$ signal is also associated with a decrease in $[^{125}I]$TSH binding and a decreased ability of TSH to increase cAMP levels. Whereas down-regulation by TSH and its cAMP signal requires the presence of insulin/IGF-I, down-regulation by the $Ca^{++}$ signal does not (93). Down-regulation by $Ca^{++}$ is transcriptional (93); a physiologic agent that appears to use the $Ca^{++}$ signal is bFGF (94). Basic FGF is recognized as an autocrine factor regulating the growth and differentiated response of target tissues affected by gonadotropins as well as TSH (88–91). Basic FGF not only decreases TSHR mRNA levels, it also decreases cAMP signal generation and cAMP signal action, for example, thyroid peroxidase gene expression (94).

Nevertheless, the bFGF action is associated with increased thymidine incorporation and thyroid cell growth (89, 90, 94, 121, 122), despite its action to decrease cAMP signal generation and transduction. This appears to be explained by the ability of bFGF to activate such protooncogenes as c-*fos* by a $Ca^{++}$ or novel tyrosine kinase activity (94). The possibility exists that bFGF is one of the autocrine factors that accounts for observations by Van Wyk et al. concerning the complex sequential regulation of growth by the cAMP signal of TSHR and gonadotropin receptors and by insulin/IGF-I (7, 87).

# The TSH Receptor Promoter: Efficient Expression, Autoregulation, Tissue Specificity, and Multihormonal Control

Taken together, the data above suggested that all glycoprotein hormone receptors might have certain common features at a promoter level that explained efficient expression, autoregulation by cAMP, tissue specificity, and regulation by insulin/IGF-I. To pursue this, we isolated and sequenced 1.7 kb of the 5′ flanking region of the rat TSHR (38). We showed that *chloramphenicol acetyltransferase* (CAT) gene chimeras of the 1.7 kb and smaller constructs expressed significant promoter activity when transfected into rat FRTL-5 and FRT thyroid cells, but not *Buffalo rat liver* (BRL) or HeLa cells (38). Initial comparisons with the 5′ flanking region of the LH/CG and FSH receptors, as well as the human TSHR (19, 20, 28, 31, 104), revealed several surprising and common features.

In the case of the rat TSHR, the region necessary for promoter activity had multiple major transcriptional start sites between −89 and −68 bp, relative to the ATG start codon (+1), had neither a TATA nor CCAAT box, was GC rich without a GC box motif, and had features of promoters seen in housekeeping genes, such as those for hypoxanthine phosphoribosyltransferase, *3-hydroxy-3-methyl-glutaryl-coenzyme A* (HMGCoA) reductase, 3-phosphoglycerate kinase, adenosine deaminase, and adenine phosphoribosyltransferase (38). Interestingly, promoters for some growth factor receptors and oncogenes also have these features: the *epidermal growth factor* (EGF) receptor, the insulin receptor, *nerve growth factor* (NGF) receptor, H-ras, Ki-ras, and N-*myc* (38). Although a consensus sequence for the promoter in this group of genes has not been identified, computer-generated comparisons readily align the promoter region between −120 and −67 bp of the TSHR gene, the 21 bp repeat sequence of the SV40 early promoter region (123), the 36 bp proximal promoter element of the EGF receptor gene (124), and the two sequences in the promoter region of the HMGCoA reductase gene (125) (Fig. 16.8B).

## A.

```
R  -214-  TGGATGGAGAGTTGCCTAGGCAAGCGGAGCACTTGAGAGCCTCTCCTTCC
          ::: :: :::::::::: :::::::::::: :: ::::::: :::::
H   -58-  TGGGTCAAG-GTTGCCTAGGGAAGCGGAGCACTTAAGTGCCTCTTTTTCC

R  -164-  CCCTCTCCAGCGTGCTCTCCAGCGATGAGGTCACAGCCCCTTGGAGCCCT
          :: :::::::: :::: :::: :: :::::::::::::::::::::::::::
H    -9-  CCTTCTCCAGCCTCCTCCACAGTGGTGAGGTCACAGCCCCTTGGAGCCCT

R  -114-  CCTCCTTCCTCCCCTTTCCCTCCGGCACCCCGGTCTCTTTCAGCGTCAG--
          :: :::::: : ::::::: :: ::::::: ::: :: :
H   +42-  CCCTCTTCCCAC----CCCTCCCGCTCCCGGGTCTCCTTTGGCCTGGGGT

R   -66-  AAGC--AGG-GCA---CTGAGAATGTGGTGACAGCGGCGCAACGATGAAGT
          :: :: ::: ::: ::::::: :: ::::: :
H   +88-  AACCCGAGGTGCAGAGCTGAGAATGAGGCGATTTCGGAGGATGGAGAAAT

R   -22-  AGCACTGGAGGTCCCTTGGAAAATG  -+3
          ::: :: :::::: ::::::::::
H  +138-  AGCCCCGA--GTCCCGTGGAAAATG  -+160
```

## B.

| Promoter | 5'-End | 3'-End | DNA Sequence |
|---|---|---|---|
| rTSH-R Proximal Promoter | -120 bp | -67 bp | AGCCCTCCTCCTTCCTTCCCTTCCCTCCGGCACCCCGGTCTCTTTCAGCGTCAG |
| SV40 Early Promoter | 5239 bp | 5214 bp | GTTCCGCCCATTCTCCGCCCCATGGC |
| SV40 Early Promoter | 38 bp | 18 bp | TCCCGCCCCTAACTCCGCCCA |
| EGF-R Proximal Promoter | -112 bp | -77 bp | GCGTCCGCCCGAGTCCCGCCCTCGCCGCCAACGCCA |
| HMG CoA Reductase Promoter | -753 bp | -729 bp | AGCTCAGTTCCTTCCGCCCGAGGCT |
| HMG CoA Reductase Promoter | -894 bp | -869 bp | CGGTGCCCGTTCTCCGCCCGGGGTGCG |

The LH/CG and FSH receptor promoters also have multiple transcriptional start sites and no TATA or CCAAT box (19, 20, 28, 31, 104). Although the putative FSH promoter is not G + C-rich, the 5' flanking region of the rat LH/CG receptor gene has a G + C-rich sequence with a 71% G + C content and 4 GC box motifs in the 200 bp immediately upstream of the ATG initiation codon (19, 20, 104).

In thyroid cells, TSH and its cAMP signal decreased promoter activity in cells transfected with the TSHR-CAT chimeric constructs (38). Similar results have been described in the case of the LH/CG receptor promoter (104), although a recent report additionally describes nontranscriptional autoregulation activity by the ligand and its cAMP signal (105). In the case of the rat TSHR promoter, deletion analyses indicated that the minimal region exhibiting promoter activity, thyroid specificity, and negative regulation by TSH/cAMP was located between −195 and −39 bp of the 5' flanking region of the TSHR gene (38). This minimal promoter region of the TSHR is highly conserved in rat and human TSHR genes (28, 38). Thus, the sequence of the human TSHR (28) matching −204 to −67 bp in the rat TSHR gene exhibits a 79% homology (Fig. 16.8A), has multiple transcriptional initiation sites, and has polypyrimidine (sense strand) and polypurine (antisense strand) stretches able to form a triple helix structure (see below).

Within the minimal promoter of the TSHR (Fig. 16.8B), a sequence exists, TGAGGTCA, that is homologous to the previously characterized CRE, TGACGTCA (126). CRE is cAMP responsive (126, 127), interacts with a multiplicity of transacting factors (127), and can act as an inducible enhancer (128). In the TSHR, the CRE-like sequence acts as a constitutive enhancer in thyroid cells, since in the absence of forskolin, it can

---

FIGURE 16.8. A: Alignment of the 5' flanking sequences of the rat (R) and human (H) TSHR genes. Numbering of the human TSHR gene is that used in reference 28; the rat gene is numbered as in reference 38. The ATG initiation codon is boxed in both cases. Nucleotide identity is indicated by double dots; gaps in the sequence, indicated by dashes, are inserted to allow for maximal alignment using PCGENE. Overall homology is 66%. Major transcriptional initiation sites are indicated by thick arrows (rat) and triangles (human). Sequence homology between the rat TSHR promoter region and the other promoter sequences. B: Several promoter sequences, such as two sequences of 21-bp repeats in SV40 early promoter, EGF receptor promoter, and two sequences of HMG CoA reductase promoter, were aligned with the rat TSHR promoter region. Matched nucleotides are depicted by dots. The major transcriptional initiation sites in the rat TSHR gene are indicated by arrowheads. The alignment was done by computer using PCGENE programs.

increase the activity of an SV40 promoter-driven CAT gene in FRT thyroid cells and since mutations of the CRE-like sequence to a consensus CRE or AP1 site increased enhancer activity, whereas a nonpalindromic mutation decreased enhancer activity (129). It is, nevertheless, a cAMP-responsive element that is not thyroid specific in its functional expression. Thus, when ligated to the SV40 promoter-driven CAT gene, forskolin increases its activity in BRL cells. The identification of the CRE-like site as a functional CRE without thyroid specificity and its action as a constitutive enhancer are supported by footprint and gel-shift data (129). Thus, both types of analyses indicate that the CRE-like region can form protein-DNA complexes with purified CREB-BR, AP1, and ATF2-BR as well as multiple CRE-binding proteins in BRL, FRT, or FRTL-5 thyroid cell extracts. Gel-shift analyses using mutant sequences of the CRE-like site show, however, that all the CRE-binding proteins in the tissue extracts, including CREB, form complexes with the CRE in a manner that parallels their effect on constitutive enhancer function in thyroid cells (129). There is, therefore, tissue-specific regulation of the CRE activity.

The promoter region between −139 and −113 bp of the minimal TSHR promoter, which contains the CRE-like sequence, is 100% conserved in human and rat TSHR genes (Fig. 16.8B). In addition, like the TSHR promoter, there is a CRE or AP1 site in the putative promoters of the LH/CG and FSH receptors (19, 20, 31, 104). In the case of the TSH and FSH receptors, this site is in the region of one of four SP1 sites in the LH/CG receptor promoter. One possibility, therefore, is that the CRE or AP1 site functions as a scaffold for a putative tethering factor to recruit TFIID complex, as is the case for SP1 in the tethering model proposed by Pugh and Tjian (130). Alternatively, it is possible that the highly conserved CRE or AP1 sites of the TSH and gonadotropin receptor promoters may be enhancer elements regulating promoter elements important for initiation of gene transcription. Precedent for this possibility exists. Thus, a protein designated CELF, a member of the CCAAT/enhancer binding protein gene family, has been shown to constitutively activate transcription of the preprotachykinin-A gene through the CRE region (131).

The CRE site is not a primary element involved in cAMP-signaled negative or positive regulation of TSHR gene expression, although it is the critical element for efficient TSHR gene expression (129, 132). Rather, elements surrounding the CRE appear to be sites of interaction for factors acting as positive or negative regulators of TSHR gene expression (95, 129, 132). As noted above, thyroid autoantigen Ku70, a DNA and TSH binding protein (8, 96, 97), is a transcriptional activator of the TSHR gene (95). Thus, TSHR promoter activity increased when cotransfected in thyroid cells with 1.7 kb, 199 bp, and 146 bp, but not 131 bp, TSHR promoter-CAT chimeras (95). The 146, but not the 131 bp, construct includes the CRE site (Fig. 16.8B). Since, however, pKu70 increased TSHR promoter activity when cotransfected with the 146 bp nonpalindromic CRE mutant, the recognition element must be different from the canonical CRE. TSH/cAMP increases Ku70 mRNA levels within 1 h, consistent with the early increase in TSHR mRNA levels (95). The Ku70 protein is a ubiquitous protein seen in many growing tissues (8).

Many groups have identified *cis*-acting, negative transcriptional regulatory elements in mammalian genes and have associated them as important components controlling the growth and differentiation function of cells (133). Negative regulation by forskolin is evident in the 131 bp chimeric CAT construct of the TSHR promoter (129), which lacks the CRE site (Fig. 16.8B). In addition, negative regulation is demonstrated with a mutant of the 146 bp TSHR promoter CAT construct, which has a nonpalindromic sequence that exhibits no specific interaction with CRE binding nuclear proteins in FRTL-5 thyroid, FRT thyroid, or BRL extracts (129). The negative regulation is, however, thyroid and promoter specific, whereas the CRE activity is not. Negative regulation appears, therefore, to be a function of the homologous promoter and may well involve nucleotides between −131 and −114 bp.

As noted above, negative regulation of gene and promoter expression by cAMP are features of MHC class I and II genes, both of which are involved in thyroid autoimmunity (110, 111, 134). cAMP can induce repression of the MHC class II promoter in lymphocytes (134). Like the TSHR, this promoter has a CRE, but the CRE is not involved in the repressive action of cAMP. Instead, repression by cAMP involves more 5' elements; that is, the S and X1 regions (134). Inspection of the TSHR promoter 3' to the CRE (Fig. 16.8B) reveals the presence of a sequence, 5'-AGCCCTCC-3', between −120 and −113 bp that has only a one-base mismatch (underlined) with the S region in the MHC class II promoter (134). The cAMP signal of TSH can induce negative regulation of the MHC class I promoter in FRT and FRTL-5 thyroid cells (110, 111). The promoter region responsible for efficient expression of MHC class I contains a CRE-like site (110, 111) that acts as an enhancer, as is the case for the TSHR and the class II gene (134); negative regulation by cAMP involves elements other than the CRE (Saji et al., in preparation).

A decanucleotide *tandem repeat* (TR) sequence flanking the CRE between −162 and −140 bp (Fig. 16.8B), also represses CRE activity (129, 132); this does not, however, directly involve a TSH/cAMP-mediated action (129, 132). Repression by the TR was first evidenced in studies of CAT activity of rat thyroid cells transfected with chimeric constructs of 5' deletion mutants of the minimal promoter. Repression by the TR was confirmed in studies showing that the decanucleotides of the TR have no significant effect on heterologous SV40-promoter activity, but, rather, inhibit the constitutive enhancer activity of the CRE when it is ligated to the SV40 promoter (132). Mutagenesis of the decanucleotides in the TR indicate that each has repressor activity and they act additively; the mechanism by which each acts is, however, different and involves nuclear factors that are not thyroid specific (132).

Thus, DNAase I footprinting shows (132) that nuclear extracts from BRL, as well as rat thyroid cells, protect a region that includes the CRE and the 3' decanucleotide of the TR, $-148$ to $-124$ bp (Fig. 16.8B). Gel mobility shift analyses reveal, however, that separate groups of nuclear proteins interact with the CRE and the 3' decanucleotide (132). Whereas nearly the same region footprinted by the nuclear extracts is protected by purified CRE binding proteins, nuclear proteins interacting with the 3' decanucleotide protect a smaller region, $-148$ to $-135$ bp (Fig. 16.8B). Consistent with these overlapping footprints, mutations of the CRE site affect nuclear protein interactions with the 3' decanucleotide, and conversely, mutations in the 3' decanucleotide influence the interaction of CRE binding proteins with the CRE. In sum, nuclear proteins interacting with the 3' decanucleotide inhibit constitutive enhancer activity of the CRE because they compete with CRE binding proteins for the CRE site (Fig. 16.9A).

In contrast to these results, DNAaseI footprinting and gel mobility shift analyses associate the repressor action of the 5' decanucleotide with the interaction of its coding strand with a single-strand binding protein present in BRL, as well as thyroid cell nuclear extracts (132). We suggest that the single-strand binding protein contributes to triple helix DNA formation (see below). This would not prevent the interaction of the 3' decanucleotide with its specific nuclear factors, since they bind to single- as well as double-strand DNA. This may, however, prevent CRE binding proteins from interacting with the CRE, since they interact only with double-strand DNA. The region of the TR in the human TSHR gene functions in the same manner, despite a nonidentical sequence (Fig. 16.8B). Thus, it represses constitutive enhancer activity of the CRE and its 5' portion acts as a binding site for a single-strand binding protein.

The 3' decanucleotide is not an example of a silencer (133) or the type of negative regulation of the CRE wherein the proteins bind to a common DNA element; that is, an isoform protein, such as CREM, derived by alternative splicing, possessing DNA binding domains, but lacking functional activation domains (127). Repression is not an example of quenching, which involves protein-protein interactions between sequence-specific DNA binding proteins that negatively regulate transcription by interacting with nonoverlapping, but adjacent or closely linked, sequences (133). Rather, repression resembles cross-coupling, as exemplified by the situation where two distinct classes of transcription factors can recognize a common regulatory sequence in the human osteocalcin gene (135). There are contiguous sequence elements that interact with different nuclear proteins with overlapping footprints (Fig. 16.9A). In essence, a competition exists between the two groups of nuclear proteins (Fig. 16.9A); cross-coupling becomes, therefore, a novel means to regulate the CRE.

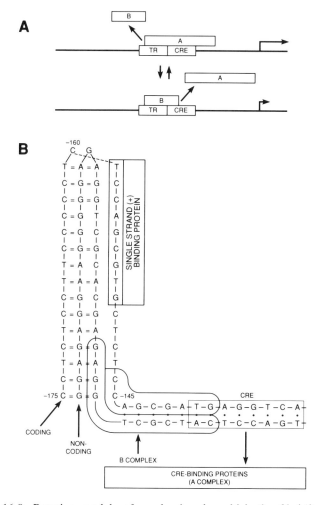

FIGURE 16.9. Putative models of mechanism by which the 3' (*A*) and 5' (*B*) decanucleotides of the TR modulate constitutive CRE-enhancer activity of the TSHR promoter. In *A*, R is repressor, A activator, TR the 3' decanucleotide, and CRE the CRE-like element of the TSHR. In *B*, the CT-rich domain of the TySHR is shown with its mirror image regions located with respect to the 5' and 3' decanucleotides of the TR. We hypothesize that the single-strand binding protein that interacts with the coding strand of the 5' decanucleotide induces triple helix formation as suggested by others. We propose that this continues to allow the nuclear proteins to interact with the 3' decanucleotide, particularly with the opposite or noncoding strand. This results in enhanced complex formation with the 3' decanucleotide, as well as decreased complex formation between CRE binding proteins and CRE. The result is decreased constitutive enhancer activity of the CRE and additive action of each decanucleotide in the TR. This model does not exclude the possibility that the nuclear proteins interacting with the 3' decanucleotide can act independently of the 5' decanucleotide. Thus, complex formation can occur with double-stranded as well as single-stranded forms of the 3' decanucleotide, and this complex can act independently as a repressor.

In the case of the 5' decanucleotide of the TR, its repressor activity is in the polypyrimidine/polypurine-rich domain between −175 and −145 bp (Fig. 16.8B). It exhibits S1 nuclease hypersensitivity (132); similar S1 nuclease hypersensitive, CT-rich domains have been identified in several housekeeping-type genes, c-*myc*, EGF receptor, insulin receptor, and Ki-*ras* (132); that is, in several of the genes identified (i) as having similar GC-rich promoter elements with multiple transcription start sites evidenced in the TSHR and LH/CG receptor promoter and (ii) as being functionally linked to glycoprotein hormone-dependent growth and function. The structure of all these CT-rich elements consists of perfect or nearly perfect mirror-image or direct repeats (132); the 31-nucleotide CT-rich element in the coding strand of the rat TSHR has this feature (Figs. 16.8B and 16.9B). The purine/pyrimidine strand asymmetry in these promoters is believed to be important for triplex formation, due to the nature of the triplex base pairing—that is, C-G-C or G-G-C (132). Evidence exists that the binding of specific nuclear factors to these sites is associated with formation of the C-G-C or G-G-C triplex in vivo.

The rat TSHR genomic sequence between −177 and −132 bp is not identical to the human TSHR promoter; there is no obvious TR (Fig. 16.8B). However, there is a 78% homology with the sequence of the human and rat TSHR genes, and the polypyrimidine and polypurine stretches are conserved, as is the completely identical CRE-like site, TGAGGTCA. The TR region of the human TSHR has the same CRE repressor activity and has the same single-strand binding protein site (132). These results suggest that the 5' flanking sequence in the human and rat TSHR gene function similarly and that an overt TR is not a marker of this activity.

The minimal TSHR promoter exhibits tissue-specific expression. Although most genes with promoters of the housekeeping type are expressed in a wide variety of cells, some, such as the NGF receptor (136), are expressed in a tissue-specific manner. Analysis of the minimal promoter 5' to the TR and CRE showed that the −199 bp chimeric CAT promoter construct expressed fourfold higher activity in FRTL-5 thyroid than BRL cells (Fig. 16.1) and expressed higher activity than FRT cells with no TTF-I (112–115). There was a specific element footprinted between −188 and −176 bp (Fig. 16.8B); this region has a sequence structurally and functionally homologous to a TTF-1 binding site (Shimura, Ikuyama, Shimura, Kohn, in preparation). TTF-1, as noted earlier, is the thyroid-specific transcription factor that is required for tissue-specific expression of both the thyroglobulin (112–115) and the thyroid peroxidase genes (116–118). TTF-1 is important for TSHR gene expression; thus, TSHR gene expression is lost in oncogene-transfected cells in association with the loss of TTF-I, as well as TSH-dependent growth, function, and thyroglobulin gene expression (119, 137). In addition, expression of TTF-1 during fetal development precedes and appears to be

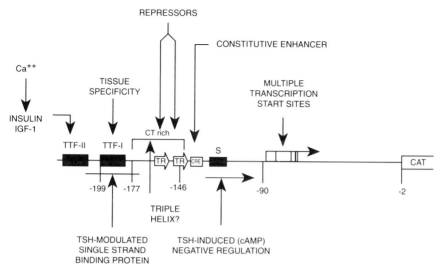

FIGURE 16.10. Diagramatic outline of the TSHR promoter and the location of each critical element thus far defined. We note each decanucleotide of the tandem repeat (TR), the CRE-like site, the polypurine/polypyrimidine domain whose coding strand is CT-rich, and GC-rich promoter with multiple major start sites between −89 to −68 bp. Also noted are the TTF-I- and TTF-II-like sites, as well as regions interacting with single-strand binding proteins.

required for expression of the TSHR, as well as the thyroglobulin and thyroid peroxidase genes (115).

In the thyroglobulin gene, TTF-I is associated with a second factor, TTF-II (138). This factor and element is linked to insulin/IGF-I regulation of thyroglobulin gene expression (138). Just upstream of the TTF-I site of the TSHR is a TTF-II–like site that is insulin sensitive. Preliminary evidence indicates that nuclear factors interacting with TTF-I and/or TTF-II sites in the TSHR may be regulated by a TSH, cAMP-regulated, single-strand binding protein different from that interacting with the TR. In addition, evidence has accumulated that negative regulation of the TSHR gene by $Ca^{++}$ operates through TTF-II (Okajima et al., in preparation).

In sum, a CRE site acts as a constitutive enhancer of promoter activity (Fig. 16.10). Flanking elements regulating the CRE respond to insulin/ IGF-I and include a thyroid specificity determinant (TTF-I) that regulates thyroid-specific gene expression. Flanking elements also account for auto-regulation by cAMP and are linked to MHC gene autoregulation by cAMP. TTF-I not only regulates thyroid-specific TSHR gene expression, it is the factor mediating expression of other thyroid-specific genes after being phosphorylated by TSHR-derived cAMP. TSHR and TTF-I thus form an autoregulated system important for tissue-specific function.

The cAMP and $PIP_2$ signals generated or modulated by TSHR and other hormone receptors—that is, the $P_1$ purinergic receptor and bFGF—autoregulate TTF-I and TSHR gene expression. Elements and factors important for cAMP- and $PIP_2$-signaled autoregulation are not identical, explaining their additive action. Insulin/IGF-I influence autoregulation of TTF-I and TSHR genes differently. Ligand-modulated single-strand binding proteins link TTF-I, constitutive CRE enhancer activity, and promoter function and may involve triplex formation.

We suggest these features may be important to understand common and divergent regulation of TSHR and gonadotropin receptor gene expression. At the least, they may offer opportunities to perform experiments involving exchanges or cross-competition of TSH and gonadotropin receptor elements. We hope this will help define common or different factors related to transcriptional regulation of the receptor genes that can explain TSH and gonadotropin regulation of growth and differentiated function.

## References

1. Pierce JG, Faith MR, Giudice LC, Reeve JR. Structure and structure-function relationships in glycoprotein hormones: molecular and cellular aspects. Ciba Found Symp 1976;41:225–49.
2. Ryan RJ, Charlesworth MC, McCormick DJ, Milius RP, Keutmann HT. The glycoprotein hormones: recent studies of structure function relationships. FASEB J 1988;2:2261–9.
3. Lustbader JW, Birkin S, Pileggi NF, et al. Crystallization and characterization of human chorionic gonadotropin in chemically deglycosylated enzymatically desialylated states. Biochemistry 1989;28:9239–43.
4. Ward DN, Bousfield GR, Mar AO. Chemical reduction-oxidation of the glycoprotein hormones disulfide bonds. In: Bellet D, Bidart JM, eds. Structure-function relationships of gonadotropins. New York: Raven Press, 1990:1–19.
5. Combarnous Y. Molecular basis for the specificity of binding of glycoprotein hormones to their receptors. Endocr Rev 1992;13:670–91.
6. Adashi EY, Resnick CE, Svoboda ME, Van Wyk JJ. Somatomedin-C synergizes with follicle stimulating hormone in the acquisition of progestin biosynthetic capacity by cultured rat granulosa cells. Endocrinology 1985;116:2135–46.
7. Takahasi S-I, Conti M, Van Wyk JJ. Thyrotropin potentiation of insulin like growth factor-I dependent deoxyribonucleic acid synthesis in FRTL-5 cells: mediation by autocrine amplification factors. Endocrinology 1990;126:736–45.
8. Kohn LD, Saji M, Akamizu T, et al. Receptors of the thyroid gland: the thyrotropin receptor is only the first violinist of a symphony orchestra. In: Ekholm R, Kohn LD, Wollman S, eds. Control of the thyroid: regulation of its normal growth and function. New York: Plenum Press, 1990:151–210.

9. Kohn LD. Relationships in the structure and function of receptors for glycoprotein hormones, bacterial toxins, and interferon. In: Cuatrecasas P, Greaves MF, eds. Receptors and recognition; series A, vol 5. London: Chapman and Hall, 1978:133–212.

10. Ascoli M, Segaloff DL. On the structure of the luteinizing hormone/chorionic gonadotropin receptor. Endocr Rev 1989;10:27–44.

11. Moncayo H, Moncayo R, Benz R, Wolf A, Lauritzen CH. Ovarian failure and autoimmunity. J Clin Invest 1989;84:1857–65.

12. Ambesi-Impiombato FS. Fast-growing thyroid cell strain. U.S. Patent no. 4,608,341, 1986.

13. Kohn LD, Valente WA, Grollman EF, Aloj SM, Vitti P. Clinical determination and/or quantification of thyrotropin and a variety of thyroid stimulatory or inhibitory factors performed in vitro with an improved thyroid cell line FRTL-5. U.S. Patent no. 4,609,622, 1986.

14. Vitti P, Rotella CM, Valente WA, et al. Characterization of the optimal stimulatory effects of Graves' monoclonal and serum immunoglobulin G on adenosine 3',5'-monophosphate production in FRTL-5 thyroid cells: a potential clinical assay. J Clin Endocrinol Metab 1983;57:782–91.

15. Fenzi GF, Vitti P, Marcocci C, Chiovato L, Macchia E. TSH receptor antibodies affecting thyroid cell function. In: Pinchera A, Ingbar SH, McKenzie JM, Fenzi GF, eds. Thyroid autoimmunity. New York: Plenum Press, 1987:83–90.

16. McFarland KC, Sprengel R, Phillips HS, et al. Lutropin-chorionic gonadotropin receptor: an unusual member of the G-protein-coupled receptor family. Science 1989;245:494–9.

17. Loosfelt H, Misrahi M, Atger M, et al. Cloning and sequencing of porcine LH-CG receptor cDNA: variants lacking transmembrane domain. Science 1989;245:525–8.

18. Minegishi T, Nakamura K, Takakura Y, et al. Cloning and sequencing of human LH/hCG receptor cDNA. Biochem Biophys Res Commun 1990; 172:1049–54.

19. Koo YB, Ji I, Slaughter RG, Ji TH. Structure of the luteinizing hormone receptor gene and multiple exons of the coding sequence. Endocrinology 1991;128:2297–308.

20. Tsai-Morris CH, Buczko E, Wang W, Xie X-Z, Dufau M. Structural organization of the rat luteinizing hormone (LH) receptor gene. J Biol Chem 1991;266:11355–9.

21. Ji I, Ji TH. Exons 1–10 of the rat LH receptor encode a high affinity bonding site and exxon 11 encodes G protein modulation and a potential second hormone binding site. Endocrinology 1991;128:2648–50.

22. Parmentier M, Libert F, Maenhaut C, et al. Molecular cloning of the thyrotropin receptor. Science 1989;246:1620–2.

23. Nagayama Y, Kaufman KD, Set P, Rapoport B. Molecular cloning, sequence and functional expression of the cDNA for the human thyrotropin receptor. Biochem Biophys Res Commun 1989;165:1184–90.

24. Libert F, Lefort A, Gerard C, et al. Cloning, sequencing and expression of the human thyrotropin receptor: evidence for binding of autoantibodies. Biochem Biophys Res Commun 1989;165:1250–5.

25. Misrahi M, Loosfelt H, Atger M, Sar S, Guiochon-Mantel A, Milgrom E. Cloning, sequencing, and expression of human TSH receptor. Biochem Biophys Res Commun 1990;166:394–403.
26. Frazier AL, Robbins LS, Stork PJ, Sprengel R, Segaloff DL, Cone RD. Isolation of TSH and LH/CG receptor cDNAs from human thyroid: regulation by tissue specific splicing. Mol Endocrinol 1990;4:1264–76.
27. Akamizu T, Ikuyama S, Saji M, et al. Cloning, chromosomal assignment, and regulation of the rat thyrotropin receptor: expression of the gene is regulated by thyrotropin, agents that increase cAMP levels, and thyroid autoantibodies. Proc Natl Acad Sci USA 1990;87:5677–81.
28. Gross B, Misrahi M, Sar S, Milgrom E. Composite structure of the human thyrotropin receptor gene. Biochem Biophys Res Commun 1991;177:679–87.
29. Sprengel R, Braun T, Nikolics K, Segaloff DL, Seeburg PH. The testicular receptor for follicle stimulating hormone: structure and functional expression of cloned cDNA. Mol Endocrinol 1990;4:525–30.
30. Minegishi T, Nakamura K, Takakura Y, Ibuki Y, Igarashi M. Cloning and sequencing of human FSH receptor cDNA. Biochem Biophys Res Commun 1991;175:1125–30.
31. Heckert LL, Daley IJ, Griswold MD. Structural organization of the follicle-stimulating hormone receptor gene. Mol Endocrinol 1992;6:70–80.
32. Winand RJ, Kohn LD. Thyrotropin effects on thyroid cells in culture: effects of trypsin on the thyrotropin receptor and on thyrotropin-mediated cyclic 3':5'-AMP changes. J Biol Chem 1975;250:6534–40.
33. Xie YB, Wang HY, Segaloff DL. Extracellular domain of lutropin choriogonadotropin receptor expressed in transfected cells binds choriogonadotropin with high affinity. J Biol Chem 265:21411–4.
34. Lefkowitz RJ, Caron MG. Adrenergic receptors: models for the study of receptors coupled to guanine nucleotide regulatory proteins. J Biol Chem 1988;263:4993–6.
35. Frazer CM. Molecular biology of adrenergic receptors: model systems for the study of G-protein-mediated signal transduction. Blood Vessels 1991;28:93–103.
36. Johnson GL. The G protein family and their interaction with receptors. Endocr Rev 1989;10:317–31.
37. Rousseau-Merck MF, Misrahi M, Loosfelt H, Atger M, Milgrom E, Berger R. Assignment of the human thyroid stimulating hormone receptor (TSHR) gene to chromosome 14q31. Genomics 1990;8:233–6.
38. Ikuyama S, Niller HH, Shimura H, Akamizu T, Kohn LD. Characterization of the 5'-flanking region of the rat thyrotropin receptor gene. Mol Endocrinol 1992;6:793–804.
39. Van Sande J, Raspe E, Perret J, et al. Thyrotropin activates both the cyclic AMP and PIP$_2$ cascades in CHO cells expressing the human cDNA of TSH receptor. Mol Cell Endocrinol 1990;74:R1–6.
40. Kosugi S, Okajima F, Ban T, Hidaka A, Shenker A, Kohn LD. Mutation of alanine 623 in the 3rd cytoplasmic loop of the rat TSH receptor results in a loss in the phosphoinositide but not cAMP signal induced by TSH and receptor autoantibodies. J Biol Chem 1992;267:24153–6.

41. Kosugi S, Akamizu T, Takai S, Prabhakar B, Kohn LD. The extra cellular domain of the TSH receptor has an immunogenic epitope reactive with Graves' sera but unrelated to receptor function as well as epitopes having different roles for high affinity TSH binding and the activity of thyroid stimulating antibodies. Thyroid 1991;1:321–30.
42. Kosugi S, Ban T, Akamizu T, Kohn LD. Site directed mutagenesis of a portion of the extracellular domain of the rat thyrotropin receptor important in autoimmune thyroid disease and nonhomologous with gonadotropin receptors: relationship of functional and immunogenic domains. J Biol Chem 1991;266:19413–8.
43. Kosugi S, Ban T, Akamizu T, Kohn LD. Further characterization of a high-affinity thyrotropin binding site on the rat thyrotropin receptor which is an epitope for blocking antibodies from idiopathic myxedema patients but not thyroid stimulating antibodies from Graves' patients. Biochem Biophys Res Commun 1991;180:1118–24.
44. Tahara K, Ban T, Minegishi T, Kohn LD. Immunoglobulins from Graves' disease patients interact with different sites on TSH receptor/LH-CG receptor chimeras than either TSH or immunoglobulins from idiopathic myxedema patients. Biochem Biophys Res Commun 1991;179:70–7.
45. Ban T, Kosugi S, Kohn LD. A specific antibody to the thyrotropin (TSH) receptor identifies multiple receptor forms in membranes of cells transfected with wild type receptor cDNA: characterization of their relevance to receptor synthesis, processing, and structure. Endocrinology 1992;131:815–29.
46. Kosugi S, Ban T, Akamizu T, Kohn LD. Identification of separate determinants on the thyrotropin receptor reactive with Graves' thyroid stimulating antibodies and with thyroid stimulating blocking antibodies in idiopathic myxedema: these determinants have no homologous sequence on gonadotropin receptors. Mol Endocrinol 1992;6:168–80.
47. Kohn LD, Kosugi S, Ban T, et al. Molecular basis for the autoreactivity against thyroid stimulating hormone receptor. Int Rev Immunol 1992;9:135–65.
48. Kosugi S, Ban T, Akamizu T, Kohn LD. Role of cysteine residues in the extracellular domain and exoplasmic loops of the transmembrane domain of the TSH receptor: effect of mutation to serine on TSH receptor activity and response to thyroid stimulating autoantibodies. Biochem Biophys Res Commun 1992;189:1754–62.
49. Kosugi S, Ban T, Kohn LD. Identification of thyroid stimulating antibody-specific interaction sites in the N-terminal region of the thyrotropin receptor. Mol Endocrinol 1993,7.114–30.
50. Yavin E, Yavin Z, Schneider MD, Kohn LD. Monoclonal antibodies to the thyrotropin receptor: implications for receptor structure and the action of autoantibodies in Graves' disease. Proc Natl Acad Sci USA 1981;78:3180–4.
51. Valente WA, Yavin Z, Yavin E, et al. Monoclonal antibodies to the thyrotropin receptor: the identification of blocking and stimulating antibodies. J Endocrinol Invest 1982;5:293–301.
52. Valente WA, Vitti P, Yavin Z, et al. Monoclonal antibodies to the thyrotropin receptor: stimulating and blocking antibodies derived from the lymphocytes of patients with Graves' disease. Proc Natl Acad Sci USA 1982;79:6680–4.

53. Ealey PA, Kohn LD, Ekins RP, Marshall NJ. Characterization of mono-clonal antibodies derived from lymphocytes from Graves' disease patients in a cytochemical bioassay for thyroid stimulators. J Clin Endocrinol Metab 1984;58:909–14.

54. Kohn LD, Alvarez F, Marcocci C, et al. Monoclonal antibody studies defining the origin and properties of autoantibodies in Graves' disease. In: Schwartz R, Rose N, eds. Autoimmunity: experimental and clinical aspects. Ann NY Acad Sci 1986;475:157–73.

55. Wadsworth HL, Chazenbalk GD, Nagayama Y, Russo D, Rapoport B. An insertion in the human thyrotropin receptor critical for high affinity hormone binding. Science 1990;249:1423–5.

56. Wadsworth HL, Russo D, Nagayama Y, Chazenbalk GD, Rapoport B. Studies on the role of amino acids 38–45 in the expression of a functional thyrotropin receptor. Mol Endocrinol 1992;6:394–8.

57. Endo T, Ikeda M, Onaya T. Thyroid stimulating activity of rabbit antibodies toward the human TSH receptor peptide. Biochem Biophys Res Commun 1991;177:145–50.

58. Ohmori M, Endo T, Ikeda M, Onaya T. Role of N-terminal region of the thyrotropin (TSH) receptor in signal transduction for TSH or thyroid stimu-lating antibody. Biochem Biophys Res Commun 1991;178:733–8.

59. Ohmori M, Endo T, Ikeda M, Onaya T. Immunization with human thyro-tropin receptor peptide induces an increase in thyroid hormone in rabbits. J Endocrinol 1992;135:479–84.

60. Takai O, Desai RK, Seetharamaiah GS, et al. Prokaryotic expression of the thyrotropin receptor and identification of the immunogenic region of the protein using synthetic peptides. Biochem Biophys Res Commun 1991;179: 319–26.

61. Dallas J, Desai RK, Seetharamaiah GS, Fan J-L, Prabhakar BS. Polyclonal antibodies to an immunodominant epitope (AA 357–372) on the human thyrotropin receptor (TSHr) inhibit 125I-uptake by FRTL-5 cells [Abstract]. Proc 66th meet Am Thyroid Assoc, 1992.

62. Seetharamaiah GS, Desai RK, Dallas JS, Tahara K, Kohn LD, Prabhakar BS. Induction of TSH binding inhibitory immunoglobulins with the extra-cellular domain of the human thyrotropin receptor produced using a baculo-virus expression system. Autoimmunity 1993.

63. Atassi MZ, Manshouri T, Sakata S. Localization and synthesis of the hor-mone binding regions of the human thyrotropin receptor. Proc Natl Acad Sci USA 1991;88:613–7.

64. Mori T, Sugowa H, Piraphatdist T, Inoue D, Enomoto T, Imura H. A synthetic oligopeptide from human thyrotropin receptor sequence binds to Graves' immunoglobulin and inhibits thyroid stimulating antibody activity but lacks interactions with TSH. Biochem Biophys Res Commun 1991;178: 165–72.

65. Kohn LD, Aloj SM, Beguinot F, et al. Molecular interactions at the cell surface: role of glycoconjugates and membrane lipids in receptor recognition processes. In: Shepard J, Anderson VE, Eaton J, eds. Membranes and genetic diseases. New York: Alan R. Liss, 1982:55–83.

66. Beguinot F, Formisano S, Rotella CM, Kohn LD, Aloj SM. Structural changes caused by thyrotropin in thyroid cells and in liposomes containing

reconstituted thyrotropin receptor. Biochem Biophys Res Commun 1983; 110:48–54.

67. Dixon AF, Sigal IS, Candelore MR, et al. Structural features required for ligand binding to the β-adrenergic receptor. EMBO J 1987;6:3269–75.

68. Dohlman HG, Caron MG, DeBlasi A, Frielle T, Lefkowitz RJ. Role of extracellular disulfide-bonded cysteines in the ligand binding function of the β$_2$-adrenergic receptor. Biochemistry 1990;29:2335–42.

69. Thompson EB. Single receptors, dual second messengers. Mol Endocrinol 1992;6:501.

70. Gudermann T, Birnbaumer M, Birnbaumer L. Evidence for dual coupling of the murine luteinizing hormone receptor to adenylyl cyclase and phospho-inositide breakdown and Ca$^{2+}$ mobilization. J Biol Chem 1992;267:4479–88.

71. Kjelsberg MA, Cotecchia S, Ostrowski J, Caron MG, Lefkowitz RJ. Constitutive activation of the α$_{1B}$-adrenergic receptor by all amino acid substitutions at a single site. J Biol Chem 1992;267:1430–3.

72. Sho K, Okajima F, Majid MA, Kondo Y. Reciprocal modulation of thyrotropin actions by P1-purinergic agonists in FRTL-5 cells. J Biol Chem 1991;266:12180–4.

73. Tate RL, Schwartz HI, Holmes JM, Kohn LD, Winand RJ. Thyrotropin receptors in thyroid plasma membranes: characteristics of thyrotropin binding and solubilization of thyrotropin receptor activity by trypsin digestion. J Biol Chem 1975;250:6509–15.

74. Okajima F, Sato K, Nazarea M, Sho K, Kondo Y. A permissive role of pertussis toxin substrate G-protein in P2-purinergic stimulation of phospho-inositide turnover and arachidonate release in FRTL-5 thyroid cells. J Biol Chem 1989;264:13029–37.

75. Okajima F, Sato K, Sho K, Kondo Y. Stimulation of adenosine receptor enhances α$_1$-adrenergic receptor-mediated activation of phospholipase C and Ca$^{2+}$ mobilization in a pertussis toxin-sensitive manner in FRTL-5 thyroid cells. FEBS Lett 1989;248:145–9.

76. Corda D, Sekura RD, Kohn LD. Thyrotropin effect on the availability of Ni regulatory protein in FRTL-5 rat thyroid cells to ADP-ribosylation by pertussis toxin. Eur J Biochem 1987;166:475–81.

77. Ribeiro-Meto F, Birnbaumer L, Field JB. Incubation of bovine thyroid slices with thyrotropin is associated with a decrease in the ability of pertussis toxin to adenosine diphosphate-ribosylate guanine nucleotide regulatory components. Mol Endocrinol 1987;1:482–90.

78. Braun T, Dchofield PR, Sprengel R. Amino-terminal leucine-rich repeats in gonadotropin receptors determine hormone specificity. EMBO J 1991;10:1885–90.

79. Morris JC, McCormick DJ, Ryan RJ. Inhibition of thyrotropin binding to receptor by synthetic human thyrotropin β-peptides. J Biol Chem 1990;265:1881–4.

80. Taub R, Hsu J-C, Garsky VM, Hill BL, Erlanger BF, Kohn LD. The hypervariable regions of two monoclonal anti-idiotypic antibodies against the TSH receptor have sequences similar to TSH and these peptides inhibit TSH-stimulated cAMP production in FRTL-5 thyroid cells. J Biol Chem 1992;267:5977–84.

81. Epstein M, Ross EM, De Wolf MJS, Fridkin M, Kohn LD. Glycoprotein hormone-cholera toxin relationships: implications for role of ADP-ribosylation in cyclase activation. In: Smulson M, Sugimura T, eds. Novel ADP-ribosylations of regulatory enzymes and proteins. New York: Elsevier North Holland, 1980:369–79.

82. Nagayama Y, Russo D, Wadsworth HL, Chazenbalk GD, Rapoport B. Eleven amino acids (Lys-201 to Lys-211) and 9 amino acids (Gly-222 to Leu-230) in the human thyrotropin receptor are involved in ligand binding. J Biol Chem 1991;266:14926–30.

83. Morris JC, Bergert ER. Structure function studies of the human TSH receptor: inhibition of binding of labeled TSH to receptor by hTSHr peptides [Abstract]. Proc 74th annu meet Endocr Soc, San Antonio, TX, 1992.

84. Tahara K, Grollman EF, Saji M, Kohn LD. Regulation of prostaglandin synthesis by thyrotropin, insulin, or insulin-like growth factor-I, and serum in rat FRTL-5 thyroid cells. J Biol Chem 1991;266:440–8.

85. Takada K, Amino N, Tada H, Miyai K. Relationship between proliferation and cell cycle dependent $Ca^{++}$ influx induced by a combination of thyrotropin and insulin-like growth factor I in rat thyroid cells. J Clin Invest 1990;86:1548–55.

86. Isozaki O, Kohn LD. Control of c-*fos* and c-*myc* protooncogene induction in rat thyroid cells in culture. Mol Endocrinol 1987;1:839–48.

87. Adashi EY, Resnick CE, D'Ercole AJ, Svoboda ME, Van Wyk JJ. Insulin-like growth factors as intraovarian regulators of granulosa cell growth and function. Endocr Rev 1985;6:400–20.

88. Adashi EY, Resnick CE, Croft CS, May JV, Gospodarowicz D. Basic fibroblast growth factor as a regulator of ovarian granulosa cell differentiation: a novel non-mitogenic role. Mol Cell Endocrinol 1988;55:7–14.

89. Emoto N, Isozaki O, Arai M, et al. Identification and characterization of basic fibroblast growth factor in porcine thyroids. Endocrinology 1991;128:58–64.

90. Logan A, Black EG, Gonzalez A-M, Buscaglia M, Sheppard MC. Basic fibroblast growth factor: an autocrine mitogen of rat thyroid follicular cells. Endocrinology 1992;130:2363–72.

91. Mullaney BP, Skinner MK. Basic fibroblast growth factor (bFGF) gene expression and protein production during pubertal development of the seminiferous tubule: follicle-stimulating hormone-induced sertoli cell bFGF expression. Endocrinology 1992;131:2928–34.

92. Saji M, Akamizu T, Sanchez M, et al. Regulation of thyrotropin receptor gene expression in rat FRTL-5 thyroid cells. Endocrinology 1992;130:520–33.

93. Saji M, Ikuyama S, Akamizu T, Kohn L. Increases in cytosolic $Ca^{++}$ down regulate thyrotropin gene expression by a mechanism different from the cAMP signal. Biochem Biophys Res Commun 1981;176:94–101.

94. Isozaki O, Emoto T, Tsushima T, et al. Opposite regulation of DNA synthesis and iodide uptake in rat thyroid cells by basic fibroblast growth factor: correlation with opposite regulation of c-*fos* and thyrotropin receptor gene expression. Endocrinology 1992;128:2723–32.

95. Ikuyama S, Shimura H, Iguchi H, et al. Thyroid autoantigen p70Ku is a transcriptional activator of the TSH receptor gene. In: Nagataki S, ed. Hashimoto's disease anniversary meeting, Fukuoka, Japan, 1992.

96. Chan JYC, Lerman MI, Prabhakar BS, et al. Cloning and characterization of a cDNA that codes for a novel human thyroid autoantigen. J Biol Chem 1989;264:3651–4.
97. Allaway GP, Vivino AA, Kohn LD, Notkins AL, Prabhakar BS. Expression of a 70 kDA human autoantigen using a baculovirus vector: association with nuclear matrix and high affinity binding to DNA. Biochem Biophys Res Commun 1990;168:747–55.
98. LaPolt PS, Oikawa M, Jia XC, Dargan C, Hsueh AJW. Gonadotropin-induced up- and down-regulation of rat ovarian LH receptor message levels during follicular growth, ovulation, and luteinization. Endocrinology 1990; 1126:3277–9.
99. Segaloff D, Wang H, Richards JS. Hormone regulation of luteinizing hormone/chorionic gonadotropin receptor mRNA in rat ovarian cells during follicular development and luteinization. Mol Endocrinol 1990;4:1856–65.
100. Wang H, Segaloff D, Ascoli M. Regulation of gonadotropin receptor mRNA. J Biol Chem 1991;266:780–5.
101. Hoffman YM, Peegel H, Sprock MJE, Zhang QY, Menon KMJ. Evidence that human chorionic gonadotropin/luteinizing hormone receptor down regulation involves decreased levels of receptor messenger ribonucleic acid. Endocrinology 1991;128:388–93.
102. LaPolt PS, Tilly JL, Aihara T, Nishiori K, Hsueh AJ. Gonadotropin-induced up- and down-regulation of ovarian follicle-stimulating hormone (FSH) receptor gene expression in immature rats: effect of pregnant mare's serum gonadotropin, human chorionic gonadotropin, and recombinant FSH. Endocrinology 1992;130:1289–95.
103. Tilly JL, LaPolt PS, Hsueh AJ. Hormonal regulation of follicle-stimulating hormone receptor message ribonucleic acid levels in cultured rat granulosa cells. Endocrinology 1992;130:1296–302.
104. Wang H, Nelson S, Ascoli M, Segaloff DL. The 5′-flanking region of the rat luteinizing hormone/chorionic gonadotropin receptor gene confers Leydig cell expression and negative regulation of gene transcription by 3′,5′-cyclic adenosine monophosphate. Mol Endocrinol 1992;6:320–6.
105. Lu DL, Peegel H, Mosier SM, Menon KMJ. Loss of lutropin/chorionic gonadotropin receptor messenger ribonucleic acid during ligand-induced down regulation occurs post transcriptionally. Endocrinology 1993;132: 235–40.
106. Bernath VA, Muro AF, Vitullo AD, Bley MA, Baranao JL, Kornbliht AR. Cyclic AMP inhibits fibronectin gene expression in a newly developed granulosa cell line by a mechanism that suppresses cAMP-responsive element-dependent transcriptional activation. J Biol Chem 1990;265:18219–26.
107. Tominaga T, Yamasita S, Izumi M, Nagataki S. Regulation of human TSH receptor. In: Proc int sym hormone action and its disorders. Tokyo: Intractable Diseases Research Foundation, 1991:44–8.
108. Huber GK, Concepcion ES, Graves PN, Davies TF. Positive regulation of human thyrotropin receptor mRNA by thyrotropin. J Clin Endocrinol Metab 1991;72:1394–6.
109. Santisteban P, Kohn LD, Di Lauro R. Thyroglobulin gene expression is regulated by insulin and IGF-I as well as thyrotropin in FRTL-5 thyroid cells. J Biol Chem 1987;262:4048–52.

110. Saji M, Moriarty J, Ban T, Kohn LD, Singer DS. Hormonal regulation of MHC class I genes in rat thyroid FRTL-5 cells: TSH induces a cAMP-mediated decrease in class I expression. Proc Natl Acad Sci USA 1992;89: 1944–8.

111. Saji M, Moriarty J, Ban T, Singer DS, Kohn LD. MHC class I expression in rat thyroid cells is regulated by hormones, methimazole, and iodide, as well as interferon. J Clin Endocrinol Metab 1992;75:871–8.

112. Musti AM, Ursini MV, Avvedimento VE, Zimarino V, Di Lauro R. A cell type specific factor recognizes the rat thyroglobulin promoter. Nucleic Acids Res 1987;15:8149–66.

113. Civitareale D, Lonigro R, Sinclair AJ, Di Lauro R. A thyroid specific nuclear protein essential for tissue-specific expression of the thyroglobulin promoter. EMBO J 1989;8:2537–42.

114. Guazzi S, Price M, Felice MD, Damante G, Mattei M-G, Di Lauro R. Thyroid nuclear factor I (TTF-I) contains a homeodomain and displays a novel DNA binding specificity. EMBO J 1990;9:3631–9.

115. Lazzaro D, Price M, De Felice M, Di Lauro R. The transcription factor TTF-1 is expressed at the onset of thyroid and lung morphogenesis and in restricted regions of the foetal brain. Development 1991;113:1093–104.

116. Kikkawa F, Gonzalez FJ, Kimura S. Characterization of a thyroid-specific enhancer located 5.5 kilobase pairs upstream of the human thyroid peroxidase gene. Mol Cell Biol 1990;10:6216–24.

117. Mizuno K, Gonzalez FJ, Kimura S. Thyroid-specific enhancer binding protein (T/EBP): cDNA cloning, functional characterization, and structural identity with thyroid transcription factor TTF-I. Mol Cell Biol 1991;11: 4927–33.

118. Francis-Lang H, Price M, Polycarpou-Schwarz, DiLauro R. Cell-type specific expression of the rat thyroperoxidase promoter indicates common mechanisms for thyroid-specific gene expression. Mol Cell Biol 1992;12:576–88.

119. Avvedimento EV, Obici S, Sanchez M, Gallo A, Musti A, Gottesman ME. Reactivation of thyroglobulin gene expression in transformed thyroid cells by 5-azacytidine. Cell 1989;58:1135–42.

120. Gallo A, Benusiglio E, Bonapace M, et al. v-Ras and protein kinase C dedifferentiate thyroid cells by down regulating nuclear cAMP dependent kinase A. Genes Dev 1992;6:1621–30.

121. Pang X-P, Hershman JM. Differential effects of growth factors on [$^3$H]thymidine incorporation and [$^{125}$I]iodine uptake in FRTL-5 rat thyroid cells. Proc Soc Exp Biol Med 1990;194:240–4.

122. Black EG, Logan A, Davis JRE, Sheppard MC. Basic fibroblast growth factor affects DNA synthesis and cell function and activates multiple signalling pathways in rat thyroid FRTL-5 and pituitary GH3 cells. J Endocrinol 1991;127:39–46.

123. Benoist C, Chambon P. In vivo sequence requirements for the SV40 early promoter gene. Nature 1981;290:304–10.

124. Hudson LG, Thompson KL, Xu J, Gill GN. Identification and characterization of a regulated promoter element in the EGF receptor gene. Proc Natl Acad Sci USA 1990;87:7536–40.

125. Reynolds GA, Basu SK, Osborne TF, et al. HMGCoA reductase: a negatively regulated gene with unusual promoter and 5'-untranslated regions. Cell 1984;38:275–85.
126. Montminy MR, Sevario KA, Wagner JA, Mandel G, Goodman RH. Identification of a cyclic-AMP-response element within the rat somatostatin gene. Proc Natl Acad Sci USA 1986;83:6682–6.
127. Habener JF. Cyclic AMP response element binding proteins: a cornucopia of transcription factors. Mol Endocrinol 1990;4:1087–94.
128. Kanei-Ishii C, Ishii S. Dual enhancer activities of the cyclic-AMP responsive element with cell type and promoter specificity. Nucleic Acids Res 1989;17:1521–36.
129. Ikuyama S, Shimura H, Hoeffler JP, Kohn LD. Role of the cyclic AMP response element in efficient expression of the rat thyrotropin receptor. Mol Endocrinol 1992;6:1701–15.
130. Pugh BF, Tjian R. Transcription from a TATA-less promoter requires a multisubunit TFIID complex. Genes Dev 1991;5:1935–45.
131. Kageyama R, Sasai Y, Nakanishi S. Molecular characterization of transcription factors that bind to the cAMP responsive region of the substance P precursor gene. J Biol Chem 1991;266:15525–31.
132. Shimura H, Ikuyama S, Shimura Y, Kohn LD. The cAMP response element in the rat thyrotropin receptor promoter: regulation by each decanucleotide of a flanking tandem repeat uses different, additive, and novel mechanisms. J Biol Chem 1993.
133. Levine M, Manley JL. Transcriptional repression of eukaryotic promoters. Cell 1989;59:405–8.
134. Ivashkiv LB, Glimcher LH. Repression of class II major histocompatibility genes by cyclic AMP is mediated by conserved promoter elements. J Exp Med 1991;174:1583–92.
135. Schule R, Umesono K, Mangelsdorf DJ, Bolando J, Pike JW, Evans RM. Jun-Fos and receptors for vitamins A and D recognize a common response element in the human osteocalcin gene. Cell 1990;61:497–504.
136. Sehgal A, Patil N, Chao M. A constitutive promoter directs expression of the nerve growth factor gene. Mol Cell Biol 1988;8:3160–7.
137. Berlingieri MT, Akamizu T, Fusco A, et al. Thyrotropin receptor gene expression in oncogene-transfected rat thyroid cells: correlation between transformation, loss of thyrotropin-dependent growth, and loss of thyrotropin receptor gene expression. Biochem Biophys Res Commun 1990;173:172–8.
138. Santisteban P, Accbron A, Polycarpou-Schwarz M, Di Lauro R. Insulin and insulin-like growth factor I regulate a thyroid-specific nuclear protein that binds to the thyroglobulin promoter. Mol Endocrinol 1992;6:1310–7.
139. Chazenbalk GD, Nagayama Y, Russo D, Wadsworth HL, Rapoport B. Functional analysis of the cytoplasmic domains of the human thyrotropin receptor by site-directed mutagenesis. J Biol Chem 1990;265:20970–5.
140. Desai RK, Dallas JS, Gupta MK, et al. Dual mechanism of perturbation of thyrotropin mediated activation of thyroid cells by antibodies to the thyrotropin receptor (TSHR) and TSHR derived peptides. J Clin Endocrinol Metab 1993.

141. Tahara K, Yamamoto K, Yoshida S. Demonstration of two types of thyroid stimulating antibodies with different epitopes on the human thyrotropin receptor [Abstract]. 66th meet Am Thyroid Assoc. Thyroid 1992;2:S-72.
142. Kosugi S, Okajima F, Ban T, Hidaka A, Shenker A, Kohn LD. Substitutions of different regions of the 3rd cytoplasmic loop of the TSH receptor have selective effects on phosphoinositide and cAMP signal generation induced by TSH and TSH receptor autoantibodies, as well as on constitutive cAMP levels. Mol Endocrinol 1993.
143. Hidaka A, Okajima F, Ban T, Kosugi S, Kondo Y, Kohn LD. Receptor crosstalk can optimize assays for autoantibodies to the thyrotropin receptor: effect of phenylisopropyladenosine on cAMP and inositol phosphate levels in rat FRTL-5 thyroid cells. J Clin Endocrinol Metab 1993.

# 17

# Studies on the Structure and Function of Gonadotropins

S.N. Venkateswara Rao and William R. Moyle

Information on the structure of a protein is useful for understanding its function. Indeed, shortly after the primary sequences of the gonadotropins were determined, investigators from several laboratories began studies to learn about their tertiary structures (1–8). Initially, these were based on chemical modifications and the sensitivities of the hormones to protease and glycosidase digestions (9–11). Partial reduction studies led to the assignments of the disulfide bonds (12, 13). The advent of hybridoma technology was soon followed by extensive immunological characterization of the hormones and the development of extensive panels of monoclonal antibodies (14–20). More recently, mutagenesis has been used to prepare analogs that permitted the assignment of monoclonal antibody binding sites and identification of the regions of the hormone β-subunits that control their receptor binding specificities and their biological activities (21–28).

  Like others, several years ago we began efforts to develop models of hCG and hFSH which would account for their endocrine, immunological, and chemical properties (discussed in 3). Initially, we identified regions of hCG and hFSH that controlled receptor binding specificity and/or that participated in antibody binding sites. Chimeras of hCG and LHβ-subunits, hCG and hFSHβ-subunits, hCG and bovine LHβ-subunits, and human and bovine α-subunits were found to be useful for identifying the binding sites of several anti-hCG and anti-hFSH monoclonal antibodies (27, 28, unpublished observations). Chimeras of hCG and hFSHβ-subunits were useful for identifying portions of the molecules that controlled receptor binding specificity. These data provided information on the relative locations of residues on the surface of the hormones and limited the number of patterns that account for the ways that these gonadotropins fold. Here we illustrate a model prepared using a low-tech strategy that is consistent with all existing data and that provides a useful means for rationalizing the interactions of gonadotropins and their receptors.

## Methods

Models were prepared manually (i.e., by bending #12 electrical wire) and by computation. To construct the wire model, we marked pieces of wire into 92 and 114 equal segments representing the α- and β-subunits, respectively. The scale of the model was chosen such that 2 cm corresponded to 3.8 Å, the approximate distance between Cα carbon atoms measured from reported crystal structures. The wires were bent to accommodate the disulfide bond assignments reported by Bahl (12, 13). The distances between the α carbons of the cysteines were kept at 3.5 cm, corresponding to the approximate distance of 6–6.5 Å between cysteines in disulfide bonds of known proteins. The wires were also bent to accommodate the locations of residues with epitope maps (27, unpublished observations). The maximal distances between the Cα carbon atoms of different amino acid sequences thought to form discontiguous epitopes were limited to 7 cm. This corresponded to approximately 13 Å, the distance of these residues in the crystal structures of antibody-lysozyme complexes (29). The minimum distances between residues in epitopes of antibodies that could bind to hCG or hFSH at the same time were kept greater than 8 cm.

Based on the binding site of antibody A109, we assumed that residues 73–75 were located adjacent to residues 11–17. The binding site of antibody A110 suggested that residues 17–26 were also likely to be near residue 83. To account for the observation that antibody A201 binds to hCG at the same time as A109, we placed Cα carbon atoms of residues 64–68 more than 8 cm from those of residues 11–26. However, since antibody A112 can block the binding of antibodies A109, A110, and A201, residues 64–68 were placed nearer residues 73–75 than 11–26. The binding sites of antibody A407 suggested that the Cα carbon atoms of residues 1–6 were located near that of residue 81 and distant from all other antibody binding sites. Similarly, the binding sites of A105 suggested that residues 50–53 were located distant from all other antibody binding sites.

Based on the binding sites of B101, B107, B109, and B204, we assumed that residues 8–10, 47–51, and 91–92 were within a 7 cm radius. Likewise, based on the binding sites of B105, B108, and B111, we placed residue 74 within 7 cm from residues 77 and 114 and kept residue 77 more than 8 cm from residue 114. The orientation of the subunits was established by noting that several anti-α-subunit antibodies competed for binding to hCG with anti-β-subunit antibodies (e.g., A102 and B101) and that bovine LHα-subunit residue Lys49 can be crosslinked to β-subunit residue

Asp111. We assumed that the antibody binding sites, charged residues, and nonconserved regions were at the surfaces of the α- and β-subunits. The remaining hydrophobic residues were positioned away from the surface, and the wires were bent to keep the molecules compact (i.e., holes were not permitted).

To learn if the wire model would be consistent with the physical properties of amino acid residues, we analyzed it quantitatively using the molecular modeling program Sybyl (Tripos Associates, St. Louis, MO). The locations of the Cα carbon atoms were digitized by placing the model on top of and in front of large sections of graph paper. Next, spotlights placed approximately 10 feet above and in front of the model were used to illuminate it in an otherwise darkened room. The positions of the x, y, and z coordinates corresponding to the locations of the marks on the wires (i.e., the locations of the Cα carbons) were read by touching the marks with a pencil and noting the positions of the shadows on the graph paper made by the pencil tip. These Cα coordinates were arranged in the Protein Data Bank format and entered into the program Sybyl. The remainder of the protein backbone and the amino acid side chains were added with the Sybyl BIOPOLYMER "Build" and "Edit"modules, using the options "alphas→backbone" and "add side chains," respectively. Major steric problems caused by interactions of amino acid side chains were removed manually, and the model was subjected to energy minimization and dynamics to overcome unfavorable contacts.

## Results and Discussion

Our long-standing efforts to decipher the function of the glycoprotein hormones have been hampered by the lack of a molecular structure of these hormones. Rational design of hormone analogs needed to study hormone function requires an understanding of hormone structure. hCG was crystallized four years ago (30, 31), and a structure should be available soon. In the interim we have been preparing models to help in the design of hormone analogs. Our first models were built from wire because they enabled us to visualize relationships between portions of the molecule easily with a minimum investment, because they could easily be manipulated by hand, and because we anticipated that they would give us about as realistic a structure as other more involved modeling procedures. Relatively few hours were required to bend the wires into a shape that would explain all the data. Thus, we anticipate that the assumptions underlying the model based on the crystal structures of other proteins (e.g., the relative distances between residues in antibody binding sites) can successfully be combined with the epitope maps for hCG to give a plausible structure. Subsequently, we have used distance geometry algorithms to prepare models of hCG using computers. These do not suffer

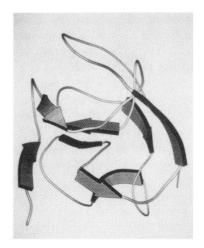

FIGURE 17.1. Folding pattern of the α-subunit (upper left panel), β-subunit (upper right panel), and α-β heterodimer (bottom panel).

from the subjectivity of wire models and can be used to evaluate several different parameters quantitatively. Both types of modeling procedures give similar overall structures that we anticipate will be crude approximations of the actual folding patterns of the hormones. Their accuracy will be known only once crystal or NMR structures have been solved.

Given their small sizes, there are only a few folding patterns consistent with a compact protein that will account for the disulfide bonds and the antibody binding sites in the α- and β-subunits. Only one of these was consistent with the data for the heterodimer (see Fig. 17.1 for a typical example). After minimization, the energy of the molecule was approximately −2 kcal. While this value is not nearly as negative as would be expected for a native protein, it indicated that all the bad contacts could be removed from the compact structure derived from the folding pattern

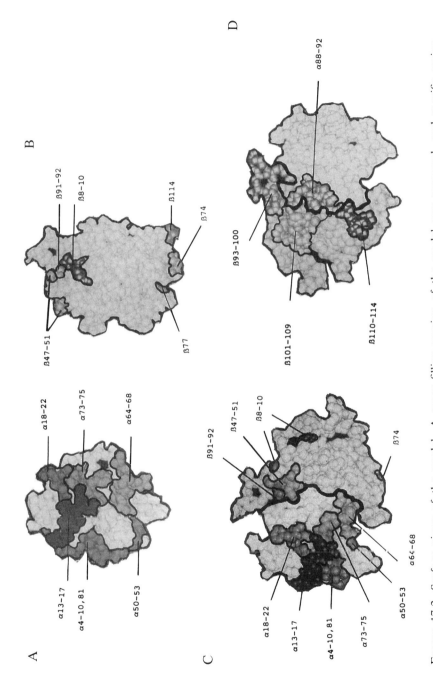

FIGURE 17.2. Surface views of the model. A space-filling version of the model was prepared, and specific amino acid residues were labeled as illustrated. A: A view of the α-subunit is shown. B: A view of the β-subunit is shown. C: A view of the α-β heterodimer is shown. D: A view of the hormone receptor interface is shown.

in the wire model. The major deficiencies in the model were that (i) the locations of the φ-ψ angles have not been optimized and (ii) hydrogen bonds have not been considered.

When epitope maps were reconstructed from the folding patterns made by bending the wires and subsequent energy minimization, they illustrated the relative positions of the antibody binding sites and, as expected, were consistent with the original data (Fig. 17.2). In addition, they were consistent with many of the chemical modifications that have been reported. This included the accessibility of the tyrosine residues to iodination (7).

While the disulfide bonds add considerable restrictions on the possible shapes that the subunits could assume, the addition of the immunological constraints limits the folding patterns considerably. This is particularly evident in the case of the α-subunit, where immunological constraints served to limit the locations of the three large intracysteine loops that account for the bulk of the α-subunit.

The model suggests that nearly all the portions of the gonadotropins that have the greatest influence on hormone activity are located in a common region near the junction of the α- and β-subunits. Figure 17.2D illustrates the view of the hormone that we anticipate is seen by the receptor. This includes the regions of the β-subunit that influence receptor binding specificity (i.e., residues 93–114) (28) and those of the α-subunit that influence binding affinity and efficacy (7, 26, 32). In addition, this portion of the hormones includes Asn52 of the α-subunit. The N-linked oligosaccharide at this site has been shown to have the greatest influence of all the oligosaccharides on hormone efficacy (21). Surprisingly, the region of the β-subunit containing the loop formed by residues Cys38–Cys57 that had been shown to stimulate steroidogenesis (33) was located on a different face of the protein (i.e., near the subunit interface just over the top of the view illustrated in Fig. 17.2D). Thus, the portion of the protein that interacts with the receptor might include much more than the surface shown in the view in Figure 17.2D. However, since the Cys38–Cys57 loop did not influence hormone binding specificity (28), we anticipate that it does not interact directly with the LH receptor. This loop may be near the receptor interface, since antibodies that bind to this region of hCG prevent hormone binding to LH receptors (14) and since some mutations in this region influence receptor binding sufficiently to cause infertility (34).

In summary, we anticipate that a combined biochemical and immunological approach can be used to devise structural models of hCG and hFSH that account for many of their hormonal properties. These models are only as accurate as the data on which they are based. While these data are currently very limited, we are encouraged to have found that they can significantly limit the number of folding patterns that the hormones might assume. As noted before, the validity of this model will be verified only after the crystal structure of the gonadotropins has been determined.

## References

1. Gordon WL, Ward DN. Structural aspects of luteinizing hormone actions. In: Ascoli M, ed. Luteinizing hormone action and receptors. Boca Raton, FL: CRC Press, 1985:173–97.
2. Liu W, Ward DN. The purification and chemistry of pituitary glycoprotein hormones. Pharmacol Ther 1975;1:545–70.
3. Ryan RJ, Keutmann HT, Charlesworth MC, et al. Structure-function relationships of gonadotropins. Recent Prog Horm Res 1987;43:383–429.
4. Ryan RJ, Charlesworth MC, McCormick DJ, Milius RP, Keutmann HT. The glycoprotein hormones: recent studies of structure-function relationships. FASEB J 1988;2:2661–9.
5. Puett D. Human choriogonadotropin. Bioessays 1986;4:70–5.
6. Strickland TW, Puett D. Contribution of subunits to the function of luteinizing hormone/human chorionic gonadotropin recombinants. Endocrinology 1981;109:1933–42.
7. Pierce JG, Parsons TF. Glycoprotein hormones: structure and function. Annu Rev Biochem 1981;50:465–95.
8. Weare JA, Reichert LE Jr. Studies with carbodiimide-crosslinked derivatives of bovine lutropin, II. Location of the crosslink and implication for interaction with the receptors in testes. J Biol Chem 1979;254:6972–9.
9. Birken S, Kolks MA, Amr S, Nisula B, Puett D. Structural and functional studies of the tryptic core of the human chorionic gonadotropin beta-subunit. Endocrinology 1987;121:657–66.
10. Birken S, Gawinowicz Kolks MA, Amr S, Nisula B, Puett D. Tryptic digestion of the α-subunit of human choriogonadotropin. J Biol Chem 1986;261: 10719–27.
11. Moyle WR, Bahl OP, Marz L. Role of the carbohydrate of human choriogonadotropin in the mechanism of hormone action. J Biol Chem 1975; 250:9163–9.
12. Mise T, Bahl OP. Assignment of disulfide bonds in the β-subunit of human chorionic gonadotropin. J Biol Chem 1981;256:6587–92.
13. Mise T, Bahl OP. Assignment of disulfide bonds in the α-subunit of human chorionic gonadotropin. J Biol Chem 1980,255.8516–22.
14. Moyle WR, Ehrlich PH, Canfield RE. Use of monoclonal antibodies to hCG subunits to examine the orientation of hCG in the hormone-receptor complex. Proc Natl Acad Sci USA 1982;79:2245–9.
15. Moyle WR, Pressey A, Dean Emig D, et al. Detection of conformational changes in human chorionic gonadotropin upon binding to rat gonadal receptors. J Biol Chem 1987;262:16920–6.
16. Ehrlich PH, Moustafa ZA, Krichevsky A, Birken S, Armstrong EG, Canfield RE. Characterization and relative orientation of epitopes for monoclonal antibodies and antisera to human chorionic gonadotropin. Am J Reprod Immunol Microbiol 1985;8:48–54.

17. Bidart JM, Troalen F, Bohuon CJ, Hennen G, Bellet DH. Immunochemical mapping of a specific domain on human choriogonadotropin using anti-protein and anti-peptide monoclonal antibodies. J Biol Chem 1987;262:15483–9.
18. Bidart JM, Troalen F, Salesse R, Bousfield GR, Bohuon CJ, Bellet DH. Topographic antigenic determinants recognized by monoclonal antibodies on human choriogonadotropin beta-subunit. J Biol Chem 1987;262:8551–6.
19. Bidart JM, Troalen F, Bousfield GR, Bohuon C, Bellet D. Monoclonal antibodies directed to human and equine chorionic gonadotropins as probes for the topographic analysis of epitopes on the human alpha-subunit. Endocrinology 1989;124:923–9.
20. Schwartz S, Krude H, Merz WE, Lottersberger C, Wick G, Berger P. Epitope mapping of the receptor-bound agonistic form of human chorionic gonadotropin (hCG) in comparison to the antagonistic form (deglycosylated hCG). Biochem Biophys Res Commun 1991;178:699–706.
21. Matzuk MM, Boime I. Mutagenesis and gene transfer define site-specific roles of the gonadotropin oligosaccharides. Biol Reprod 1989;40:48–53.
22. Smith PL, Kaetzel D, Nilson J, Baenziger JU. The sialylated oligosaccharides of recombinant bovine lutropin modulate hormone bioactivity. J Biol Chem 1990;265:874–81.
23. Chen F, Puett D. Contributions of arginines-43 and -94 of human chori-ogonadotropin β to receptor binding and activation as determined by oli-gonucleotide-based mutagenesis. Biochemistry 1991;30:10171–5.
24. Chen F, Wang Y, Puett D. Role of the invariant aspartic acid 99 of human choriogonadotropin β in receptor binding and biological activity. J Biol Chem 1991;266:19357–61.
25. Chen F, Puett D. Delineation via site-directed mutagenesis of the carboxyl-terminal region of human choriogonadotropin required for subunit assembly and biological activity. J Biol Chem 1991;266:6904–8.
26. Yoo J, Ji I, Ji TH. Conversion of lysine 91 to methionine or glutamic acid in human choriogonadotropin α results in the loss of cAMP inducibility. J Biol Chem 1991;266:17741–3.
27. Moyle WR, Matzuk MM, Campbell RK, et al. Localization of residues that confer antibody binding specificity using human chorionic gonadotropin/luteinizing hormone beta subunit chimeras and mutants. J Biol Chem 1990; 265:8511–8.
28. Campbell RK, Dean Emig DM, Moyle WR. Conversion of human chori-ogonadotropin into a follitropin by protein engineering. Proc Natl Acad Sci USA 1991;88:760–4.
29. Davies DR, Sheriff S, Padlan EA. Antibody-antigen complexes. J Biol Chem 1988;263:10541–4.
30. Lustbader JW, Birken S, Pileggi NF, et al. Crystallization and characterization of human chorionic gonadotropin in chemically deglycosylated and enzymatically desialated states. Biochemistry 1989;28:9239–43.
31. Harris DC, Machin KJ, Evin GM, Morgan FJ, Isaacs NW. Preliminary X-ray diffraction analysis of human chorionic gonadotropin. J Biol Chem 1989; 264:6705–6.
32. Chen F, Wang Y, Puett D. The carboxy-terminal region of the glycoprotein hormone alpha-subunit: contributions to receptor binding and signaling in human chorionic gonadotropin. Mol Endocrinol 1992;6:914–9.

33. Keutmann HT, Charlesworth MC, Mason KA, Ostrea T, Johnson L, Ryan RJ. A receptor-binding region in human choriogonadotropin/lutropin beta subunit. Proc Natl Acad Sci USA 1987;84:2038–42.
34. Weiss J, Axelrod L, Whitcomb RW, Harris PE, Crowley WF, Jameson JL. Hypogonadism caused by a single amino acid substitution in the β subunit of luteinizing hormone. N Engl J Med 1992;326:179–83.

# 18

# Hormone Binding and Activation of the LH/CG Receptor

TAE H. JI, HUAWEI ZENG, AND INHAE JI

LH and hCG are members of the glycoprotein hormone family, which consists of noncovalently associated α- and β-subunits (1). Upon dissociation of α-β dimers, each subunit loses high-affinity hormonal activities. LH and CG interact with the LH/CG receptor. This is expressed in the gonads of both sexes at specific stages of development and, particularly, in the ovary during the ovulation cycle (2, 3).

The LH/CG receptor cDNAs have been cloned (4–7), and the amino acid sequences have been deduced. The LH/CG receptor consists of a signal peptide of 26 amino acids and the mature receptor of 674 amino acids. The molecular weight of the mature LH/CG receptor polypeptide is calculated to be 75,000, whereas the actual molecular weight is in the range of 85,000 (5, 11, 12) to 92,000 (13). These higher molecular weights are due to glycosylation of the receptor (11–13). The amino acid sequences of the rat and human LH/CG receptors are highly homologous, indicating structural and functional similarities between the human and rat receptors.

The sequences indicate that the receptors are comprised of two halves of similar sizes, the extracellular N-terminal half and the membrane-embedded C-terminal half (Fig. 18.1). The C-terminal half is considered to have seven transmembrane domains, three extracellular loops, three cytoplasmic loops, and a C-terminal tail, as does bacteriorhodopsin (14). These receptors are coupled to G protein and comprise the superfamily of G protein-coupled receptors.

## Conformational Diversity of the Extracellular N-Terminal Half of the LH/CG Receptor

The LH/CG receptor is encoded by 11 exons. The bulk of the N-terminal half of the LH/CG receptor is encoded by exons 1–10, and exon 11 encodes primarily the C-terminal half (15, 16). When exons 1–9 are aligned without gaps to increase the stringency of sequence comparison, considerable matches are found between some exons, revealing an imperfectly matching motif (Fig. 18.2). Those showing substantial similarities are: exon 2 with exons 6 and 9; exon 3 with exons 4, 6 and 8; exon 6 with exons 2, 3 and 8; exon 8 with exons 3, 6 and 7; and exon 9 with exons 3 and 5. Overall, considerable similarities exist among exons 2, 3, 6, 7, and 8. Since exon boundaries often mark functional domains of proteins and map to amino acids located at the protein surface, they could reveal a criterion for defining structural and functional domains. Indeed, all of the exon boundaries of the LH receptor have one or more ionic Lys, Arg, Glu, or Asp residues within a distance of two amino acids. Normally one would expect a similar peptide structure from peptide segments with similar sequences, particularly from repeating sequences. Furthermore, most of the exon junctions coincide with structural transitions.

## Multiple Hormone Contact Sites in the LH/CG Receptor

Several lines of evidence demonstrate the multiple contacts between hCG and its receptor. Photoaffinity labeling studies not only indicate the existence of a minimum of three hormone contact sites in the receptor, but also suggest the approximate locations of the contact sites (12, 17). Another important piece of information made available from the photoaffinity and affinity labeling studies is that there is a Lys residue(s) at or near the contact site. The three sites are located in the three distinct receptor domains of 24 kd, 28 kd, and 34 kd (12). The 24 kd and 34 kd receptor domains are heavily N-sialoglycosylated, whereas the 28 kd domain is not glycosylated (18). Therefore, the glycosylated 24 kd and 34 kd domains are logically considered to comprise the extracellular N-terminal half, and the 28 kd domain forms the bulk of the C-terminal half (Fig. 18.3). These results indicate that the hormone-contact sites are located in both the extracellular N-terminal half and the membrane-associated C-terminal half of the receptor.

This assumption is verified by studies using anti-hCG monoclonal antibodies (19), truncated mutant receptors (20–24), and synthetic receptor peptides (25). The truncated receptors (Fig. 18.4) consisting of exons 1–3 (20), 1–4 (20), and 1–10 (21–23) are capable of hormone binding, but they are incapable of generating hormone signals (Table 18.1). On the other hand, the C-terminal half is capable of hormone binding and signal generation (24). Recently, four LH/CG receptor peptides, peptide[21–38], peptide[102–115], peptide[253–272], and peptide[573–583], have been shown to inhibit hCG binding to the receptor, albeit at low affinities (25). The first three peptides represent the boundaries between exons 1–2, 4–5, and 9–10. This is consistent with the observation that exons 1–3 are capable

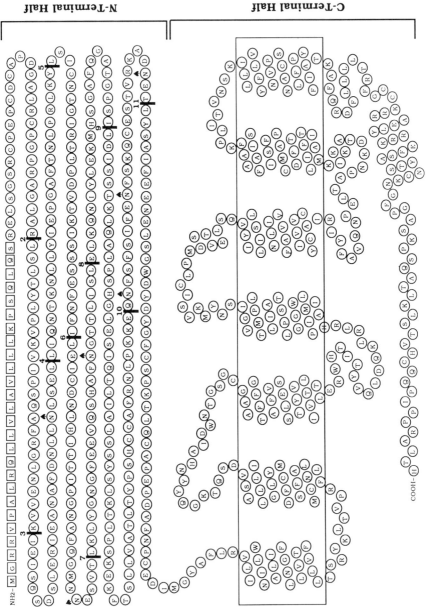

FIGURE 18.1. The LH/CG receptor. The amino acid sequence of the mature wild-type receptor is shown as circles, except the signal peptide of 26 amino acids, which are presented as squares. The starting points of exons are marked by vertical lines and exon numbers. The seven transmembrane helices are boxed and the six con-

Exon        Sequence

1           IVLAVLLIKPSQLQSREL
2           SITYLPVKVIPSQAFRGLNEVVK
3           EISQSDSLEFIEANAFDNLLNLSE
4           LIQNTKMLLYIEPGAFTNLPRIKY
5           SICNTGIRTLPDVTKISSSEFNFI
6           ELQDNLHITTIPGNAFQGMNNESVT
7           KLYGNGFEFVQSHAFNGTTLIS
8           ELKENIYLEKMHSQAFQGATGPSI
9           DISSTKLQALPSHGLESIQTLIALSSYSLKTLPSKEKFTSLLVATLTYPSHCCAFRNLPKK
            AISSYSIKTLPSKEKFTSLLVATIT
5           SICNTGIRTLPDVTKISSSEFNFI

Consensus   L----L--LPSQAF-----L    L----LKTLP---K-TS-----L

FIGURE 18.2. Alignment of amino acid sequences of exons. Amino acid sequences of exons are aligned without gaps to increase the stringency of alignment. Homologies, shown in capital letters, are based on either exact matches or the stringent definition of exchange groups (I = L = V = M, F = Y = W, G = A). The consensus sequence was determined when homology was 50% or more. Exon 9 has two tandem repeats of the imperfectly matched motif. The second of the repeats is underlined and aligned with the first of the repeats. Exon 10 was excluded because of the absence of significant homologies to other exons.

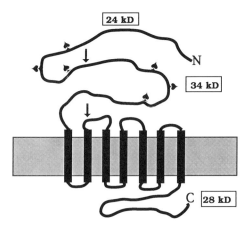

FIGURE 18.3. A schematic drawing of the LH/CG receptor. Three receptor domains of 24 kd, 34 kd, and 28 kd are tentatively assigned. The 28 kd domain corresponds to the bulk of the C-terminal half of the receptor. The 24 kd and 34 kd domains are located in the N-terminal half, but which of the two corresponds to the N-terminus is unknown.

of hormone binding and the prediction that exon boundaries are often functional sites (15). All these peptides possess one or more Lys residues or have it within a distance of several amino acids, consistent with the photoaffinity labeling result.

## Multiple-Step Interactions of hCG and the Receptor

Recent studies using mutant receptors have identified a few distinct receptor states, including free receptor, high-affinity hormone-receptor complex, low-affinity hormone-receptor complex, and low-affinity hormone-activated receptor complex. The high-affinity hCG-receptor binding is primarily due to the slow dissociation rate (27). Therefore, the low-affinity binding of some truncated receptors, receptor peptides, and hormone peptides may be due in part to presumptive fast dissociation rates. If this is true, the high-affinity binding may involve a certain (latching) mechanism to slow dissociation rates. Such latching is likely to occur in the N-terminal half of the receptor where the high-affinity binding occurs. Whether the high- and low-affinity sites are independent sites or comprise a hormone binding site is unknown. If they are independent sites, hormone binding may occur in a sequential manner. For example, initially the hormone binds to the high-affinity site and then additionally interacts with the low-affinity site where the receptor might be activated and signal generated. This is consistent with multistep binding

```
                        83   84   85   86   87   88   89   90   91   92

Natural α    NH₂-82 Amino Acids-His-Cys-Ser-Thr-Cys-Tyr-Tyr-His-Lys-Ser-COOH

αΔ92         NH₂-82 Amino Acids-His-Cys-Ser-Thr-Cys-Tyr-Tyr-His-Lys-COOH

αΔ91-92      NH₂-82 Amino Acids-His-Cys-Ser-Thr-Cys-Tyr-Tyr-His-COOH

αΔ89-92      NH₂-82 Amino Acids-His-Cys-Ser-Thr-Cys-Tyr-Tyr-COOH

αΔ89-92      NH₂-82 Amino Acids-His-Cys-Ser-Thr-Cys-Tyr-COOH

αΔ88-92      NH₂-82 Amino Acids-His-Cys-Ser-Thr-Cys-COOH

αX90         NH₂-82 Amino Acids-His-Cys-Ser-Thr-Cys-Tyr-Tyr--X--Lys-Ser-COOH

αX91         NH₂-82 Amino Acids-His-Cys-Ser-Thr-Cys-Tyr-Tyr-His--X--Ser-COOH


Peptide α83-92         NH₂-His-Cys-Ser-Thr-Cys-Tyr-Tyr-His-Lys-Ser-COOH
```

FIGURE 18.4. Mutant α-subunits. Two types of mutant α-subunits were prepared. The C-terminal five amino acids of the α-subunit were progressively truncated by replacing amino acid codons with a stop codon in the α cDNA. Also, α Lys$^{91}$ or α His$^{90}$ was substituted with a series of other amino acids. In addition, peptide α$^{83-92}$ was synthesized.

TABLE 18.1. Truncated LH/CG receptor.

|  | hCG binding (kd) | cAMP induction (EC$_{50}$) |
|---|---|---|
|  | nM | |
| Wild type | 0.26 | 0.05 |
| Exons 1–3 | 0.24 | ND |
| Exons 1–4 | 30 | ND |
| Exons 1–10 | 25 | ND |
| Exons 1–10 + (11) | 0.17 | ND |
| Exons 1 + 11 | 650 | 580 |
| LH-R$_{\Delta 6-336}$ | 1040 | 900 |

Note: Truncated mutant LH/CG receptors were expressed in 293 cells and assayed for hormone binding and induction of cAMP synthesis.

models that entail more than one type of hCG-receptor complex (26, 27). The fact is that hCG-receptor binding data are not consistent with a single-step hormone-receptor binding (27). It is presumptuous, however, to assume that the high-affinity and low-affinity hormone binding sites and binding itself are totally independent of each other. In fact, evidence indicates otherwise. For example, the conversion of Asp383 of trans-membrane domain 2 of the LH/CG receptor affected the high-affinity hormone binding and resulted in a significant reduction of the hormone binding affinity (28).

## Mutiple Receptor Contact Points in hCGα

Studies using photoaffinity labeling (12, 17) and hCGα and β peptides (29, 30) demonstrated that both subunits of hCG directly interact with the receptor. There is ample evidence for multiple contacts of hCGα with the receptor. α C-terminal 2–5 residues are important for bioactivity of LH and TSH (31). Chemical modifications of a number of amino acids affect receptor binding and/or bioactivity (32). Photoaffinity labeling indicates a minimum of three receptor contact points in hCGα, and each of these sites has a Lys residue (17). Two α peptides, peptide[26–46] and peptide[76–92], are capable of inhibition of hCG binding to the receptor (29).

## Receptor Binding of Mutant hCG and FSH with Truncated α C-Terminal Residues

Truncation of α C-terminal 2–5 residues results in the loss of bioactivity of LH and TSH (31). To systematically examine the roles of the α C-terminal region, we utilized a combination of three different approaches

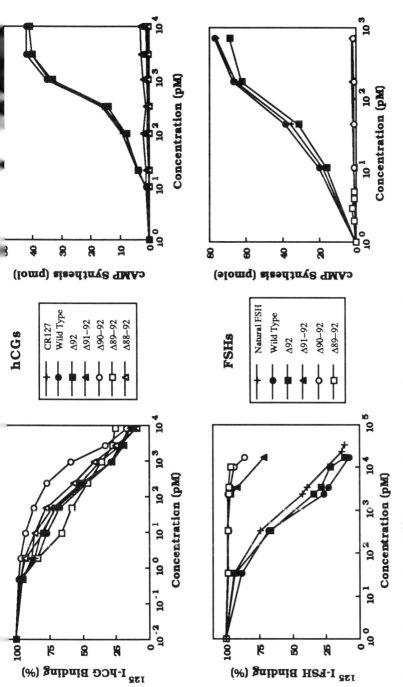

FIGURE 18.5. Inhibition of [125]I-hormone binding and induction of cAMP synthesis by deletion mutants of hCG (upper panels) and FSH (lower panels). Porcine granulosa cells possessing the LH/CG receptor were incubated with [125]I-hCG in the presence of increasing concentrations of unlabeled recombinant hCGs. After washing cells, the radioactivity associated with the cells. Likewise, receptor binding of mutant FSHs was determined with Y1 cells. For cAMP induction assay, increasing concentrations of recombinant hCGs and FSHs were incubated with MLTC/J cells or Y1 cells, respectively. After washing cells, they were lysed and centrifuged to remove pellets. The resulting supernatants were collected and assayed for intracellular cAMP concentrations.

involving deletion-mutant hormones, mutant hCGs with substitutions for a single amino acid, and synthetic peptides corresponding to different regions of α (33). A set of five deletion-mutant hCGs were generated in which the α C-terminal five amino acids were progressively truncated (Fig. 18.5). In contrast to 92 amino acids present in normal α-subunit, the number of amino acids in the α-subunits of these truncated hCGs is 91 in $hCG_{\alpha\Delta92}$, 90 in $hCG_{\alpha\Delta91-92}$, 89 in $hCG_{\alpha\Delta90-92}$, 88 in $hCG_{\alpha\Delta89-92}$, and 87 in $hCG_{\alpha\Delta88-92}$. Likewise, we have generated a set of four deletion mutants of human FSH, $FSH_{\alpha\Delta92}$, $FSH_{\alpha\Delta91-92}$, $FSH_{\alpha\Delta90-92}$, and $FSH_{\alpha\Delta89-92}$.

These deletion mutants, as opposed to single amino acid substitution mutants, make it easy to identify important amino acids because a limited number of mutants need to be prepared and the collective effect of deletions can readily be detected. In contrast, single amino acid substitution mutations may not necessarily reveal important amino acids. Once important amino acids were identified, we attempted to determine whether they contact receptor or are necessary for the hormone to assume a necessary conformation. To answer this question, peptide $\alpha^{83-92}$, was used. Then the important amino acids were individually substituted with a series of other amino acids (Fig. 18.4) to determine effects of different side chains.

Natural hCG, wild-type recombinant hCG, and all truncated recombinant hCGs were capable of competing with [125]I-hCG for receptor binding (Fig. 18.5). The $EC_{50}$ values for inhibition of [125]I-hCG binding to the receptor were similar among natural hCG, recombinant wild-type hCG, $hCG_{\alpha\Delta92}$, $hCG_{\alpha\Delta91-92}$ and $hCG_{\alpha\Delta89-92}$ (Table 18.2). $hCG_{\alpha\Delta90-92}$ and $hCG_{\alpha\Delta88-92}$, however, were less effective in receptor binding, as they showed higher $EC_{50}$ values. The results indicate that the C-terminal amino acids of hCGα play a role in receptor binding, but they are not absolutely essential. Therefore, if they are involved in contacting the receptor, the affinity is low. This is not surprising, as there is ample evidence indicating that a number of receptor contact sites exist in both subunits of the hormone (12, 17, 29, 30). Each site has a low affinity for receptor binding and is expected to contribute part of the intact hormone's high affinity toward the receptor.

Deletion of the α C-terminal residues had a dramatically different effect on FSH. Natural FSH, wild-type recombinant FSH, and $FSH_{\alpha\Delta92}$ were capable of competing with [125]I-FSH for receptor binding with similar $EC_{50}$ values (Fig. 18.5 and Table 18.3). Unlike their hCG counterparts, $FSH_{\alpha\Delta91-92}$, $FSH_{\alpha\Delta90-92}$, and $FSH_{\alpha\Delta89-92}$ lost the ability to inhibit [125]I-FSH binding by nearly 100-fold, or completely, under the same binding conditions. Since $FSH_{\alpha\Delta92}$ was fully capable of receptor binding, as was $hCG_{\alpha\Delta92}$, these results clearly demonstrate that the loss of receptor binding activity was specific. More importantly, they indicate the existence of a critical difference in the mechanisms of hormone-receptor interactions of hCG and FSH. Interestingly, the difference resides in the

TABLE 18.2. Affinities for inhibition of $^{125}$I-hCG binding and for cAMP induction.

| Hormones | Inhibition of $^{125}$I-hCG binding (EC$_{50}$) | cAMP synthesis | |
| | | EC$_{50}$ | Max level |
| | pM | pM | pmole (%) |
| --- | --- | --- | --- |
| Natural hCG | 173 ± 19 | 302 ± 27 | 41 (100) |
| Wild-type hCG | 197 ± 26 | 319 ± 30 | 42 (102) |
| hCG$_{\alpha\Delta92}$ | 263 ± 58 | 341 ± 42 | 40 (98) |
| hCG$_{\alpha\Delta91-92}$ | 251 ± 39 | NS | NS |
| hCG$_{\alpha\Delta90-92}$ | 1210 ± 163 | NS | NS |
| hCG$_{\alpha\Delta89-92}$ | 178 ± 24 | NS | NS |
| hCG$_{\alpha\Delta88-92}$ | 617 ± 57 | NS | NS |

*Note:* The data in Figure 18.4 were used to determine the EC$_{50}$ values for inhibition of $^{125}$I-hCG binding and for cAMP induction. cAMP induction of some of the truncated hCGs were not significant (NS). A student's t-test analysis indicated that the probability of chance occurrence of the variations in EC$_{50}$ values for hCG CR-127, recombinant wild-type hCG, hCG$_{\alpha\Delta92}$, hCG$_{\alpha\Delta91-92}$, and hCG$_{\alpha\Delta89-92}$ is $0.7 > P > 0.2$, an indication of no significance. In contrast, variations in the values for hCG$_{\alpha\Delta90-92}$ and hCG$_{\alpha\Delta88-92}$ from those of the rest are significant, with $0.05 > P > 0.02$.

TABLE 18.3. Affinities for inhibition of $^{125}$I-FSH binding and for cAMP induction.

| Hormones | Inhibition of $^{125}$I-FSH binding (EC$_{50}$) | cAMP synthesis | |
| | | EC$_{50}$ | Max level |
| | pM | pM | pmole (%) |
| --- | --- | --- | --- |
| Natural FSH | 1490 ± 120 | 53 ± 6 | 76 (100) |
| Wild-type FSH | 760 ± 56 | 42 ± 4 | 78 (103) |
| FSH$_{\alpha\Delta92}$ | 920 ± 58 | 47 ± 4 | 72 (95) |
| FSH$_{\alpha\Delta91-92}$ | NS | NS | NS |
| FSH$_{\alpha\Delta90-92}$ | NS | NS | NS |
| FSH$_{\alpha\Delta89-92}$ | NS | NS | NS |
| FSH$_{\alpha\Delta88-92}$ | NS | NS | NS |

*Note:* The data in Figure 18.4 were used to determine the EC$_{50}$ values for inhibition of $^{125}$I-FSH binding and for induction of cAMP synthesis. Inhibition of $^{125}$I-FSH binding and induction of cAMP synthesis by FSH$_{\alpha\Delta91-92}$, FSH$_{\alpha\Delta90-92}$, and FSH$_{\alpha\Delta89-92}$ were not significant (NS).

penultimate C-terminal Lys$^{91}$ of the α-subunit. This distinctive role of αLys$^{91}$ in hCG and FSH is in contrast to the apparent lack of an essential role of the α C-terminal Ser$^{92}$ in both hormones.

## Induction of cAMP and Progesterone Synthesis

Natural hCG, wild-type recombinant hCG, and hCG$_{\alpha\Delta92}$ were similar in their ability to induce cAMP synthesis (Fig. 18.5 and Table 18.2). In contrast, hCG$_{\alpha\Delta91-92}$ was less effective, while hCG$_{\alpha\Delta90-92}$, hCG$_{\alpha\Delta89-92}$,

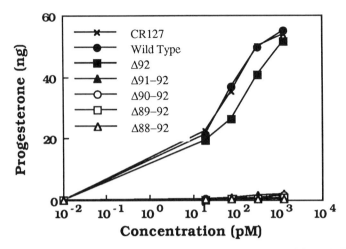

FIGURE 18.6. Induction of progesterone synthesis by mutant hCGs. MLTC/J cells were incubated with increasing concentrations of recombinant hCGs and washed. After cell lysis as described in Figure 18.5, the supernatants were collected and assayed for intracellular progesterone concentrations.

and $hCG_{\alpha\Delta88-92}$ were completely ineffective in inducing cAMP synthesis. The hormones had a similar effect on induction of progesterone synthesis in MLTC cells (Fig. 18.6). These results indicate that the α C-terminal $Ser^{92}$ and its carboxyl group are not necessary for hCG's induction of cAMP and progesterone synthesis, whereas $Lys^{91}$ and $His^{90}$ are important for this bioactivity. Truncation of FSHα had similar results on cAMP induction (Fig. 18.5 and Table 18.3).

## Activities of Peptide $hCG\alpha^{83-92}$

To define the role of the α C-terminal residues, peptide $\alpha^{83-92}$ was used for receptor binding. As shown in Figure 18.7, it binds to the LH/CG receptor, and the binding is specific since its binding can be blocked with natural hCG. The Ka was, however, $10^6$ times less than that of natural hCG. In addition, peptide $hCG\alpha^{83-92}$ induced cAMP production with an $EC_{50}$ value of 21 μM, compared to an $EC_{50}$ value of 32 pM for natural hCG. The maximal level of cAMP induction by peptide $\alpha^{83-92}$ was significantly lower than that by natural hCG. On the other hand, two control peptides, peptides $\alpha^{11-25}$ and $\alpha^{51-65}$, failed to bind to the LH/CG receptor and to induce cAMP synthesis. The sequences of peptides $\alpha^{11-25}$ and $\alpha^{51-65}$, respectively, correspond to the sequences of α amino acids 11–25, Thr-Leu-Gln-Glu-Asn-Pro-Phe-Phe-Ser-Gln-Pro-Gly-Ala-Pro-Ile, and 51–65, Lys-Asn-Ile-Thr-Ser-Glu-Ala-Thr-Cys-Cys-Val-

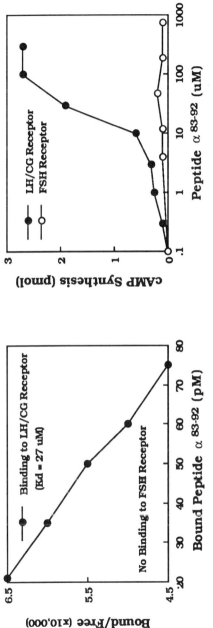

FIGURE 18.7. Activities of peptide hCGα^{83-92}. To determine receptor binding of peptide α^{83-92}, 293 cells expressing the LH/CG receptor or Y1 cells expressing the FSH receptor were incubated with increasing concentrations of ^{125}I-peptide α^{83-92}. Specific binding at a given concentration was calculated by subtraction of non-specific ^{125}I-peptide α^{83-92} binding from total ^{125}I-peptide α^{83-92} binding at that concentration. Nonspecific binding was determined by ^{125}I-hCGα^{83-92} binding in the presence of an excess amount of hCG or FSH. For cAMP induction assay, the cells were incubated with unlabeled hCGα^{83-92}

Ala-Lys-Ala-Phe. What is more important is the observation that peptide $\alpha^{83-92}$ did not bind to the FSH receptor, nor did it stimulate cAMP synthesis under the same experimental conditions (Fig. 18.7).

These data indicate the specificity of peptide $\alpha^{83-92}$ binding to the LH/CG receptor and support the significant difference in the interactions of hCG with the LH/CG receptor and FSH with the FSH receptor. In addition, our results demonstrate for the first time that peptide $\alpha^{83-92}$ is capable of not only specific binding to the LH/CG receptor but also cAMP induction, albeit with low affinity. This result is consistent with the data in Table 18.1, indicating that the $\alpha$ C-terminal residues bind the receptor with a low affinity. A similar observation was reported that a 17-mer peptide corresponding to the $\alpha$ C-terminal 76–92 inhibited hCG binding to the receptor (29).

# Substitution of $\alpha$His$^{90}$ or $\alpha$Lys$^{91}$

The results demonstrate that the $\alpha$ C-terminal region is directly involved in LH/CG receptor binding with low affinity, that this binding is sufficient for receptor activation and subsequent cAMP induction, and that $\alpha$His$^{90}$ and $\alpha$Lys$^{91}$ are involved in these processes. The next step is to determine the importance of these two amino acids. Although a study with sequentially deleted mutants implies potential roles of deleted amino acids, it is not able to define the role of each amino acid. This is because deletion of more than one amino acid could collectively, rather than individually, affect the function of the hormone. Therefore, it is necessary to examine the role of each amino acid. For this purpose, a series of substitution mutants were prepared. As seen in Figure 18.8 and Table 18.4, $\alpha$His$^{90}$ was substituted with Phe, Ser, Ala, Leu, Trp, Pro, Gln, or Arg. All of these mutant hCGs, hCG$_{\alpha Phe^{90}}$, hCG$_{\alpha Ser^{90}}$, hCG$_{\alpha Ala^{90}}$, hCG$_{\alpha Leu^{90}}$, hCG$_{\alpha Trp^{90}}$, hCG$_{\alpha Pro^{90}}$, hCG$_{\alpha Gln^{90}}$, and hCG$_{\alpha Arg^{90}}$ were capable of inhibiting $^{125}$I-hCG binding to the receptor. Their affinities, except those of hCG$_{\alpha Trp^{90}}$ and hCG$_{\alpha Pro^{90}}$, are similar to those of natural and wild-type recombinant hCGs. The receptor binding affinities of hCG$_{\alpha Trp^{90}}$ and hCG$_{\alpha Pro^{90}}$ are somewhat less. The data indicate that $\alpha$His$^{90}$ plays a minor role in high-affinity receptor binding.

Most of the mutant hCGs were able to induce cAMP synthesis with $EC_{50}$ values several times larger than those of natural and wild-type recombinant hCGs. However, hCG$_{\alpha Pro^{90}}$, hCG$_{\alpha Gln^{90}}$ and hCG$_{\alpha Arg^{90}}$ failed to stimulate cAMP synthesis at the tested concentrations up to 1 nM (Fig. 18.8 and Table 18.4). These results are in agreement with the results of the deletion mutation study and indicate the lack of a direct correlation between receptor binding and cAMP inducibility of hCG$_{\alpha His^{90}}$ substitution mutants. In addition, the data demonstrate the essential role of $\alpha$His$^{90}$ in the induction cAMP synthesis, but not in receptor binding.

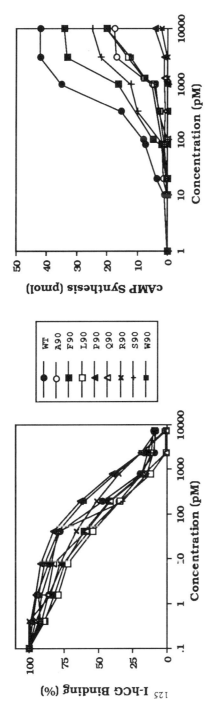

FIGURE 18.8. Inhibition of $^{125}$I-hCG binding and induction of cAMP synthesis by mutant hCGs with substitution at $\alpha$His$^{90}$. *Left panel*: Porcine granulosa cells were incubated with a constant amount of $^{125}$I-hCG in the presence of increasing concentrations of unlabeled hCG CR-127, recombinant wild-type hCG, or substitution mutant hCGs. *Right panel*: MLTC/J cells were incubated with increasing concentrations of various hCGs described above. The increase in intracellular cAMP content was determined.

TABLE 18.4. $\alpha His^{90}$ substitute mutant hCGs.

| Hormones | Inhibition of $^{125}I$-hCG binding (EC$_{50}$) pM | cAMP synthesis | |
| --- | --- | --- | --- |
| | | EC$_{50}$ pM | Max level pmole (%) |
| Natural hCG | 173 ± 19 | 302 ± 27 | 41 (100) |
| Wild-type hCG | 197 ± 26 | 319 ± 30 | 42 (102) |
| hCG$_{\alpha Phe^{90}}$ | 152 ± 17 | 1072 ± 16 | 35 (85) |
| hCG$_{\alpha Ser^{90}}$ | 150 ± 26 | 620 ± 53 | 25 (61) |
| hCG$_{\alpha Ala^{90}}$ | 67 ± 19 | 1427 ± 118 | 17 (41) |
| hCG$_{\alpha Leu^{90}}$ | 60 ± 11 | 2029 ± 227 | 21 (51) |
| hCG$_{\alpha Trp^{90}}$ | 548 ± 40 | 2182 ± 142 | 21 (51) |
| hCG$_{\alpha Pro^{90}}$ | 616 ± 39 | NS | NS |
| hCG$_{\alpha Gln^{90}}$ | 142 ± 68 | NS | NS |
| hCG$_{\alpha Arg^{90}}$ | 231 ± 29 | NS | NS |

*Note:* The data in Figure 18.8 were used to determine the EC$_{50}$ values for inhibition of $^{125}I$-hCG binding and for cAMP induction. cAMP induction of some of the mutant hCGs were not significant (NS).

Maximal levels of cAMP synthesis show an interesting pattern. hCG$_{\alpha Phe^{90}}$ produced 85% of the maximal cAMP level for natural hCG, in comparison with 61%, 51%, 51%, and 41% for hCG$_{\alpha Ser^{90}}$, hCG$_{\alpha Leu^{90}}$, hCG$_{\alpha Trp^{90}}$, and hCG$_{\alpha Ala^{90}}$, respectively. This result indicates that Phe$^{90}$ is capable of more effectively activating the receptor to induce cAMP synthesis than Ser$^{90}$, Leu$^{90}$, Trp$^{90}$, and Ala$^{90}$. A simple explanation of this observation is that Phe$^{90}$ fits better in the contact site in the receptor than do Ser$^{90}$, Leu$^{90}$, Trp$^{90}$, and Ala$^{90}$.

The effect of substitution for $\alpha Lys^{91}$ is similar to that of substitutions for $\alpha His^{90}$ (Fig. 18.9 and Table 18.5). The substitution of Arg, Ala, Met, Thr, Glu, Val, Ile, Asp, and His did not noticeably affect receptor binding. cAMP inducibility, however, was suppressed or abolished. The EC$_{50}$ value of hCG$_{\alpha Arg^{91}}$ for cAMP synthesis was similar to those of natural and wild-type recombinant hCGs, whereas the maximal level of cAMP synthesis was reduced to 83%. EC$_{50}$ values and maximal levels of cAMP induction for hCG$_{\alpha Ala^{91}}$, hCG$_{\alpha Met^{91}}$, and hCG$_{\alpha Thr^{91}}$ were not so good. On the other hand, hCG$_{\alpha Glu^{91}}$, hCG$_{\alpha Val^{91}}$, hCG$_{\alpha Ile^{91}}$, hCG$_{\alpha Asp^{91}}$, and hCG$_{\alpha His^{91}}$ were not able to induce cAMP synthesis at tested concentrations up to 10 nM.

# Inhibition by Mutant hCGs of cAMP Induction by Natural hCG

Some of the mutant hCGs were capable of high-affinity receptor binding, but incapable of cAMP induction. To validate this antagonistic nature, we examined whether the mutant hCGs were capable of suppressing natural

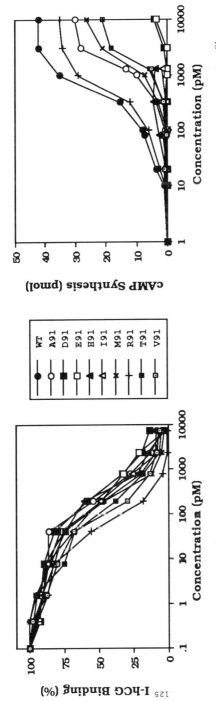

FIGURE 18.9. Inhibition of [125]I-I-CG binding and induction of cAMP synthesis by mutant hCGs with substitution at αLys[91]. *Left panel:* Porcine granulosa cells were incubated with a constant amount of [125]I-hCG in the presence of increasing concentrations of unlabeled hCG CR-127, recombinant wild-type hCG, or substitution mutant hCGs. *Right panel:* MLTC/J cells were incubated with increasing concentrations of various hCGs described above. The increase in intracellular cAMP content was determined.

TABLE 18.5. $\alpha Lys^{91}$ substitute mutant hCGs.

| Hormones | Inhibition of $^{125}$I-hCG binding ($EC_{50}$) pM | cAMP synthesis $EC_{50}$ pM | cAMP synthesis Max level pmole (%) |
|---|---|---|---|
| Natural hCG | $173 \pm 19$ | $302 \pm 27$ | 41 (100) |
| Wild-type hCG | $197 \pm 26$ | $319 \pm 30$ | 42 (102) |
| $hCG_{\alpha Arg^{91}}$ | $75 \pm 13$ | $346 \pm 47$ | 34 (83) |
| $hCG_{\alpha Ala^{91}}$ | $333 \pm 22$ | $1443 \pm 44$ | 30 (73) |
| $hCG_{\alpha Met^{91}}$ | $207 \pm 25$ | $1840 \pm 215$ | 26 (63) |
| $hCG_{\alpha Thr^{91}}$ | $159 \pm 108$ | $1954 \pm 131$ | 21 (51) |
| $hCG_{\alpha Glu^{91}}$ | $117 \pm 45$ | NS | NS |
| $hCG_{\alpha Val^{91}}$ | $139 \pm 16$ | NS | NS |
| $hCG_{\alpha Ile^{91}}$ | $232 \pm 12$ | NS | NS |
| $hCG_{\alpha Asp^{91}}$ | $426 \pm 32$ | NS | NS |
| $hCG_{\alpha His^{91}}$ | $521 \pm 68$ | NS | NS |

*Note:* The data in Figure 18.9 were used to determine the $EC_{50}$ values for inhibition of $^{125}$I-hCG binding and for cAMP induction. cAMP induction of some of the mutant hCGs were not significant (NS).

hCG's ability to induce cAMP synthesis. In addition, this test would further verify the results of the radioimmunoassay. If the concentration estimates of mutant hCGs by the radioimmunoassay are accurate, the less potent hCGs should be able to inhibit natural hCG-dependent cAMP synthesis due to the fact that they have receptor binding capabilities comparable to natural hCG and the more potent truncated hCGs. To optimize conditions, such inhibition studies are normally performed at concentrations of natural hCG that permit the maximal level of cAMP production. This, however, requires considerable quantities of the truncated hCGs, which are difficult to obtain at the moment. Therefore, we chose to perform the inhibition study at a hCG concentration below the effective concentration to induce the maximal level of cAMP synthesis.

cAMP synthesis decreased in the presence of the less potent or impotent hCGs, while it dramatically increased by 80-fold in the presence of hCG CR-127 (Fig. 18.10). In the case of $hCG_{\alpha \Delta 91-92}$, cAMP production was suppressed at concentrations below 748 pM and stimulated above this concentration, but never back to the level achieved by natural hCG. This dual effect of $hCG_{\alpha \Delta 91-92}$ was expected because while $hCG_{\alpha \Delta 91-92}$ is less potent than natural hCG, it is nonetheless capable of inducing cAMP production. At low concentrations, the primary effect of $hCG_{\alpha \Delta 91-92}$ is its competition with natural hCG for receptor binding, resulting in a decrease in cAMP production. At moderate concentrations, $hCG_{\alpha \Delta 91-92}$ should exhibit dual effects. While competing with natural hCG for receptor binding, it might also occupy receptors that could have been left unoccupied. At high concentrations, $hCG_{\alpha \Delta 91-92}$ is expected to occupy

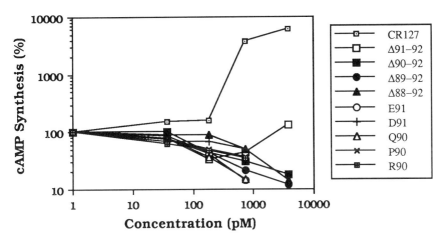

FIGURE 18.10. Suppression of cAMP synthesis by mutant hCGs. Mutant hCGs were examined to test their antagonistic activity. MLTC/J cells were incubated with 187 pM of hCG CR-127 in the presence of increasing concentrations of truncated hCGs, and intracellular cAMP concentrations were determined.

most of the receptors and stimulate cAMP production, albeit at a low rate, which will result in a net increase in cAMP production, as seen in Figure 18.10. These results indicate that $hCG_{\alpha\Delta90-92}$, $hCG_{\alpha\Delta89-92}$, $hCG_{\alpha\Delta88-92}$, $hCG_{\alpha Glu^{91}}$, $hCG_{\alpha Asp^{91}}$, $hCG_{\alpha Gln^{90}}$, $hCG_{\alpha Pro^{90}}$, and $hCG_{\alpha Arg^{90}}$ act as antagonists in cAMP induction. Furthermore, they verify the conclusion that the α C-terminal residues, particularly His$^{90}$ and Lys$^{91}$, play important roles in cAMP and probably progesterone induction.

## Conclusions

Our results demonstrate for the first time that the hCG α C-terminal region makes direct contact with the LH/CG receptor. This low-affinity contact is necessary and sufficient for hCG's ability to activate the receptor and involves αLys$^{91}$ and His$^{90}$. However, this contact is not necessary for the overall interaction of hCG and the receptor (Table 18.2 and Figs. 18.4 and 18.5). This conclusion is consistent with our observation that the LH/CG receptor can be activated for the induction of cAMP synthesis by hormone binding to a low-affinity hormone contact site(s) in the membrane-associated C-terminal half of the receptor (23, 24), while a high-affinity hormone contact site(s) of the receptor is present in the extracellular N-terminal half (21–23). We cannot, however, exclude the possibility that Lys$^{91}$ and His$^{90}$ are also involved in the formation of the necessary conformation of the hormone for receptor activation.

At the same time, it is clear that such a hypothetical structure is not required for high-affinity receptor binding.

This study of substitutions at the position of $Lys^{91}$ indicates that among the substitute amino acids, Arg is the most favorable (Fig. 18.9 and Table 18.5). The affinities of receptor binding and cAMP induction of $hCG_{\alpha Arg^{91}}$ are similar to those of wild-type recombinant hCG, but the efficacy to induce cAMP synthesis is 83% of that of wild-type hCG. Obviously, $hCG_{\alpha Arg^{91}}$ is capable of high-affinity receptor binding and receptor activation, but the activated receptor is not as efficient in stimulating G protein and adenylyl cyclase as wild-type hCG. The simplest explanation of this intriguing observation is that $Arg^{91}$ of $hCG_{\alpha Arg^{91}}$ does not perfectly interact with the receptor. The contact site in the receptor can also accept Ala, Met, and Thr, but with lower affinities. The interaction appears to favor a positive charge and a straight side chain of amino acids. Probably the contact site has a limited space that can specifically accommodate the straight chain of Lys and, as a result, may not effectively accommodate the bulkier side chain of an Arg. Consequently, Glu, Val Ile, Asp, and His substitutions failed to activate the receptor.

The result of substitutions at $\alpha^{90}$ shows the essential role of $\alpha His^{90}$ in cAMP induction. Substitutions of Phe, Ser, Ala, Leu, and Trp resulted in 2–7-fold losses of the affinity for cAMP induction, whereas the maximal level of cAMP synthesis decreased to 85%–41%. Of particular interest is Phe substitution, which showed an 85% activity, as opposed to a 51% activity for Trp substitution. The size of side chains, therefore, appears to be an important factor, while a ring structure is favored. These results, taken together with the data on substitutions at $Lys^{91}$, suggest that there is a tight contact area in the LH/CG receptor for $\alpha His^{90}$-$Lys^{91}$. In this regard, it is of interest to note that the $\alpha$ C-terminal sequence including $Tyr^{88}$-$Tyr^{89}$-$His^{90}$-$Lys^{91}$ is similar to consensus sequences for protein-protein recognition sites that consist of aromatic amino acids flanked by ionic amino acids (15). The specificity of such an interaction is supported by the fact that the $\alpha$ C-terminal $Ser^{92}$ and its carboxyl group are not necessary for receptor activation nor for receptor binding. Apparently, the exposed carboxylic group of $Lys^{91}$ in $hCG_{\alpha\Delta92}$ does not interfere with receptor activation and receptor binding, reinforcing the notion of a close-fit area for $His^{90}$-$Lys^{91}$.

Considering the common nature of the $\alpha$-subunits, it is surprising to see that sequential truncation of the $\alpha$ C-terminal 4 amino acids has a strikingly similar effect on FSH and hCG, as well as dramatically different impacts. On the one hand, both FSH and hCG are not affected by deletion of $\alpha Ser^{92}$, while on the other hand, they lose the cAMP inducibility by additional deletion of $Lys^{91}$. Therefore, the $\alpha$ penultimate C-terminal $Lys^{91}$ is equally important to FSH and hCG for cAMP induction. Despite this similarity, the degree of the essential role does not appear to be identical. The $\alpha$ penultimate C-terminal Lys[91] is necessary for high-affinity receptor binding for FSH, but not for hCG. The most conspicuous difference in the two hormones is found when Lys[91]-Ser[92] are truncated. It has no effect on receptor binding of hCG, while it results in the dramatic loss of receptor binding of FSH. In addition, deletion of Lys[91]-Ser[92] significantly, but not completely, reduces cAMP induction by hCG, while it completely abolishes cAMP induction by FSH. Another difference is the trend of changes in the receptor binding activity. On sequential deletion of the $\alpha$ C-terminal residues, FSH abruptly lost the receptor binding activity, whereas the receptor binding activity of hCG changed slightly and these changes fluctuated.

These differences between FSH and hCG are manifested by the fact that peptide $\alpha^{83-92}$ failed to bind to the FSH receptor and to induce cAMP synthesis at concentrations of up to nearly 1 mM, in contrast to its successful binding to the LH receptor and subsequent cAMP induction. It is possible that the $\alpha$ C-terminal region does not contact the FSH receptor, but circumstantial evidence suggests otherwise. For example, $\alpha$Lys[91] is essential for high-affinity receptor binding. Peptide $\alpha^{83-92}$ is capable of not only receptor binding, but also cAMP induction in the case of the LH/CG receptor. Alternatively, the $\alpha$ C-terminal region has to assume a specific structure to be recognized by the receptor, and peptide $\alpha^{83-92}$ does not have this structure. There are three half-cystines in the $\alpha$ C-terminal region (1) and two of them, Cys[82] and Cys[84], are thought to form an unusual disulfide link in LH and CG (34). If this is true for FSH, the $\alpha$ C-terminal region is highly structured upstream, while the downstream 5 residues, Tyr[88]-Tyr[89]-His[90]-Lys[91]-Ser[92], are likely to be exposed at the surface of the molecule (31) and to be somewhat flexible.

It is doubtful that the synthetic peptide $\alpha^{83-92}$ can assume a similar structure. Due to the expected steric hindrance, such a disulfide link between adjacent Cys residues would be difficult unless assisted by enzymes during posttranslational processing. The failure of the FSH receptor to interact with peptide $\alpha^{83-92}$ also indicates a specific requirement not only of the C-terminal five residues, but also of a surrounding contiguous structure. In light of the successful receptor binding and cAMP induction of the same peptide in the case of the LH/CG receptor (17), the specificity of the interaction of peptide $\alpha^{83-92}$ and the LH/CG receptor appears to be less rigorous, and the affinity is expected to be low. This is in complete agreement with the observed low binding affinity of the peptide (17). The differences in receptor binding of FSH and hCG may be attributed to distinct conformations of the $\alpha$ C-terminal region in FSH and hCG. Alternatively, it may be the result of distinct structures and mechanisms of the two receptors to recognize the $\alpha$ C-terminal region of the respective hormones.

The importance of the $\alpha$ C-terminal region is contrasted to the hCG$\beta$ C-terminal region, which could be deleted without losing hormonal activity

(35). Another interesting observation in this study is that deletion of the $\alpha$ C-terminal $Ser^{92}$ does not have a significant effect on receptor binding and induction of cAMP and progesterone. As $Ser^{92}$ is conserved among various animal species, it is expected to play some role including hormonal actions other than receptor binding and induction of cAMP and progesterone. It may be possible that $Ser^{92}$ is necessary for the hormonal action of thyrotropin.

Our results are not entirely the same as the results of earlier reports (31, 36). In those studies, progressive enzymatic cleavage of hCG$\alpha$ C-terminal residues resulted in the gradual and more extensive loss of receptor binding activity (31). The discrepancies may be attributed to the different methods used to prepare the modified hormones. In the former studies (31, 36), the hCG$\alpha$-subunit was purified and treated with commercial carboxypeptidase A. The ratio of enzyme to hCG$\alpha$ was varied to control the extent of truncation. Another study, however, showed that the extent of truncation by carboxypeptidases could not be precisely regulated, and the resulting proteolytic products were not homogeneous (36). The treated hCG$\alpha$ preparations containing the proteolytic enzyme were used for reconstitution with intact $\beta$-subunit to produce $\alpha$-$\beta$ dimers. The effect of residual carboxypeptidase A on reconstituted hCGs and receptor binding is not well understood. Reconstituted hCGs were tested for receptor binding and their ability to induce weight gains in rat male accessory glands (31). Other differences in methodology involve the precision and yield of truncation and formation of $\alpha$-$\beta$ dimers. Each recombinant truncated hCG in our study consists of a homogeneous population in which the truncation is complete. Furthermore, there are no contaminating proteolytic enzymes in our preparations. Another potentially important difference is the manner in which $\alpha$-$\beta$ dimer is formed and processed. The recombinant hCGs in our study were formed and processed in CHO cells. While in vitro reconstitution and biological activity of reconstituted hormones are well documented, the efficiency of in vitro reconstitution is 90% or less (37).

Our results indicated that hCG$_{\alpha\Delta89-92}$ and hCG$_{\alpha\Delta88-92}$ were capable of receptor binding, in contrast to the report indicating that hCG$_{\alpha\Delta90-92}$, but not hCG$_{\alpha\Delta88-92}$, were capable of receptor binding (38). It was surprising to see that the deletion of one more amino acid, $\alpha$Tyr88, resulted in the complete loss of receptor binding. Therefore, we repeated the experiment five times, and hCG$_{\alpha\Delta88-92}$ repeatedly inhibited $^{125}$I-hCG binding to the receptor. However, there is a difference in the experimental conditions of the two studies. The maximum hormone concentration in the study was 250 pM (38), in contrast to 10,000 pM in our study. This 40-fold higher concentration might have made the difference in receptor binding, particularly when the inhibition of $^{125}$I-hCG binding by hCG$_{\alpha\Delta88-92}$ was marginal (Fig. 18.5).

*Acknowledgment.* This work was supported by Grant HD-18702 from the National Institutes of Health.

# References

1. Pierce JG, Parson TF. Glycoprotein hormones: structure and function. Annu Rev Biochem 1981;50:465–95.
2. Adashi EA, Hsueh AJW. Hormonal induction of receptors during ovarian granulosa cell differentiation. Receptors 1984;1:587–626.
3. Richards JS. Maturation of ovarian follicles: actions and interactions of pituitary and ovarian hormones on follicular development. Physiol Rev 1980;60:51–89.
4. McFarland KC, Sprengel R, Phillips HS, et al. Lutropin-choriogonadotropin receptor: an unusual member of the G protein-coupled receptor family. Science 1989;245:525–8.
5. Loosfelt H, Misrahi M, Atger M, et al. Cloning and sequencing of porcine LH-hCG receptor cDNA: variant lacking transmembrane domain. Science 1989;245:525–8.
6. Minegishi T, Nakamura K, Takakura Y, et al. Cloning and sequencing of human LH/CG receptor cDNA. Biochem Biophys Res Commun 1990; 172:1049–54.
7. Jia X-C, Oikawa M, Bo M, et al. Expression of human luteinizing hormone (LH) receptor: interaction with LH and chorionic gonadotropin from human but not equine, rat, and ovine species. Mol Endocrinol 1991;5:759–68.
8. Sprengel R, Braun T, Nikolics K, Segaloff DL, Seeburg PH. The testicular receptor for follicle stimulating hormone: structure and functional expression of cloned cDNA. Mol Endocrinol 1990;4:525–30.
9. Minegishi T, Nakamura K, Takakura Y, Ibuki Y, Igarashi M. Cloning and sequencing of human FSH receptor cDNA. Biochem Biophys Res Commun 1991;175:1125–30.
10. Tilly JL, Aihara T, Nishimori K, et al. Expression of recombinant human follicle-stimulating hormone receptor: species-specific ligand binding, signal transduction, and identification of multiple ovarian messenger ribonucleic acid transcripts. Endocrinology 1992;131:799–806.
11. Kusuda S, Dufau ML. Characterization of ovarian gonadotropin receptor: monomer and associated form of the receptor. Biol Chem 1988;263:3046–9.
12. Ji I, Okada Y, Nishimura R, Ji TH. The domain structure of the choriogonadotropin receptor. In: Chin WW, Boime I, eds. Glycoprotein hormones: structure, synthesis, and biological functions. Norwell, MA: Serono Symposia, USA, 1990:355–65.
13. Kim I-C, Ascoli M, Segaloff DL. Immunoprecipitation of the lutropin/choriogonadotropin receptor from biosynthetically labeled Leydig tumor cells: a 92-kDa glycoprotein. J Biol Chem 1987;262:470–7.
14. Sprengel R, Braun T, Nikolics K, Seagaloff DL, Seeburg PH. The testicular receptor for follicle stimulating hormone: structure and functional expression of cloned cDNA. Mol Endocrinol 1990;4:525–30.

15. Koo YB, Ji I, Slaughter RG, Ji TH. Structure of the luteinizing hormone receptor gene and mutiple exons of the coding sequence. Endocrinology 1991;128:2297–308.
16. Tsai-Morris CH, Buczko E, Wang W, Xie X-Z, Dufau ML. Structural organization of the rat luteinizing hormone (LH) receptor gene. J Biol Chem 1991;266:11355–9.
17. Ji I, Ji TH. Both α and β subunit of human choriogonadotropin photoaffinity label the hormone receptor. Proc Natl Acad Sci USA 1981;78:5465–9.
18. Okada Y, Ji I, Nishimura R, Ji TH. Carbohydrates of the lutropin receptor. In: Mochizuki M, Hussa R, eds. Placental protein hormones. Amsterdam: Excerpta Medica, 1988:269–78.
19. Moyle WR, Ehrlich PH, Canfield RE. Use of monoclonal antibodies to subunits of human chorionic gonadotropin to examine the orientation of the hormone in its complex with receptor. Proc Natl Acad Sci USA 1982; 79:2245–9.
20. Koo YB, Ji I, Ji TH. Intronic polyadenylation and characterization of rat LH receptor mRNAs. J Biol Chem (in press).
21. Tsai-Morris CH, Buczko E, Wang W, Dufau ML. Intronic nature of the rat luteinizing hormone receptor gene defines a soluble receptor subspecies with hormone binding activity. J Biol Chem 1990;265:19385–8.
22. Xie Y-B, Wang H, Segaloff DL. Extracelluar domain of lutropin/chori-ogonadotropin receptor expressed in transfected cells binds choriogonado-tropin with high affinity. J Biol Chem 1990;265:21411–4.
23. Ji I, Ji TH. Exons 1–10 of the rat LH receptor encode a high affinity hormone binding site and exon 11 encodes G-protein modulation and a potential second hormone binding site. Endocrinology 1991;128:2648–50.
24. Ji I, Ji TH. Human choriogonadotropin binds to a lutropin receptor with essentially no N-terminal extension and stimulates cAMP synthesis. J Biol Chem 1991;266:13076–9.
25. Roche PC, Ryan RJ, McCormick DJ. Identification of hormone-binding regions of the luteinizing hormone/human chorionic gonadotropin receptor using synthetic peptides. Endocrinology 1992;131:268–74.
26. Moyle WR. Biochemistry of gonadotropin receptors. In: Finn CA, ed. Oxford review of reproductive biology; vol 2. Oxford, 1980:128–204.
27. Roche PC, Ryan RJ. The LH/hCG receptor. In: Ascoli M, ed. Luteinizing hormone action and receptors. New York: CRC Press, 1985:17–56.
28. Ji I, Ji TH. Asp383 in the second transmembrane domain of the lutropin receptor is important for high affinity hormone binding and cAMP production. J Biol Chem 1991;266:14953–7.
29. Charlesworth MC, McCormick DJ, Madden B, Ryan RJ. Inhibition of human choriogonadotropin binding to receptor by human choriogonadotropin α peptides. J Biol Chem 1987;262:13409–16.
30. Keutman HT, Charlesworth MC, Mason KA, Ostrea T, Johnson L, Ryan RJ. A receptor-binding region in human choringonadotropin/lutropin β subunit. Proc Natl Acad Sci USA 1987;84:2038–942.
31. Cheng K-W, Glazer AN, Pierce JG. The effects of modification of the COOH-terminal regions of bovine thyrotropin and its subunits. J Biol Chem 1973;248:7930–7.

32. Gordon WL, Ward DN. Structural aspects of luteinizing hormone actions. In: Ascoli M, ed. Luteinizing hormone action and receptors. Boca Raton, FL: CRC Press, 1985:173–98.
33. Yoo J, Zeng H, Ji I, Murdoch WJ, Ji TH. C-terminal amino acids of the α subunit play common and different roles in hCG and FSH. J Biol Chem 1993:268.
34. Mise T, Bahl OP. Assignment of disulfide bonds in the α subunit of human chorionic gonadotropin. J Biol Chem 1980;255:8516–22.
35. Chen F, Puett D. Delineation via site-directed mutagenesis of the carboxyl-terminal region of human choriogonadotropin β required for subunit assembly and biological activity. J Biol Chem 1991;266:6904–8.
36. Merz, WE. Studies of the specific role of the subunits of choriogonadotropin for biological, immunological and physical properties of the hormone: digestion of the α subunit with carboxypeptidase A. Eur J Biochem 1979; 101:541–53.
37. Parson, TF, Strckland TW, Pierce JG. Disassembly and assembly of glycoprotein hormones. Methods Enzymol 1985;109:736–49.
38. Chen F, Wang Y, Puett D. The carboxyl-terminal region of the glycoprotein hormone α-subunit: contributions to receptor binding and signaling in human chorionic gonadotropin. Mol Endocrinol 1992;6:914–9.

# Part VI

## Clinical Implications

# 19

# Use of the Free Alpha Subunit (FAS) of Glycoprotein Secreting Hormones as a Surrogate Marker of GnRH Secretion in the Human

William F. Crowley, Jr., Ann E. Taylor, Kathryn A. Martin, Randall C. Whitcomb, Joel S. Finkelstein, and Janet E. Hall

Since GnRH secretion in the human cannot be monitored directly, the pulsatile secretion of LH from the anterior pituitary has been chosen as a surrogate marker of release of endogenous GnRH into the hypophyseal-portal blood supply (1). Since GnRH is the only known stimulator of LH release, almost all studies of the neuroendocrine control of reproduction in the human has employed frequent sampling of LH to make inferences regarding the frequency and amplitude of antecedent endogenous GnRH secretion. LH has proven to be a reasonable marker of endogenous GnRH secretion because of: (i) its known one-to-one relationship with endogenous GnRH secretion as validated in several animal species in which both hypophyseal portal secretion of GnRH and pituitary output of LH can be monitored (2, 3); (ii) its rapid circulating half-life of approximately 20 min (4); (iii) the widespread availability of relatively specific radioimmunoassays for its determination; and (iv) the previously demonstrated immediate response of LH to exogenous GnRH stimulation (5).

However, an alternative marker of endogenous GnRH secretion would be desirable for two reasons. First, any such alternative marker could serve to corroborate estimates of endogenous GnRH secretion by LH monitoring since there are several circumstances in which the monitoring of pulsatile LH release yields ambiguous information. Second, certain circumstances involving rapid frequencies of GnRH secretion (such as are seen at the midcycle gonadotropin surge [6] and in women with polycystic ovarian disease [7]), low amplitude of GnRH secretion (as seen in the prepubertal state [8]), and/or high mean levels of LH (as seen in castrate or menopausal patients [9]) still represent problematic circumstances where monitoring of pulsatile pattern of LH secretion yields information that can prove problematic in providing accurate estimates of endogenous GnRH secretion.

The secreted or FAS of glycoprotein hormone fulfills many of these requirements, with the recent availability of a monoclonal antibody of relatively absolute specificity and excellent sensitivity for the uncombined form of alpha subunit (10). Its half-life of secretion is even more rapid than that of endogenous LH, providing the opportunity to increase the resolving power of its pulsatile component in several of the above-mentioned circumstances.

However, the dual control of FAS secretion by both GnRH (11) and TRF (12) has made its use in these circumstances somewhat problematic. This chapter explores the hypothesis that GnRH is responsible for the pulsatile component of FAS secretion using several disease models, as well as normals and GnRH antagonist administration, to validate this concept.

## Materials and Methods

### Patient Population

Normal men and women of reproductive age (18–40) with a normal prior medical history and no evidence of intercurrent disease were selected. Each had evidence of normal gonadal physiology, normal baseline levels of gonadal steroids and gonadotropins, and were free from intercurrent illness. Patients with polycystic ovarian disease were all documented to have ovarian enlargement by transvaginal ultrasound, hyperandrog-enemia, high LH:FSH ratios when compared to the early follicular phase of the normal menstrual cycle, and irregular menses as previously described (7). Patients with GnRH deficiency each had documented absence of secondary sexual characteristics by age 18, gonadotropin levels in the low–normal range in the face of prepubertal levels of gonadal steroids, an apulsatile pattern of LH and FAS release prior to GnRH ex-posure, normal radiographic imaging of the hypothalamic-pituitary axis, and no other evidence by baseline or stimulatory biochemical testing of other pituitary dysfunction. Moreover, each demonstrated a responsive-ness to pulsatile exogenous GnRH administration in a physiologic pattern by restoration of normal pituitary-gonadal axis activity.

### GnRH Antagonist Administration

The Nal-Glu GnRH antagonist was used as previously described (13). It was administered at 150 µg/kg sc, a dose previously determined to ablate endogenous pulsatile LH activity in normal men and women via its receptor blockade that competes with the GnRH receptor.

## Radioimmunoassays

The assay used for LH is absolutely specific for its beta subunit and cross-reacts completely with hCG as previously described (14). The FSH assay is specific for the FSH dimer as previously described (15). The monoclonal antibody assay used for determination of FAS has been previously reported (10) and was purchased from the BioAmerica Corporation. This antibody had been previously raised by Professor Om Bahl to the dissociated alpha subunit of hCG and was quite specific for the uncombined form of the alpha subunit.

## Blood Sampling Protocol

All subjects were sampled at least at 10 min intervals. In some circumstances 5 min intervals were employed where it was anticipated that rapid frequencies of endogenous GnRH secretion would be manifest.

# Results

Figure 19.1 demonstrates the excellent correlation seen with the pulsatile behavior of the FAS in a normal male. It is noteworthy that whereas LH secretion continues to fall following each pulse, FAS secretion appears to terminate more quickly than LH secretion and plateau during long interpulse intervals. As can be seen from Figure 19.2, which shows a histogram of the frequency of varying interpulse intervals (i.e., frequencies) and amplitudes of LH and FAS pulsations in normal men, there is excellent correlation ($r = 0.9$, $P < 0.001$) between the frequency of LH and FAS pulses. However, whereas the amplitudes of LH pulses span a large spectrum varying from small to large amplitude pulses, FAS pulses are considerably more homogeneous, being generally of smaller amplitude than those observed for LH. This may well be due to the fact that the nonpulsatile component of the FAS secretion—that is, the component presumably driven by TRH secretion—is a tonic secretion that elevates the mean level of FAS. This elevation of the baseline, in turn, reduces the amplitude of each subsequent GnRH-induced pulse.

Figure 19.3 demonstrates the typical findings of men and women with complete GnRH deficiency—that is, a complete absence of pulsatile LH and FAS secretion. The FAS levels in patients with complete GnRH deficiency, however, are not undetectable, but rather, approximately 50% of those observed in normal men and women. Since all of these patients are euthyroid with normal TRH stimulation testing, we concluded that this tonic component of FAS secretion is driven by endogenous TRF secretion and represents approximately 50% of the normal mean level of FAS secretion and its apulsatile component.

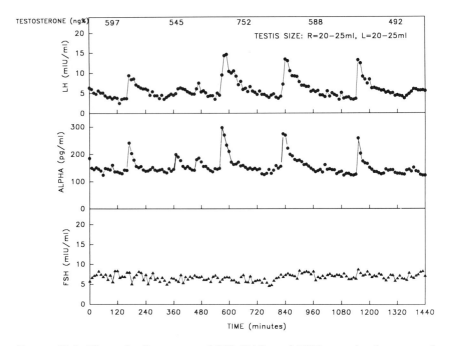

FIGURE 19.1. The pulsatile pattern of LH, FAS, and FSH secretion in a normal 30-year-old male sampled at 10 min intervals for 24 h. LH is shown in the upper panel, FAS in the middle panel, and FSH in the lower panel. Serum testosterone levels (ng/dl) and testicular size (ml) are shown. Note the excellent correlation between the pulsatile release of LH and FAS.

The fact that the endogenous pulses of FAS secretion are attributable to the pulsatile release of GnRH is shown in Figure 19.4, in which a normal female is administered a GnRH antagonist followed by a subsequent exogenous GnRH stimulation test. Prior to GnRH antagonist administration, clear-cut pulsatile LH and FAS secretion are seen with complete agreement between the two. Subsequent to GnRH antagonist administration, there is a total ablation of the pulsatile component of both LH and FAS secretion. Following readministration of a pharmacologic dose (100 μg) of exogenous GnRH, there is a prompt restoration of a single pulse of LH and FAS secretion. Consequently, the pulsatile component in normal men and women of FAS secretion is attributable to endogenous GnRH, at least as determined by its ability to be selectively and completely ablated by blockade of the GnRH receptor.

The use of the pulsatile component of FAS secretion in circumstances of particularly rapid endogenous GnRH secretion is demonstrated in Figures 19.5 and 19.6. Figure 19.5 shows a frequent sampling study in a woman with polycystic ovarian disease, demonstrating a clear cut resolu-

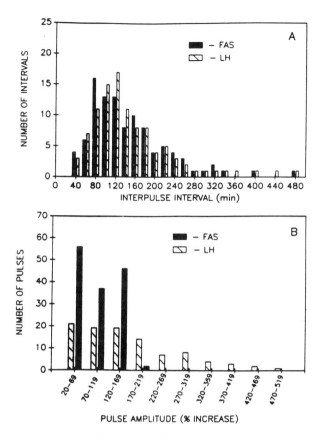

FIGURE 19.2. Frequency histogram of LH interpulse interval (*A*) and amplitudes (*B*) comparing LH and FAS pulses in 10 normal men. Note that while the spectrum of interpulse intervals of LH and FAS pulses are nearly identical, the spectrums of their amplitudes vary considerably, with LH pulses being larger and more heterogeneous in amplitude than the FAS pulses, which are smaller and more homogeneous in magnitude. Reprinted with permission from Whitcomb, O'Dea, Finkelstein, Heavern, and Crowley (11), © The Endocrine Society, 1990.

tion of FAS pulses to an equivalent or greater degree than those of LH. Similarly, Figure 19.6, representing a study taken from a normal woman at the midcycle gonadotropin surge, again shows both the excellent correlation and enhanced resolution of FAS secretion. Finally, Figure 19.7 demonstrates a male with GnRH deficiency undergoing progressive increase in the frequency of stimulation with exogenous GnRH from once to twice hourly. As can be seen at the more rapid frequency of GnRH stimulation in this GnRH-deficient patient, shown previously in Figure 19.2 to be entirely without exogenous GnRH pulses, the hourly frequency of stimulation causes a lack of complete resolution in the LH pulses,

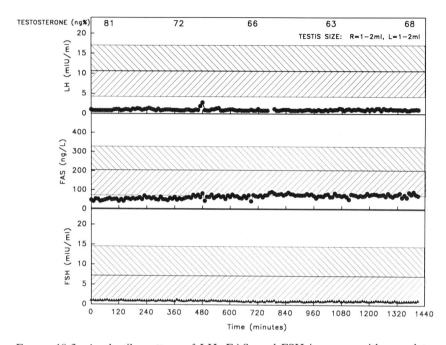

FIGURE 19.3. Apulsatile pattern of LH, FAS, and FSH in a man with complete GnRH deficiency. Exogenous GnRH administration restored gonadotropin and FAS secretion to normal in terms of both mean level and pattern of release. Note that the FAS level is not undetectable in this man, but merely half that of normal men. Since this man was euthyroid with a normal TRH stimulation test, the residual, tonic FAS secretion is attributable to the thyrotroph.

where the FAS pulses continue to provide considerably greater resolution at these higher GnRH pulse frequencies. Thus, it appears that FAS is particularly useful in those circumstances of rapid endogenous frequency of exogenous GnRH.

## Discussion

In the past, it has been demonstrated that secretion of the FAS of glycoprotein hormone has been under the dual control of TRH and GnRH (11, 12). The TRH component of FAS secretion has been demonstrated by elevated levels of FAS in hypothyroidism, with their subsequent return to normal with exogenous thyroid replacement (11). Similarly, exogenous TRF stimulation results in release of FAS in normal individuals (16). Similarly, exogenous GnRH causes release of FAS, which is elevated in castrates and returned to normal with replacement of exogenous sex steroids (12).

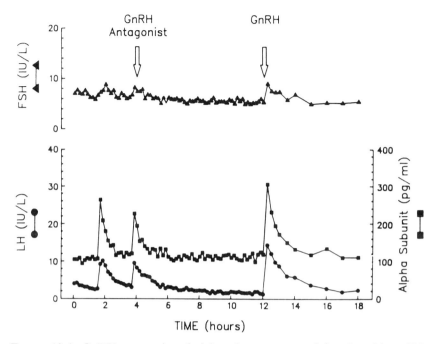

FIGURE 19.4. GnRH antagonist administration to a normal female subject. Following a 4 h baseline period of every 10 min sampling in which spontaneous LH and FAS pulses coincide, 150 µg/kg of the Nal-Glu antagonist is administered subcutaneously to a normal female volunteer in the early follicular phase of the cycle, resulting in a complete eradication of pulsatile LH and FAS secretion that is not restored until a pharmacologic dose of exogenous GnRH is administered to override this competitive blockade. Reprinted with permission from Hall, Whitcomb, Rivier, Vale, and Crowley (13), © The Endocrine Society, 1990.

However, these previous studies have not examined the relationship of the pattern of FAS secretion to its dual control. The present study has demonstrated that: (i) GnRH-deficient subjects are completely devoid of any pulses of LH and FAS, whereas their mean levels of FAS are approximately half those of normal; (ii) exogenous replacement in GnRH-deficient subjects restores both pulsatile LH and FAS secretion; (iii) there is an excellent concordance of LH and FAS in normal subjects with intact gonads (17); and (iv) administration of a GnRH antagonist results in complete ablation of LH and FAS pulses that can only be restored by exogenous GnRH override of this blockade. Thus, it appears that the GnRH-driven component of FAS secretion is pulsatile in its behavior, whereas the TRF-driven component is tonic in nature. Moreover, these two components would appear to contribute roughly equal amounts to the mean circulating levels of this hormone in normal individuals.

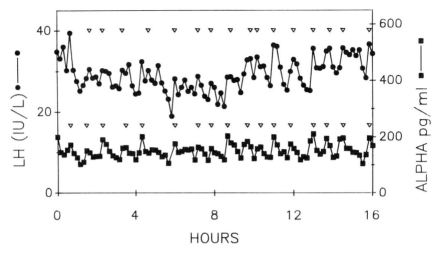

FIGURE 19.5. The monitoring of pulsatile FAS secretion is particularly useful in circumstances in which LH and FAS pulses are occurring at a rapid frequency, such as in this woman with polycystic ovarian disease. Note the excellent agreement of pulsatile LH and FAS in this circumstance. Reprinted with permission from Hall JE, Taylor AE, Martin KA, and Crowley WF. Neuroendocrine investigation of polycystic ovary syndrome: New approaches. In: *Polycystic Ovary Syndrome*. A Dunaif, JR Givens, FP Haseltine, and GR Merriam (eds). Blackwell Scientific Publications: Cambridge, MA, 1992.

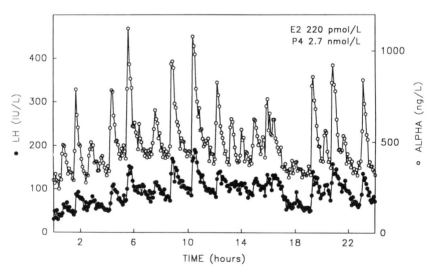

FIGURE 19.6. LH and FSH are sampled at 5 min intervals in this normal female, who was studied at the midcycle surge as determined prospectively by ultrasonic evaluation of her ovary and retrospectively by daily blood samples for gonadotropins and sex steroids throughout the menstrual cycle. It is apparent that excellent correlation of LH and FAS secretion is manifest and that the pattern of FAS secretion is clearer than that of LH during this physiologic period.

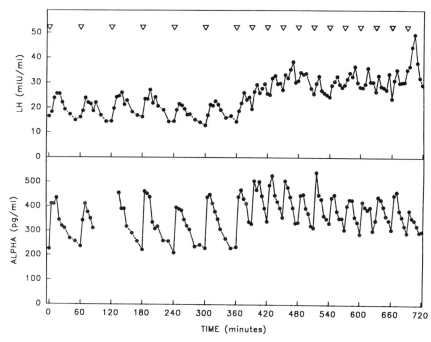

FIGURE 19.7. As the frequency of exogenous GnRH administration is increased from every 60 min to every 30 min in this GnRH-deficient male, pulsatile LH secretion is lost, whereas pulsatile FAS remains apparent. Open triangles indicate the timing of administration of a physiologic dose of GnRH.

Having validated the use of pulsatile FAS secretion as a surrogate marker for exogenous GnRH, it then becomes very desirable to take advantage of its more rapid half-life than LH. In circumstances where rapid GnRH frequency is anticipated, such as the midcycle gonadotropin surge in women with polycystic ovarian disease, monitoring of pulsatile component of FAS enhances resolution of estimates of GnRH pulse frequency. Moreover, in patients who are castrate or in menopausal women where the mean levels of LH level are high and there may well be alterations of the half-life of circulating hormones resulting in longer half-lives, the use of FAS secretion represents a most attractive alternative marker for monitoring endogenous GnRH secretion.

There are a few drawbacks to using FAS as a surrogate monitor of GnRH secretion that are worthy of mention. First, because of its more rapid half-life, it demands a sampling interval of at least 10 min, and enhanced resolution can be obtained with 5 min sampling in circumstances of very rapid pulse frequency. Clearly, these types of studies are thereby restricted to an investigational setting. Second, the lack of widespread availability of assays specific for the FAS are a major rate-limiting feature of its measurement. The antibody employed for these studies, although available commercially, is expensive and requires large numbers of assays. On the other hand, the assay is relatively robust, can be easily run, once established, by a knowledgeable technician, and provides the reasonable intra-assay precisions required of these studies. Third, the patients under study must clearly be euthyroid, since any elevation of the TRH-driven component of FAS secretion will interfere with the monitoring of its pulsatile behavior, as mentioned above. However, given these limitations, the attractive availability of a surrogate marker of endogenous GnRH secretion in the form of FAS monitoring is noteworthy.

*Acknowledgment.* Supported in part by NIH Grants HD-15080, HD-15788, HD-29164, and RR-M0101066.

## *References*

1. Crowley WF, Filicori M, Spratt DI, Santoro NF. The physiology of gonadotropin-releasing hormone (GnRH) secretion in men and women. Recent Prog Horm Res 1985;41:473–531.
2. Clarke IJ, Cummins JT. The temporal relationship between gonadotropin releasing hormone (GnRH) and luteinizing horomone (LH) secretion in ovariectomized ewes. Endocrinology 1983;111:1737–9.
3. Levine JE, Ramirez VD. Luteinizing hormone-releasing hormone release during the rat estrous cycle and after ovariectomy, as estimated with push-pull cannulae. Endocrinology 1982;111:1439–48.
4. Yen SSC, Llerena O, Little B, Pearson OH. Disappearance rates of endogenous luteinizing hormone and chorionic gonadotropin in man. J Clin Endocrinol Metab 1968;28:1763–7.
5. Conn PM, Crowley WF. Gonadotropin releasing hormone and its analogues. N Engl J Med 1991;324:93–103.
6. Filicori M, Santoro N, Merriam GR, Crowley WF. Characterization of the physiological pattern of episodic gonadotropin secretion throughout the human menstrual cycle. J Clin Endocrinol Metab 1986;62:1136–44.
7. Waldstreicher J, Santoro NF, Hall JE, Filicori M, Crowley WF. Hyperfunction of the hypothalamic-pituitary axis in women with polycystic ovarian disease: indirect evidence for partial gonadotroph desensitization. J Clin Endocrinol Metab 1988;66:165–72.
8. Styne DM, Kaplan SL, Grumbach MM. Plasma glycoprotein hormone alpha-subunit in the neonate and in prepubertal and pubertal children: effects of luteinizing hormone-releasing hormone. J Clin Endocrinol Metab 1980;50:450–5.
9. Yen SSC, Tsai CC, Naftolin F, Vandenberg G, Ajabor L. Pulsatile patterns of gonadotropin release in subjects with and without ovarian function. J Clin Endocrinol Metab 1972;34:671–5.
10. Whitcomb RW, Sangha JS, Schneyer AL, Crowley WF. Improved measurement of free alpha subunit of glycoprotein hormones by assay with use of a monoclonal antibody. Clin Chem 1988;34:2022–5.

11. Whitcomb RW, O'Dea LStL, Finkelstein JS, Heavern DM, Crowley WF. Utility of free alpha-subunit as an alternative neuroendocrine marker of gonadotropin-releasing hormone (GnRH) stimulation of the gonadotroph in the human: evidence from normal and GnRH-deficient men. J Clin Endocrinol Metab 1990;70:1654–61.
12. Kourides IA, Weintraub BD, Ridgway EC, Maloof F. Pituitary secretion of free alpha and beta subunits of human thyrotropin in patients with thyroid disorders. J Clin Endocrinol Metab 1975;40:872–85.
13. Hall JE, Whitcomb RW, Rivier J, Vale W, Crowley WF. Differential regulation of LH, FSH, and free alpha subunit secretion from the gonadotroph by GnRH: evidence from the use of two GnRH antagonists. J Clin Endocrinol Metab 1990;70:328–35.
14. Filicori M, Butler JP, Crowley WF. Neuroendocrine regulation of the corpus luteum in the human. Evidence for pulsatile progesterone secretion. J Clin Invest 1984;73:1638–47.
15. Landy H, Boepple PA, Mansfield MJ, Crigler JF, Crwaford JD, Crowley WF. Renal clearance of free alpha subunit (FAS) decreased during pituitary desensitization. Proc 71st annu meet Endocr Soc, 1989:989.
16. Kagen C, McNeilly AS. Changes in circulating levels of LH, FSH, LH beta- and alpha-subunit after gonadotropin-releasing hormone and of TSH, LH beta- and alpha-subunit after thyrotropin-releasing hormone. J Clin Endocrinol Metab 1975;41:466–70.
17. Winters SJ, Troen P. Pulsatile secretion of immunoreactive alpha-subunit secretion in man. J Clin Endocrinol Metab 1985;60:344–8.

# 20

# Free hCGβ and Nongonadal Neoplasms

D. BELLET, N. ROUAS, I. MARCILLAC, F. HOUSSEAU, P. COTTU,
F. TROALEN, AND J.-M. BIDART

The production of *human chorionic gonadotropin* (hCG) and its free *alpha subunit* (hCGα) and *beta subunit* (hCGβ) by nongonadal neoplasms is controversial. Following the initial report (1), numerous studies have provided indirect evidence for the production of hCG, hCG-like substances, or free subunits by nongonadal cancers (2–8). Indeed, between 1973 and 1985, 105 studies reported immunoreactive hCG in sera of nontrophoblastic tumor patients; between 1962 and 1986, 107 investigations showed the immunocytochemical localization of hCG in various tissue (9, 10). However, these results were questionable, since most investigations used polyclonal and monoclonal antibodies lacking the required specificity. One of the major problems was that most antibodies cross-reacted with *human luteinizing hormone* (hLH) (6, 11, 12). More recently developed assays based on monoclonal antibodies continued to display low, but significant, cross-reactivity with hLH (13–17). More importantly, most assays were unable to distinguish between the intact hCG molecule and the free hCGβ-subunit. Indeed, the observed cross-reactivity of intact hCG varied between 3% and 100% in radioimmunoassays used for the detection of free hCGβ. It was much lower, but still significant (0.2% to 4%), with most *immunoradiometric assays* (m-IRMAs) for free hCGβ based on monoclonal antibodies (13–15, 18–20). Finally, the nomenclature for hCGβ assays is ambiguous, since the same designation was given to at least two types of assays with widely different specificities, including those measuring only the free hCGβ-subunit and those measuring both intact hCG and the free hCGβ-subunit (21, 22). Similarly, the nomenclature for hCG assays may be misinterpreted, since these assays also measure the free hCGβ-subunit in addition to hCG (23).

Recently, m-IRMAs entirely specific for intact hCG and the free hCGα- or hCGβ-subunit enabled investigators to reevaluate the production of hCG and free subunits by normal individuals, pregnant women, and patients with gestational trophoblastic diseases (24–29). These assays have provided direct evidence that hCG is secreted by the normal human pituitary and is present at low levels (<1000 pg/ml) in sera of normal adults (25, 26, 30–32). A significant rise in serum levels of the free hCGα-subunit (as high as 3000 pg/ml) was found in 97.5% of normal individuals and nonmalignant disease controls (32). Using a highly sensitive assay for free hCGβ (5.9 pg/ml), it was found that free hCGβ was present at low levels (<100 pg/ml) in sera of men and nonpregnant women without evidence of cancer (16). Thus far, hCG and free hCGβ in serum have been used mainly as tumor markers for gestational trophoblastic disease and testicular cancer (19, 27, 33).

We took advantage of the specificity of m-IRMAs developed in our laboratory to assess the production of hCG and free subunits by nongonadal and nontrophoblastic malignancies. It was found that isolated production of the free hCGβ-subunit was associated with tumors having a poor prognosis. Moreover, this free subunit appeared to represent a biological marker, as well as a potential target for specific immunotherapy of such malignancies.

# Production of hCG and Its Free Subunits by Nongonadal Neoplasms

We used four different assays entirely specific for the molecules of interest, namely hCG, free hCGα, free hCGβ, and free hCGβ with the *carboxyl-terminal portion* (CTP). The assay for hCG was described previously and is based on two monoclonal antipeptide antibodies, designated FB12 and FB09, which are directed against the CTP and serve as capture antibodies (29). Monoclonal antibody HT13 directed against the hCGα-subunit and labeled with $^{125}$I served as a radiolabeled indicator probe (Fig. 20.1A). The assay designated m-IRMA 9-12-13 had a lower limit of sensitivity of 50 pg/ml for hCG and demonstrated no cross-reaction with the free α- and β-subunits of either hCG or hLH, nor with intact hLH, human follicle stimulating hormone or human thyroid stimulating hormone.

To measure the free hCGα-subunit, a specific assay (m-IRMA 20-13) was used as described previously (24). This assay is based on monoclonal antibody AHT20, which captures the free α-subunit of hCG on a solid phase support and uses the $^{125}$I-labeled HT13 described above as a radiolabeled detecting antibody (Fig. 20.1B). The assay had a lower specificity of 50 pg/ml and showed no cross-reactivity with the free hCGβ-subunit or with other intact glycoprotein hormones.

Two different assays based on monoclonal antibody FBT11 were used for measurement of the free hCGβ-subunit, as described previously (24, 32). The first assay (m-IRMA 11-9) utilized FBT11 as the capture anti-

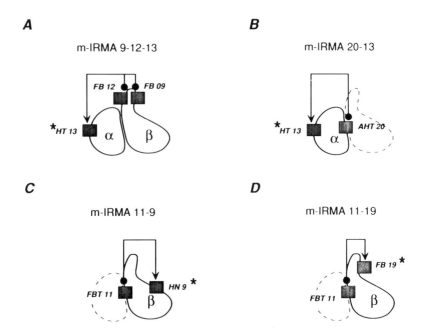

FIGURE 20.1. Localization of antigenic regions recognized by monoclonal anti-bodies within hCGα-subunits (α) and hCGβ-subunits (β) and schematic repre-sentation of m-IRMAs utilized to measure either intact α-β dimeric hCG (*A*), free hCGα (*B*) or free hCGβ (*C, D*). * = antibody used as radiolabeled indicator in a given assay.

body and a monoclonal antibody directed against the central portion of hCGβ (HN9) as radiolabeled indicator antibody (Fig. 20.1C). This assay had a sensitivity limit of 100 pg/ml of highly purified hCGβ, with no cross-reactivity with hCGα or with hCG or hLH. It displayed a cross-reactivity of less than 0.2% with the *free beta chain of LH* (hLHβ). In order to avoid potential cross-reactivity with the free hLHβ-subunit, we developed a second assay (m-IRMA 11-19) by using a monoclonal antipeptide anti-body designed FB19, instead of antibody HN9, as the radiolabeled in-dicator probe. This antibody is directed to the 110–116 amino acid carboxyl-terminal portion of hCGβ and does not bind to the corresponding region of hLHβ. Consequently, the m-IRMA 11-19 based on FBT11 as capture antibody and FB19 as tracer was entirely specific for hCGβ, with no cross-reactivity with hLHβ. Moreover, this assay measured any free hCGβ-subunit containing the carboxyl-terminal portion of hCGβ (Fig. 20.1D). Its sensitivity was 100 pg/ml of highly purified hCGβ.

We studied the presence of either hCG or its free subunits in sera of 661 patients with nontrophoblastic and nongonadal tumors from various primary sites. Serum samples were stored at −20 °C before analysis. Since

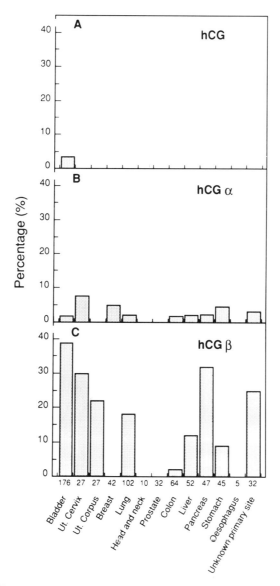

FIGURE 20.2. Percentage of cancer patients with serum values of hCG (A), free hCGβ (B) or free hCGβ (C) higher than values found in sera of healthy subjects and patients with benign diseases. Ut = uterine. Free hCGβ levels were determined with m-IRMA 11-9.

serum levels for the free hCGβ-subunit, hCG, and the free hCGα-subunit of up to 100 pg/ml, 1000 pg/ml, and 3000 pg/ml, respectively, were found in normal subjects and nonmalignant disease controls, we determined the percentage of patients with a given malignancy and a serum level of hCG, free hCGα, or free hCGβ higher than that found in those control groups. Results presented in Figure 20.2 clearly indicate that both an hCG serum level >1000 pg/ml and a free hCGα level >3000 pg/ml are marginally observed in patients with nongonadal and nontrophoblastic malignancies (Figs. 20.2A and 20.2B). In striking contrast, as many as 39% of those patients had serum levels of free hCGβ >100 pg/ml (Fig. 20.2C). In such patients, the presence of free hCGβ was most frequently detected in sera of individuals with bladder (39%), pancreatic (32%), and cervical carcinomas (30%). Interestingly, 25% of patients with a cancer of an unknown primary site had free hCGβ levels >100 pg/ml in their sera.

These findings are at odds with previous observations describing elevated serum levels of hCG immunoreactivity in patients with nontrophoblastic neoplasms, such as cancer of the stomach (23%), liver (17%), lung (19%–33%), and urinary bladder (30%) (34–36), as well as adenocarcinoma (33%–50%) and endocrine carcinoma of the pancreas (69%) (3, 34, 37). Elevated levels of serum hCG have also been reported in 10%–18% of patients with nontrophoblastic gynecological cancer (38, 39). Present results clearly indicate that hCG immunoreactivity observed in those patients was due in part to the production of intact hCG by the pituitary (30, 31, 40), particularly in females over 45 with gynecological cancers; indeed, serum levels related to a pituitary origin have been found to increase with age (26, 27). Alternatively, the hCG reactivity observed in those patients might be due to the presence, in serum, of free hCGβ produced by their tumors and detected by assays cross-reacting with this free subunit. Moreover, previous results suggested that free hCGβ is rarely, if ever, produced by nongonadal and nontrophoblastic malignancies and that free hCGβ secretion occurs only in association with synthesis of high levels of intact hCG (13, 41, 42).

In the present investigation, 134 patients with various tumors had elevated free hCGβ levels. Among these patients, 138 had either isolated secretion of free hCGβ or a predominant production of this free subunit in excess of hCG. This pattern of secretion contrasted with that observed in pregnant women, who displayed predominant secretion of hCG that greatly exceeded that of free hCGβ (24).

There is accumulating evidence that the isolated production of free hCGβ may be associated with aggressive nongonadal and nontrophoblastic malignancies. For example, we found that isolated production of the free hCGβ-subunit was associated with tumors having poor prognosis, such as cancers of the lung, pancreas, and liver. This observation is in agreement with prior documented cases in which the presence of free hCGβ was associated with aggressive malignancies, as measured by a duration of survival of two weeks to three months (5, 43). Moreover, it is interesting to note that the half-life of the free hCGβ-subunit was previously found to be 10- to 30-fold less than that of intact hCG and varied between 40 min and 4 h (44, 45). It is likely that the free hCGβ-subunit may be rapidly cleared by the kidney, and therefore a low serum level of free hCGβ may indicate the high secreting activity of some tumors; further studies are required to clarify this point.

The histological origin of the hCGβ-producing cells in these tumors deserves comment. Indeed, the existence of nongonadal tumors with histopathologic similarities to choriocarcinoma has been known for a number of decades. In 1904, a Russian pathologist reported a case of "chorioepithelioma" arising from a bladder tumor in a postmenopausal woman that had the histological features of a syncytiotrophoblast (46).

Several hypotheses have been proposed to explain the origin of hCGβ-producing cells, including the following: (i) displaced totipotential or gonadal cells (extragonadal choriocarcinomas); (ii) metastasis from an intrauterine or gonadal lesion; and (iii) evolution from a somatic cell that underwent morphologic and functional transformation (metaplasia) into a cell functionally similar to the trophoblast. Since the pioneering work of Pick (47) in 1926, most authors favor the concept of a trophoblastic metaplasia within carcinomatous tissues (48–52). Indeed, the pathway of cytotrophoblast differentiation ensures that hCGα mRNA accumulates before hCGβ mRNA, resulting in increasing hCG production as hCGβ synthesis is initiated (15). Our results are in agreement with this pathway; most gonadal and extragonadal choriocarcinomas, as well as metastatic choriocarcinomas, produce both hCG and the free hCGβ-subunit and thus differ from nongonadal cells predominantly producing the free hCGβ-subunit (15, 19, 21). In addition, it has been shown that in the early stages of implantation, as well as in trophoblastic disease, the origin of free hCGβ production may be due to poorly differentiated trophoblastic tissue (19, 53). These observations suggest that the origin of nongonadal malignant cells involves metaplasia of carcinomatous tissue into a tissue similar to that of poorly differentiated trophoblasts.

Collectively, these results indicate that the presence of elevated serum levels of free hCGβ is not a rare occurence, but rather, may be relatively common, particularly in patients with bladder tumors. As this free subunit might represent a unique biological marker of such cancers, we decided to further study the production of free hCGβ by bladder cancers.

# Production of hCG and Its Free Subunits by Bladder Cancers

The finding that bladder tumors produce the free hCGβ-subunit was not totally unexpected, since in vivo production of either hCG or free hCGβ by such neoplasms had been reported previously (35, 48, 54–58). Moreover, in vitro studies performed on culture media derived from urothelial cell lines strongly suggested that the material excreted by these cells consisted principally of the free hCGβ-subunit or fragments thereof (59, 60). In order to identify precisely the molecular form(s) excreted by bladder cancers and responsible for the detection of hCG or hCGβ immunoreactivity in the serum, we measured the serum level of either the intact hCG or the free hCGβ-subunit in samples collected from 176 patients with bladder cancers at various stages. As described above, only 6 patients (3%–5%) had serum hCG levels higher than those found in healthy individuals or benign disease controls, whereas 69 patients (39%) had free hCGβ >100 pg/ml. Thus, it is clear that the free hCGβ-subunit is the major molecular form produced by bladder cancers. Moreover, the percentage of patients with a free hCGβ-producing tumor increased with the stage of the disease: At the time of first determination, 6.5% of patients with $T_a$, 14% of those with $T_2$ and 20%–25% of those with $T_3$ tumors had serum free hCGβ >100 pg/ml, while 42% of those with $T_4$ tumors and 56% of patients with metastases showed a rise in the free hCGβ serum level (Fig. 20.3). These results are in accordance with previous studies demonstrating that serum hCGβ levels were substantially elevated in 76% of patients with widespread metastatic disease (54). However, using a sensitive and totally specific assay for the free hCGβ-subunit, 6.5%–42% of patients in whom disease was limited to the renal pelvis ($T_a$-$T_4$) had serum levels >100 pg/ml, while a detectable level of hCGβ was previously found in only 4% of patients (54).

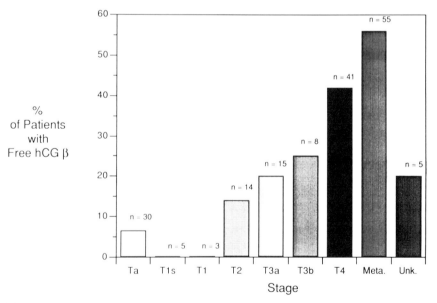

FIGURE 20.3. Percentage of cancer patients with bladder cancers at various stages and free hCGβ serum levels >100 pg/ml.

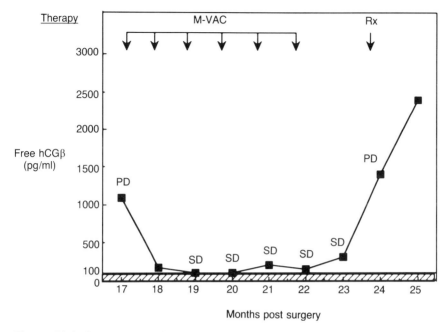

FIGURE 20.4. Serum levels of free hCGβ-subunit during the follow-up of a patient with transitional cell carcinoma of the bladder (stage T3a). M-VAC = chemotherapy with Methotrexate, Velbe, Adriamycine, or CDDP. Rx = radiotherapy; PD = progressive disease; SD = stable disease. Hatching indicates the limit of detection of the assay (m-IRMA 11-9).

During patient follow-up, the percentage of patients with free hCGβ serum levels >100 pg/ml increased to 28%, 45%, and 70% for patients with $T_3b$, $T_4$, and metastatic diseases, respectively. Figure 20.4 and Figure 20.5 show the free hCGβ serum levels observed during the follow-up of two patients taken as representative. Indeed, preliminary results observed in 30 patients with bladder cancer at previous stages suggest a significant correlation between free hCGβ serum levels and regression, stability, or progression of the disease.

As noted above for nongonadal tumors of various origins, hCGβ expression in patients with bladder cancers was associated with poor prognosis. Since the initial report of Djewitzki in 1904, most reports were case studies of advanced-stage disease (46, 58, 61). The most recent studies, primarily immunohistochemical, have identified local tumors that expressed hCGβ. All were poorly differentiated and invasive (55). In addition, such expression correlates with those local tumors that do not respond to radiotherapy (62, 63). Similarly, chemotherapy gives disappointing results in transitional cell carcinoma of the bladder with trophoblastic metaplasia (64). Since nonspecific immunotherapy

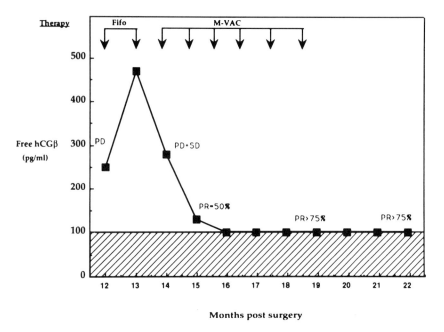

FIGURE 20.5. Serum levels of free hCGβ-subunit during the follow-up of a patient with undifferentiated transitional cell carcinoma of the bladder (stage T3b). Symbols are as described in the legend to Figure 20.4. Fifo = chemotherapy with 5-FU and Ifosfamide; PR = partial response.

(Bacille Calmette-Guérin therapy) was found to alter the progression of transitional-cell carcinoma of the bladder (65–67), we sought to determine whether specific immunotherapy aimed at eradicating free hCGβ-producing cells might be an efficient adjunct for treating aggressive bladder tumors. As a first step, we thus studied the immune response to hCG and its free subunits in patients with bladder cancer.

## Immune Response to hCG and Its Free Subunits in Patients with Bladder Cancer

Bladder cancer is of particular interest in terms of antitumor immunity, since intravesical immunotherapy with *Bacille Calmette-Guérin* (BCG) mycobacterium has been shown to be effective in the treatment of in situ carcinoma of the bladder and can reduce tumor recurrence in transitional cell carcinoma of the bladder (66, 68). Several animal and clinical studies have strongly suggested an immunological pathway through which BCG exerts its antitumor effects (69–71). Interestingly, the increased presence of intense infiltrates of T cells and macrophages was observed in human bladder biopsies after intravesical BCG treatment (72). Moreover, the neoexpression of *major histocompatibility complex* (MHC) class II antigens by the urothelium was induced after intravesical BCG therapy. It is likely that such neoexpression is related to T cell interferon-γ production in the local anti-BCG response (72). Increased levels of cytokines interleukin-1, interleukin-2, and tumor necrosis factor have also been detected in urine directly following intravesical BCG instillations (73). Finally, internalization of BCG by human bladder tumor cells was demonstrated and may play a role in antigen presentation to the immune system (74).

Taken together, these data demonstrate that bladder tumor cells are the target of nonspecific immunological mechanisms. However, a major obstacle to successful immunotherapy of malignancy has been a lack of well-characterized tumor antigens that can be utilized as specific targets for autochthonous T cell responses. As production of the free hCGβ-subunit was found to be highly specific to patients with aggressive malignancies, this subunit can be viewed as a potential target for immunotherapy in patients with productive tumors and, in particular, with bladder cancers. Such immunotherapy is based on the stimulation of tumor antigen-specific T cells involved in in vivo antitumor immune responses. In general, T cells recognize peptide fragments of protein antigens bound to MHC molecules. In light of these considerations, we examined whether T cells specific for free hCGβ, as well as hCG and free hCGα, were present in patients with bladder cancer and whether peptides derived from hCG subunits were recognized by such T cells. As a prelude to investigations in humans, we had previously described immune recognition of hCG and its subunits in a murine model. Collectively, the results observed in our model showed that the α/β-subunit association played a critical role in hCG processing (75, 76) and that receptor-mediated uptake of hCG drastically enhanced its presentation to T cells (unpublished observations).

TABLE 20.1. PBMCs proliferative response to hCG and its subunits in cancer patients and pregnant women.

| Primary site | Histology | Total no. | No. with cellular proliferative response |
|---|---|---|---|
| Bladder | Transitional cell carcinoma | 8 | 2 |
| Testis | Seminoma choriocarinoma | 2 | 0 |
| Controls (pregnant women) | | 21 | 0 |

*Note:* PBMCs ($1 \times 10^{-5}$ cells per well) were stimulated in the absence or presence of $3\,\mu M$ of hCG or its free subunits. Cells were pulsed on day 6 with $1\,\mu Ci$ $^3$H-thymidine and collected the next day for counting the amount of incorporated label.

In order to study the cellular immune response to hCG and its subunits in humans, we tested the in vitro proliferative response of *peripheral blood mononuclear cells* (PBMCs) from either cancer patients or pregnant women as controls. In those experiments, purified hCG, free subunits, and peptides covering the entire sequence of hCGα and hCGβ were used to induce a proliferative response (77). Twenty-one pregnant women were tested and did not display a proliferative response. Among eight patients with bladder cancer and two patients with testicular cancer,

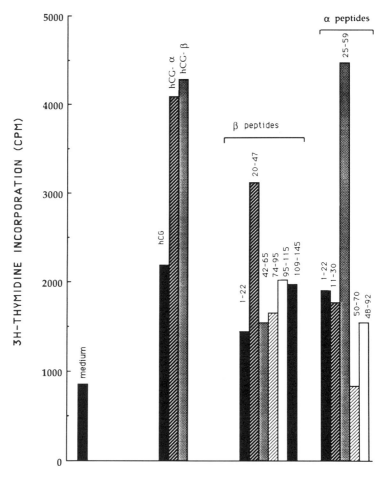

FIGURE 20.6. Proliferative response of PMBCs from a patient with a bladder cancer. PBMCs ($1 \times 10^{-5}$ cells per well) were stimulated in the absence (medium) or presence of either $3\,\mu$M of hCG, hCGβ, and hCGα or $25\,\mu$M of synthetic peptides encompassing sequences of the α- and β-subunits. Cells were pulsed on day 6 with $1\,\mu$Ci $^3$H-thymidine and collected the next day for counting of the amount of incorporated label.

two patients with bladder cancer (CI.2 and CI.6) demonstrated an in vitro T cell response to hCG and its free subunits (Table 20.1). We further studied the in vitro proliferative response of patient CI.2 using peptides analogous to various portions of hCG subunits as stimulators. Results presented in Figure 20.6 show that peptides hCGβ (20–47) and hCGα (25–59) are the most potent in inducing T cell proliferation. It should be noted that among tested patients with bladder cancer, patient CI.2 exhibited one of the highest serum levels of free hCGβ (13.7 ng/ml). Taken together, these results indicate that memory T cell responses of patients CI.2 and CI.6 were due to prior in vivo stimulation by peptides processed from hCG subunits. The finding that hCG-derived peptides are immunogenic in man suggests that immunotherapy may be targeted against specific proteins secreted by neoplastic cells. For example, T cells derived by specific stimulation with corresponding synthetic peptides might retain functional competence in vivo and could thus be useful for therapeutic purposes.

## Conclusion

Secretion of the free hCGβ-subunit was strongly correlated with aggressive nontrophoblastic neoplasms, such as tumors of the pancreas, lung, and bladder. It is noteworthy that these nongonadal tumors are generally resistant to low-risk chemotherapy and often require high-risk cytotoxic protocols (21). It is important to determine whether such treatment is warranted for patients with free hCGβ-subunit-producing neoplasms, including tumors of an unknown primary site (78, 79). Moreover, the potential interest of specific immunotherapy directed toward free hCGβ-producing tumor cells and using T cell stimulating peptides must be evaluated. Finally, the utilization of sensitive assays capable of detecting pg amounts of the free hCGβ-subunit may facilitate the follow-up of patients undergoing surgery, radiation, chemotherapy, or immunotherapy.

## *References*

1. Reeves RL, Tesluk H, Harrison CE. Precocious puberty associated with hepatoma. J Clin Endocrinol Metab 1959;19:1651–60.
2. Borkowski A, Muquardt C. Human chorionic gonadotropin in the plasma of normal nonpregnant subjects. N Engl J Med 1979;301:298–302.
3. Kahn CR, Rosen SW, Weintraub BD, Fajans SS, Gorden P. Ectopic production of chorionic gonadotropin and its subunits by islet-cell tumors. N Engl J Med 1977;297:565–9.
4. Rosen SW, Weintraub BD. Ectopic production of the isolated alpha subunit of the glycoprotein hormones. N Engl J Med 1974;290:1441–7.
5. Weintraub BD, Rosen WS. Ectopic production of the isolated beta subunit of human chorionic gonadotropin. J Clin Invest 1973;52:3135–42.

6. Tormey DC, Waalkes TP, Simon RM. Biological markers in breast carcinoma. Cancer 1977;39:2391–6.
7. Gailani S, Chu TM, Nussbaum A, Ostrander M, Christoff N. Human chorionic gonadotrophins (hCG) in nontrophoblastic neoplasms. Cancer 1976;38:1684–6.
8. Blackman MR, Weintraub BD, Rosen SW, Kourides IA, Steinwascher K, Gail MH. Human placental and pituitary glycoprotein hormones and their subunits as tumor markers: a quantitative assessment. J Natl Cancer Inst 1980;65:81–93.
9. Hussa RO. Clinical applications of hCG tests: tumors. In: Hussa RO, ed. The clinical marker hCG. New York: Praeger, 1987:119–36.
10. Hussa RO. Subunits of hCG. In: Hussa RO, ed. The clinical marker hCG. New York: Praeger, 1987:161–78.
11. Papapetrou PD, Sakarelou NP, Braouzi H, Fessas PH. Ectopic production of human chorionic gonadotropin (hCG) by neoplasms: the value of measurements of immunoreactive hCG in the urine as a screening procedure. Cancer 1980;45:2583–92.
12. Monteiro JCMP, Ferguson KM, McKinna JA, Greening WP, Neville AM. Ectopic production of human chorionic gonadotrophin-like material by breast cancer. Cancer 1984;53:957–62.
13. Saller B, Clara R, Spöttl G, Siddle K, Mann K. Testicular cancer secretes intact human choriogonadotropin (hCG) and its free β subunit: evidence that hCG (+hCG-β) assays are the most reliable in diagnosis and follow-up. Clin Chem 1990;36:234–9.
14. Hay D, Murphy JR. Evaluating human choriogonadotropin heterogeneity in nongestational malignancy with monoclonal immunoassays. Oncology 1987;44:174–9.
15. Hay DL. Histological origins of discordant chorionic gonadotropin secretion in malignancy. J Clin Endocrinol Metab 1988;66:557–64.
16. Alfthan H, Haglund C, Dabek J, Ulf-Häkan S. Concentrations of human choriogonadotropin, its β-subunit, and the core fragment of the β-subunit in serum and urine of men and nonpregnant women. Clin Chem 1992;38:1981–7.
17. Alfthan H, Haglund C, Roberts P, Ulf-Häkan S. Elevation of free β subunit of human choriogonadotropin and core β fragment of human choriogonadotropin in the serum and urine of patients with malignant pancreatic and biliary disease. Cancer Res 1992:4628–33.
18. Hussa RO. Tissue localization of hCG. In: Hussa RO, ed. The clinical marker hCG. New York: Praeger, 1987;52:151–60.
19. Fan C, Goto S, Furuhashi Y, Tomoda Y. Radioimmunoassay of the serum free β-subunit of human chorionic gonadotropin in trophoblastic disease. J Clin Endocrinol Metab 1987;64:313–8.
20. Mann K, Siddle K. Evidence for free beta-subunit secretion in so-called human chorionic gonadotropin-positive seminoma. Cancer 1988;62:2378–82.
21. Case records of the Massachusetts General Hospital (case 34-1983). N Engl J Med 1983;309:477–87.
22. Lindstedt G, Lundberg PA, Hedman LA. Circulating choriogonadotropin β subunit in a patient with primary amenorrhea and embryonal ovarian carcinoma. Clin Chim Acta 1980;104:195–200.

23. Cole LA, Kardana A. Discordant results in human chorionic gonadotropin assays. Clin Chem 1992;38:263–70.
24. Ozturk M, Bellet D, Manil L, Hennen G, Frydman R, Wands JR. Physiological studies of human chorionic gonadotropin (hCG), αhCG, and βhCG as measured by specific monoclonal immunoradiometric assays. Endocrinology 1987;120:549–58.
25. Griffin J, Odell WD. Ultrasensitive immunoradiometric assay for chorionic gonadotropin which does not cross-react with luteinizing hormone nor free β chain of hCG and which detects hCG in blood of non-pregnant humans. J Immunol Methods 1987;103:275–83.
26. Odell WD, Griffin J. Pulsatile secretion of human chorionic gonadotropin in normal adults. N Engl J Med 1987;317:1688–91.
27. Ozturk M, Berkowitz R, Goldstein D, Bellet D, Wands JR. Differential production of human chorionic gonadotropin and free subunits in gestational trophoblastic disease. Am J Obstet Gynecol 1988;158:193–8.
28. Stenman U-H, Alfthan H, Myllynen L, Seppälä M. Ultrarapid and highly sensitive time-resolved fluoroimmunometric assay for chorionic gonadotropin. Lancet 1983;2:647–9.
29. Bellet DH, Ozturk M, Bidart JM, Bohuon CJ, Wands JR. Sensitive and specific assay for human chorionic gonadotropin (hCG) based on anti-peptide and anti-hCG antibodies: construction and clinical implications. J Clin Endocrinol Metab 1986;63:1319–27.
30. Odell WD, Griffin J, Bashey HM, Snyder PJ. Secretion of chorionic gonadotropin by cultured human pituitary cells. J Clin Endocrinol Metab 1990;71:1318–21.
31. Hammond E, Griffin J, Odell WD. A chorionic gonadotropin-secreting human pituitary cell. J Clin Endocrinol Metab 1991;72:747–4.
32. Marcillac I, Troalen F, Bidart J-M, et al. Free human chorionic gonadotropin beta subunit in gonadal and nongonadal neoplasms. Cancer Res 1992;52:3901–7.
33. Madersbacher S, Klieber R, Mann K, et al. Free α-subunit, free β-subunit of human chorionic gonadotropin (hCG), and intact hCG in sera of healthy individuals and testicular cancer patients. Clin Chem 1992;38:370–6.
34. Braunstein GD, Vaaitukaitis JL, Carbone PP, Ross GT. Ectopic production of human chorionic gonadotrophin by neoplasms. Ann Med 1973;78:39–45.
35. Dexeus F, Logothetis C, Hossan E, Samuel ML. Carcinoembryonic antigen and beta-human chorionic gonadotropin as serum markers for advanced urothclial malignancics. J Urol 1986;136:403 7.
36. Gropp C, Havemann K, Scheuer A. Ectopic hormones in lung cancer patients at diagnosis and during therapy. Cancer 1980;46:347–54.
37. Oberg K, Wide L. hCG and hCG subunits as tumour markers in patients with endocrine pancreatic tumours and carcinoids. Acta Endocrinol 1981;98:256–60.
38. Rutanen E-M, Seppälä M. The hCG-β subunit radioimmunoassay in nontrophoblastic gynecological tumors. Cancer 1978;41:692–6.
39. Carenza L, di Gregorio R, Mocci C, Moro M, Pala A. Ectopic human chorionic gonadotropin: gynecological tumors and nonmalignant conditions. Gynecol Oncol 1980;10:32–9.

40. Hoermann R, Spoettl G, Moncayo R, Mann K. Evidence for the presence of human chorionic gonadotropin (hCG) and free β-subunit of hCG in the human pituitary. J Clin Endocrinol Metab 1990;71:179–86.

41. Heitz PU, von Herbay G, Klöppel G. The expression of subunits of human chorionic gonadotropin (hCG) by nontrophoblastic, nonendocrine, and endocrine tumors. Am J Clin Pathol 1987;88:467–72.

42. Papapetrou PD, Nicopoulou SCh. The origin of a human chorionic gonadotropin β-subunit-core fragment excreted in the urine of patients with cancer. Acta Endocrinol 1986;112:415–22.

43. Nagelberg SB, Marmorstein B, Khazaeli MB, Rosen SW. Isolated ectopic production of the free beta subunit of chorionic gonadotropin by an epidermoid carcinoma of unknown primary site. Cancer 1985;55:1924–30.

44. Wehmann RE, Nisula BC. Metabolic and renal clearance rates of purified human chorionic gonadotropin. J Clin Invest 1981;68:184–94.

45. Wehmann RE, Nisula BC. Renal clearance rates of the subunits of human chorionic gonadotropin in man. J Clin Endocrinol Metab 1980;50:674–9.

46. Djewitzki WSt. Uber einen fall von chorionepithelioma der harnblase. Virchows Arch Anat Physiol Klin Med 1904;178:451–64.

47. Pick L. Uber die chorionepitheliomätnlich metastasierende form des magencarcinoms. Klin Wochenschr 1926;5:1728–9.

48. Vahlensieck W, Riede U, Wimmer B, Ihling C. Beta-human chorionic gonadotropin-positive extragonadal germ cell neoplasia of the renal pelvis. Cancer 1991;67:3146–9.

49. Park CH, Reid JD. Adenocarcinoma of the colon with choriocarcinoma in its metastases. Cancer 1980;46:570–5.

50. Obe JA, Rosen N, Koss LG. Primary choriocarcinoma of the urinary bladder. Cancer 1983;52:1405–9.

51. Kubosawa H, Nagao K, Kondo Y, Ishige H, Inaba N. Coexistence of adenocarcinoma and choriocarcinoma in the sigmoid colon. Cancer 1984;54: 866–8.

52. Campo E, Algaba F, Palacin A, Germa R, Sole-Balcells FJ, Cardesa A. Placental proteins in high-grade urothelial neoplasms. An immunohistochemical study of human chorionic gonadotropin, human placental lactogen, and pregnancy-specific beta-1-glycoprotein. Cancer 1989;63:2497–504.

53. Hay DL. Discordant and variable production of human chorionic gonadotropin and its free α- and β-subunits in early pregnancy. J Clin Endocrinol Metab 1985;61:1195–200.

54. Iles RK, Jenkins BJ, Oliver RTD, Blandy JP, Chard T. Beta human chorionic gonadotrophin in serum and urine. A marker for metastatic urothelial cancer. Br J Urol 1989;64:241–4.

55. Iles RK, Chard T. Human chorionic gonadotropin expression by bladder cancers: biology and clinical potential. J Urol 1991;145:453–8.

56. McLoughlin J, Pepera T, Bridger J, Williams G. Serum and urinary levels of beta human chorionic gonadotrophin in patients with transitional cell carcinoma. Br J Cancer 1991;63:822–4.

57. Fetissof F, Bellet D, Guilloteau D, Maillot O. Hormone gonadotrophine chorionique et carcinome à cellules transitionnelles de la vessie. Ann Pathol 1988;8:276–80.

58. Iles R. β-hCG expression by bladder cancers. Br J Cancer 1992;65:305.

59. Iles RK, Oliver RTD, Kitau M, Walker C, Chard T. In vitro secretion of human chorionic gonadotropin by bladder tumour cells. Br J Cancer 1987; 55:623–6.
60. Iles RK, Purkis PE, Whitehead PC, Oliver RTD, Leigh I, Chard T. Expression of beta human chorionic gonadotrophin by non-trophoblastic non-endocrine "normal" and malignant epithelial cells. Br J Cancer 1990;61:663–6.
61. Young RH, Eble JN. Unusual forms of carcinoma of the urinary bladder. Hum Pathol 1991;22:948–65.
62. Martin JE, Jenkins BJ, Zuk RJ, et al. Human chorionic gonadotropin expression and histological findings as predictors of response to radiotherapy in carinoma of the bladder. Virchows Arch [A] 1989;414:273–7.
63. Oliver RTD, Stephenson C, Collino CE, et al. Clinicopathological significance of immunoreactive beta-hCG production by bladder cancer. Mol Biother 1988;1:43–5.
64. Burry AF, Munn SR, Arnold EP, et al. Trophoblastic metaplasia in urothelial carcinoma of the bladder. Br J Urol 1986;58:143–6.
65. Herr WH, Laudone VP, Badalament RA, et al. Bacillus Calmette-Guérin therapy alters the progression of superficial bladder cancer. J Clin Oncol 1988;6:1450–5.
66. Lamm DL, Blumenstein BA, Crawford ED, Montie JE, Scardino P, Grossman HB, et al. A randomized trial of intravesical doxorubicin and immunotherapy with Bacille Calmette-Guérin for transitional-cell carcinoma of the bladder. N Engl J Med 1991;325:1205–9.
67. Cookson MS, Sarosdy MF. Management of stage T1 superficial bladder cancer with intravesical Bacillus Calmette-Guérin therapy. J Urol 1992;148: 797–801.
68. Brosmann SA. Experience with Bacillus Calmette-Guérin in patients with superficial bladder cancer. J Urol 1982;128:27–30.
69. Ratlif TL, Gillen DP, Catalona WJ. Requirement of a thymus dependent immune response for BCG-mediated antitumor activity. J Urol 1987;137: 155–8.
70. Becich MJ, Carroll S, Ratliff TL. Internalization of Bacille Calmette-Guérin. J Urol 1991;145:1316–24.
71. Cornel EB, Van Moorselaar JA, Van Stratum P, Debruyne MJ, Schalken JA. Antitumor effects of Bacillus Calmette-Guérin in a syngeneic rat bladder tumor model system RT323. J Urol 1993;149:179–82.
72. Prescott S, James K, Hargreave TB, Chisholm GD, Smyth JF. Intravesical Evans strain BCG therapy: quantitative immunohistochemical analysis of the immune response within the bladder wall. J Urol 1992;147:16536–42.
73. Böhle A, Nowc CH, Ulmer AJ, et al. Elevations of cytokines interleukin-1, interleukin-2 and tumor necrosis factor in the urine of patients after intravesical Bacillus Calmette-Guérin immunotherapy. J Urol 1990;144:59–63.
74. Kuroda K, Brown EJ, Telle WB, Russel DG, Ratliff TL. Characterization of the internalization of Bacillus Calmette-Guérin by human bladder tumor cells. J Clin Invest 1993;91:69–76.
75. Rouas N, Christophe S, Bellet D, Troalen F, Guillet JG, Bidart J-M. Immune recognition of a molecule naturally presented as a monomeric or an oligo-

meric structure: the model of the human chorionic gonadotropin α subunit. Mol Immunol 1992;29:883–93.

76. Rouas N, Christophe S, Housseau F, Bellet D, Guillet JG, Bidart J-M. Influence of protein quaternary structure on antigen processing. J Immunol 1993;150:782–92.

77. Gedde-Dahl T III, Spurkland A, Eriksen JA, Thorsby E, Gaudernack G. Memory T cells of a patient with follicular thyroid carcinoma recognize peptides derived from mutated p21 ras (Gln-Leu 61). Int Immunol 1992;11: 1331–7.

78. Richardson RL, Schoumacher RA, Fer MF. The unrecognized extragonadal germ cell cancer syndrome. Ann Intern Med 1981;94:181–6.

79. Greco A, Vaughn WK, Hainsworth JD. Advanced poorly differentiated carcinoma of unknown primary site: recognition of a treatable syndrome. Ann Intern Med 1986;104:547–53.

# 21

# New Clinical Applications of hCG Assays

R.E. Canfield, J.F. O'Connor, Y. Chen, S. Birken,
M.C. Hatch, K. Friedman, C. Matera, and A.C. Kelly

This chapter focuses on some clinical applications of *human chorionic gonadotropin* (hCG) assays, especially to detect early pregnancy loss. The collection and storage of urine samples for assay is an easy and painless procedure, so urinary measurements of hCG, rather than serum measurements, have been utilized for most epidemiologic studies of human reproduction.

Measurements to detect the presence of urinary hCG have been made for over five decades to diagnose pregnancy and also to make the diagnosis of choriocarcinoma (1). When these measurements evolved from biological assays to those with an immunological basis, the sensitivity became much greater, but there were limits imposed by the cross-reactivity with *human luteinizing hormone* (hLH) since its structure is so similar to hCG. An important advance in the history of hCG measurement occurred when Vaitukaitis, Braunstein, and Ross described the rabbit SB-6 antiserum (raised against one of our preparations of the hCG beta subunit), which exhibited a distinct preference for hCG over hLH (2). Utilizing this antiserum, these investigators were able to establish that the epitope, which was principally involved in hCG recognition by this antiserum, is frequently expressed on molecules secreted by malignant tumors (3). More recently, many monoclonal antibodies that bind at different sites on hCG have been developed, and we demonstrated in a previous Serono Symposia, USA publication how these could be utilized to distinguish among various hCG fragments (4).

## Assays for Different hCG Forms

Figure 21.1 is a cartoon depicting the alpha and beta subunits of hCG, joined to form the native hormone. Hypothetical antibody binding sites are drawn in such a way as to illustrate the different regions of the beta

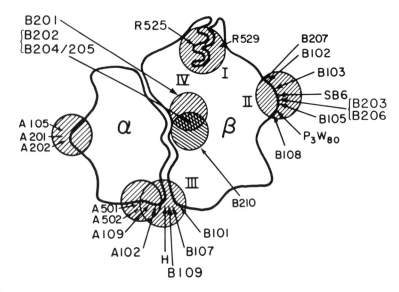

FIGURE 21.1. A cartoon that schematically depicts the intact hCG alpha-beta heterodimer of hCG. The four different beta subunit epitope binding regions, indicated by Roman numerals, are discussed in the text. Additional epitope regions indicated are for the hCG free alpha subunit and another set of alpha-directed antibodies that exhibit a preference for the intact hCG molecule.

subunit that give rise to hCG-specific antibodies. The locations of the antibody binding sites are not precisely known, but we infer that each of these regions contains a structure unique to the hCG beta subunit, since some of the antibodies react much more with hCG than with hLH. Region I is composed of the unique carboxyterminal sequence of hCG. Region II is conformationally dependent and is recognized as a part of the whole hormone, or in the free beta subunit, or in the beta core fragment. Region III reactivity requires that the alpha and beta subunits be joined to form the native hormone—that is, it appears to be specific for the whole hormone. Previous reports of the development of human antibodies to hCG appear to involve this region. Region IV is not reactive in the whole hormone but is exposed when the subunits are separated; thus, we infer that it is located at the interface between the subunits. This latter binding region is. sufficiently extensive to give rise to different monoclonal antibodies that can bind to the region at the same time; for example, both B210 and B201 have been shown to bind to the beta core fragment at the same time. The evidence for these different binding regions has been detailed elsewhere (5, 6) and comparable findings have emerged from the work of Bellet et al. (7) and Berger et al. (8, 9).

Given these antibody specificities, it is possible to devise two-site immunoassays that recognize a particular hCG molecular form, and these

TABLE 21.1. Formats of hCG assays.

| Capture antibody | Detection antibody | Antigen(s) measured | Sensitivity (fmole/ml) |
|---|---|---|---|
| B-109/B-204 combination | B-108* | Intact hCG<br>hCG free beta subunit<br>hCG beta fragment | 1.1 |
| B-109 | B-108* | Intact hCG | 1.2 |
| B-201 | CTP-104* | hCG free beta subunit | 9.0 |
| B-210 | B-108* | hCG beta fragment | 1.0 |
| A-105 | A-501* | hCG free alpha subunit | 2.1 |

* Radiolabeled detection antibody.
*Note:* 1.0 fmole hCG = 36.7 pg.

assay formats are summarized in Table 21.1. Illustrations showing their application to construct assays for specific fragments were published earlier (10).

## hCG Metabolism

If we are to measure this hormone accurately, it is necessary to take into account the metabolic events that may occur prior to excretion in the urine. Several regions of the hCG beta subunit in the native hormone appear to be vulnerable to early proteolysis. Some reports have described the early proteolytic removal of the beta carboxyterminus in vivo (11, 12), and in separate studies, we found that following limited proteolysis with trypsin, this region is readily removed from the whole hormone (13).

Recently, we and others have found that the region between half-cystines 38 and 57 in the beta subunit also appears to be susceptible to early proteolysis. When the enzyme leukocyte elastase is used to digest hCG, a cleavage occurs between residues 44 and 45 in the beta subunit (14). Also, a so-called nicked form of hCG has been isolated from choriocarcinoma urine, as well as normal pregnancy urine, where the proteolytic nicking occurs between residues 47 and 48 of the beta subunit (15, 16). Thus, we infer that both the lysine at residue 115 and the region of the beta subunit between residues 44 and 49 are more accessible for enzyme attack than other portions of the molecule, since the primary structure contains many other sequences known to be susceptible to cleavage by these enzymes.

While proteolytic cleavage at Lys-115 does not appear to lead to hormone dissociation, genetic engineering experiments have suggested that the absence of the carboxyterminal region of the hCG beta subunit does lead to reduced biological activity of the hormone, presumably due to a shortened lifetime in the circulation (17). Additional support for the role of this region of hCG beta for prolonging hormone circulatory lifetime was shown by the enhanced bioactivity of a recombinant expressed form of FSH with this peptide added to its carboxyterminus (18). The so-called nicking between residues 47 and 48 has been reported to promote alpha and beta subunit dissociation and markedly reduce biological activity (19). It is unclear whether the loss of biological activity is due to subunit dissociation or to a reduction in the action of the whole hormone caused by the nicking.

Thus, it would appear that an early event in hormone metabolism is proteolytic cleavage of the beta subunit; further cleavages are known to give rise to an hCG beta core fragment, which has been isolated and characterized (20, 21). Little is known about the metabolic fate of the alpha subunit. Although it has a smaller molecular weight than the free beta subunit, little is found in pregnancy urine specimens that contain substantial amounts of the hCG beta core fragment, so it is likely that the free alpha subunit is also cleaved to much smaller, immunologically nonreactive fragments. While it has been demonstrated that hCG is degraded during urinary excretion, it is possible that some of this pro-teolytic processing can also occur during hCG synthesis, since tropho-blastic cell culture media has also been shown to contain the hCG beta core fragment (22).

While the proteolytic events and subunit dissociation are known to occur, one must also consider the possibility of metabolic degradation associated with alterations to the carbohydrate side chains. Removal of a terminal sialic acid, for example, will expose the penultimate galactose, which will result in removal of the hormone from the blood after it binds to the asialoglycoprotein receptor (Ashwell receptor) in the liver. This may account for the fact that only a fraction of the hCG released into the circulation is ever excreted in any form in the urine.

We have measured the levels of the whole hormone, the subunits and the beta core fragment in matched blood and urine specimens at various stages of pregnancy. These results are shown in Figure 21.2.

Human CG and hLH have very similar structures, with half-cystines in homologous locations, suggesting that they form homologous disulfide bridges. Both hormones act at the same receptor. Thus, it may be reasonable to infer that hLH structure is very similar to hCG and their metabolism may follow the same pathway, leading to an hLH beta core fragment. Thus far, such a fragment has not been isolated from urine at the time of ovulation or associated with the menopause, but this type of fragment seems likely to be the end product of hLH metabolism. The recent finding that piituitary extracts contain an hLH fragment, similar to the hCG beta core fragment, lends weight to this hypothesis (23).

The stability of these various molecular forms of hCG must be tested, as loss of stability could lead either to loss of immunoreactivity or to further degradation or subunit dissociation. This line of experimentation has been pursued by others (24), and our findings are basically in agree-

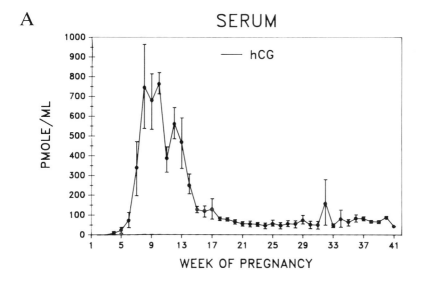

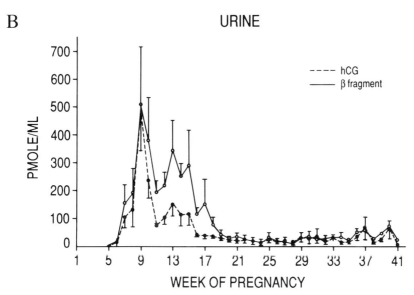

FIGURE 21.2. *A:* A composite profile of hCG values in serum (mean ± SEM) from seven normal term pregnancies. *B:* Corresponding urine values (mean ± SEM) for hCG and hCG beta core fragment in the same subjects collected at the same time.

ment with published reports, namely, that hCG urine specimens can be thawed and refrozen many times without loss of immunological reactivity or dissociation. Intact hCG is stable at room temperature for at least two days. After five days dissociation into free subunits occurs. Incubation at 37° results in dissociation at two days or before. A combination assay exploiting two capture antibodies and one detection antibody will measure native hormone (using B109 capture), as well as free beta subunit and beta core fragment (using B204 as capture). The antibody-antigen complexes are quantified by antibody B108. When a measurement of total hCG immunoreactivity is required, the use of this assay should obviate potential problems associated with thermal dissociation of the subunits.

## The Mikveh Study

In pursuit of a clinical research model to study the urinary excretion of hLH and hCG in women with normal reproductive function, we have entered into a collaboration with the Orthodox Jewish community in upper Manhattan, whose religious practices place particular emphasis on reproductive function. Throughout a normal menstrual cycle, volunteers collected twice-daily (morning and evening) urine specimens that were frozen until assay. One of our goals was to study the metabolic fragments of hLH that appear in the urine at the time of ovulation (future report), and the other was to study the renal clearance of hCG early in pregnancy, since the religious practices of these women result in a high rate of conception.

In the pioneering studies of Marshal et al. (25) and later work of Norman et al. (26), it was shown that, with considerable individual variation, the concentration of hCG in the blood of pregnant subjects was roughly equal to the concentration in the urine. While these studies dealt with the later stages of pregnancy, we sought to enroll these volunteers shortly after conception to measure whether the renal clearance of hCG was the same during the first week of pregnancy as later in the first trimester when the levels are quite high. Our earlier studies had shown that urinary secretion at very low levels was likely to be proportional to the level in blood (27), but more recently, the question of a renal threshold has arisen, possibly leading at times early in pregnancy to blood concentrations greatly in excess of those in the urine. Six of the twenty women became pregnant during the month of study and agreed to a day of hospitalization in the Clinical Research Center for the purpose of measuring the urinary clearance of hCG and its beta core fragment. They then returned for similar measurements later in the first trimester. Our findings enhanced the earlier results, indicating that urine levels of hCG reflect the concentrations in blood even at the very earliest stages of pregnancy.

# Early Pregnancy Loss

It has been appreciated for over a decade that a fertilized ovum can implant briefly, only to be lost as a viable pregnancy. The inference that this occurs has been drawn from data that depict a small rise in urinary hCG concentration (i.e., in the maternal circulation as well), followed by a fall at the onset of menses. The estimates of the frequency with which this occurs vary widely from 8% to 57% of pregnancies (28–30). We collaborated in a study with Dr. Allan Wilcox and his colleagues in which we found that 22% of pregnancies in normal women terminated in this fashion (31).

In the course of reviewing the existing data on this subject, we have developed the belief that many epidemiologic results may be in error because of insufficient sampling or the nature of the assay for hCG. For example, in developing an algorithm that analyzes urinary hCG concentration variations through time in order to detect early pregnancy loss, it is important to know that the "noise" level in urine specimens from the subject prior to implantation is low and also to be sure that the assay is sensitive and specific for a rise and fall in hCG at the time of the expected menses.

We have participated in a variety of epidemiologic research studies, either to perform the assays or to provide supplies of monoclonal antibodies. Some of the recent studies are summarized in Table 21.2. Based on our experience, we conclude that an immunoradiometric assay that detects both the native hCG hormone and its dissociated and partially degraded beta subunit forms is the most useful for this application. For example, in the studies conducted with Dr. Nancy Ellish, this combination assay was approximately 35% more effective at detecting urinary hCG excretion indicative of early pregnancy loss than was the assay that detected only the intact native hormone.

If studies of early pregnancy loss, especially to detect the effect of environmental hazards, are to be placed on a sound scientific basis, it is important to focus attention on the nature of the sampling and the assays for hCG and possibly also to develop a primate model in which to correlate this event with exposure to environmental toxins.

# Cancer

Measurements of hCG are well recognized to be of value in the management of patients with choriocarcinoma and nonseminomatous testicular cancer (1). In addition, Braunstein et al. have shown that a wide variety of other neoplasms may produce detectable levels of hCG (32). With the development of both more sensitive and fragment-specific assays, we have sought to make measurements in urine specimens from populations with

TABLE 21.2. EPL studies.

| Investigator/affiliation | Assay type | Sampling protocol | hCG assay | EPL definition sensitivity |
|---|---|---|---|---|
| Wittaker/Newcastle, UK | hCG RIA | Serum assay, day 25 of cycle | 4 mIU/ml | 16 mIU/ml hCG |
| Miller/Southampton London, UK | hCG RIA | Start day 21 of cycle, alternate days, urine | 1 ug/L | 1 day > 5 ug/L OR, 2 day > 2 ug/L hCG |
| Edmonds/Southampton, UK | SB-6 RIA | Start day 21 of cycle, alternate days, until menses or pregnancy | 10 IU/L | >56 IU/L hCG |
| Wilcox/NIEHS | B-101 R-525* IRMA | 15-day window including 1st 5 days of menses, urine | 0.01 ng/ml | 3 consecutive days above 0.025 ng/ml hCG |
| Hembree/Columbia | B-101 R-525* IRMA | Daily–entire cycle, urine (donor insemination) | 0.01 ng/ml | 3 consecutive days above 0.025 ng/ml hCG |
| Landrigan/Marcus/Mt. Sinai, NY | B-109/B-204 B-108* IRMA | First 2 days of menses, urine | 0.05 ng/ml | — |
| Ellish/NYS/CDC | B-109/B-204 B-108* IRMA | Days 9, 10, 11 post midcycle, urine | 0.05 ng/ml | Any 2 of 3 days above 0.25 ng/ml hCG |
| Lippman/McGill Univ. | — | Day 21 thru 1st 3 days of menses, urine | — | — |
| Gray/Johns Hopkins | B-109/B-204 B-108* IRMA | 12-day window, urine | 0.05 ng/ml | 2 consecutive days 0.25 ng/ml hCG |
| Swan/Lasley/Cal. Dept. of Health, Univ. Cal.-Davis | B-109/B-204 B108* IRMA to screen<br>B-109 B108* IRMA to confirm | 15-day window including 1st 5 days of menses, urine | 0.05 ng/ml | 0.25 ng hCG/mg creatinine any 3 of 4 days during the window |

* Radiolabeled detection antibody.

TABLE 21.3. Sensitivity and specificity of intact hCG and hCG beta fragment assays for the detection of lung cancer.

| Sex | N | Assay | fmol/mg creatinine cutoff point | Sensitivity | Specificity |
|---|---|---|---|---|---|
| Female | 43 | Intact hCG | 4.68 | 52.9 | 77.8 |
| | | Beta fragment | 3.51 | 76.5 | 86.1 |
| Male | 34 | Intact hCG | 1.63 | 65.1 | 57.2 |
| | | Beta fragment | 1.53 | 69.8 | 68.5 |

lung cancer and with ovarian cancer. We are particularly attracted to the hypothesis that hCG, synthesized ectopically by a malignant cell, may be degraded to the beta core fragment by intracellular processes, leading to an altered ratio of beta core fragment to native hCG in the urine specimens of patients with malignancy.

Our results for the lung cancer study are listed in Table 21.3. In both lung and ovarian (not shown) cancer it would appear that the tumor-bearing patient is, indeed, excreting an increased amount of the hCG beta core fragment in comparison with age- and sex-matched controls. However, the data overlap with nonmalignant controls so much that this is not a reliable marker for either disease.

Our early findings have led us to the conclusion that any strategy for usefully measuring hCG or its fragments as produced by neoplasms requires that one be able to circumvent the increases in the pituitary synthesis of glycoprotein hormones that occurs at the menopause. While hLH secretion is greatly increased in this state of ovarian failure, numerous reports have documented that hCG secretion also occurs in the pituitary (33–35). While some of the pituitary hCG carbohydrate side chains may terminate in sulfate, rather than sialic acid, the low levels of hCG in menopausal urine specimens make this difficult to determine by chemical analysis. It is likely that pituitary synthesis of hCG or its fragments may be suppressed in the menopause either by estrogen administration or by GnRH inhibitors. All these avenues of clinical research deserve to be explored.

*Acknowledgment.* This work was supported by Grants HD-15454, HD-28632, and RR-00645.

## References

1. Hussa RO. The clinical marker hCG. New York: Praeger, 1987.
2. Vaitukaitis JL, Braunstein GD, Ross GT. A radioimmunoassay which specifically measures human chorionic gonadotropin in the presence of human luteinizing hormone. Am J Obstet Gynecol 1992;113:751–8.

290    R.E. Canfield et al.

3. Vaitukaitis JL. Immunologic and physical characterization of human chorionic gonadotropin (hCG) produced by tumors. J Clin Endocrinol Metab 1973;37: 505–14.
4. Canfield RE, O'Connor JF, Chen Y, Krichevsky A, Birken S, Wilcox AJ. The clinical utility of the measurement of urinary hCG and its fragments. In: Bellet D, Bidart JM, eds. Structure-function relationships of the gonadotropins. New York: Raven Press, 1989:297–311.
5. Ehrlich PH, Moustafa ZA, Krichevsky A, Birken S, Armstrong EG, Canfield RE. Characterization and relative orientation of epitopes for monoclonal antibodies and antisera to human chorionic gonadotropin. Am J Reprod Immunol Microbiol 1985;8:48–54.
6. Krichevsky A, Armstrong EG, Schlatterer J, et al. Preparation and characterization of antibodies to the urinary fragment of the human chorionic gonadotropin beta subunit. Endocrinology 1988;123:584–93.
7. Bidart JM, Troalen F, Lazar V, et al. Monoclonal antibodies to the free beta subunit of human chorionic gonadotropin define three distinct antigenic domains. Endocrinology 1992;131:1832–40.
8. Schwarz S, Krude S, Wick G, Berger P. Twelve of fourteen surface epitopes of receptor-bound human chorionic gonadotropin being antibody-inaccessible suggest an extensive involvement of the long extracellular domain of the hCG receptor. Mol Cell Endocrinol 1991;82:71–9.
9. Schwarz S, Krude H, Klieber R, et al. Number and topography of epitopes of human chorionic gonadotropin (hCG) are shared by desialylated and deglycosylated hCG. Mol Cell Endocrinol 1991;80:33–40.
10. O'Connor JF, Schlatterer JP, Birkin S, et al. Development of highly sensitive immunoassays to measure human chorionic gonadotropin, its beta subunit, and beta core fragment in the urine: application to malignancies. Cancer Res 1988;48:1361–6.
11. Amr S, Wehmann RE, Birken S, Canfield RE, Nisula BC. Characterization of a carboxyterminal peptide fragment of the human choriogonadotropin beta subunit excreted in the urine of a woman with choriocarcinoma. J Clin Invest 1983;71:329–39.
12. Amr S, Rosa C, Birken S, Canfield R, Nisula B. Carboxyterminal peptide fragments of the beta subunit are urinary products of the metabolism of desialylated human choriogonadotropin. J Clin Invest 1985;76:350–6.
13. Birken S, Canfield R, Agosto G, Lewis J. Preparation and characterization of an improved beta COOH-terminal immunogen for generation of specific and sensitive antisera to human chorionic gonadotropin. Endocrinology 1982; 110:1555–63.
14. Birken S, Gawinowicz MA, Kardana A, Cole LA. The heterogeneity of human chorionic gonadotropin, II. Characteristics and origins of nicks in hCG reference standards. Endocrinology 1991;129:1551–8.
15. Bidart JM, Puisieux A, Troalen F, Foghietti MJ, Bohuon C, Bellet D. Characterization of the cleavage product in the human choriogonadropic beta-subunit. Biochem Biophys Res Commun 1988;154:626–32.
16. Kardana A, Elliot MM, Gawinowicz MA, Birken S, Cole LA. The heterogeneity of human chorionic gonadotropin (hCG), I. Characterization of peptide heterogeneity in 13 individual preparations of hCG. Endocrinology 1991;129:1541–50.

17. Matzuk MM, Hsueh AJ, Lapolt P, Tsafriri A, Keene JF, Boime I. The biological role of the carboxyl-terminal extension of human chorionic gonadotropin beta-subunit. Endocrinology 1990;126:376–83.
18. LaPolt PS, Nishimori K, Farer FA, Perlas E, Boime I, Hsueh AJ. Enhanced stimulation of follicle maturation and ovulatory potential by long acting follicle-stimulating hormone agonists with extended carboxyl-terminal peptides. Endocrinology 1992;131:2514–20.
19. Cole LA, Kardana A, Park S-Y, Braunstein GD. The deactivation of hCG by nicking and dissociation. J Clin Endocrinol Metab 1993;76:704–14.
20. Birken S, Armstrong EG, Gawinowicz-Kolks MA, et al. Structure of the human chorionic gonadotropin beta-subunit fragment from pregnancy urine. Endocrinology 1988;123:572–83.
21. Blithe DL, Akar AH, Wehmann RE, Nisula BC. Purification of beta-core fragment from pregnancy urine and demonstration that its carbohydrate moieties differ from those of native human chorionic gonadotropin beta. Endocrinology 1988;122:173–80.
22. Cole LA, Birken S. Origin and occurrence of human chorionic gonadotropin beta-subunit core fragment. Mol Endocrinol 1988;2:825–30.
23. Birken S, Chen Y, Gawinowicz MA, Canfield RE, Hartree AS. The structure and significance of hLH beta core fragment purified from human pituitary extracts. Endocrinology 1993.
24. deMedeiros SF, Amato F, Norman RJ. Stability of immunoreactive beta-core fragment of hCG. Obstet Gynecol 1991;77:53–9.
25. Marshall JR, Hammond CB, Ross GT, Jacobson A, Rayford P, Odell WD. Plasma and urinary chorionic gonadotropin during early human pregnancy. Obstet Gynecol 1968;32:760.
26. Norman RJ, Buck RH, Rom L, Joubert SM. Blood or urine measurement of human chorionic gonadotropin for detection of ectopic pregnancy? A comparative study of quantitative and qualitative methods in both fluids. Obstet Gynecol 1988;71:315–8.
27. Armstrong EG, Ehrlich PH, Birken S, et al. Use of a highly sensitive and specific immunoradiometric assay for detection of human chorionic gonadotropin in urine of normal, nonpregnant and pregnant individuals. J Clin Endocrinol Metab 1984;59:867–73.
28. Wittaker PG, Taylor A, Lind T. Unsuspected pregnancy loss in healthy women. Lancet 1983;1:1126–7.
29. Miller JF, Williamson E, Glue J, Gordon YB. Fetal loss after implantation. Lancet 1980;2:554–6.
30. Edmonds DK, Lindsay KS, Miller JF, Williamson E, Wood PJ. Early embryonic mortality in women. Fertil Steril 1982;38:447–53.
31. Wilcox AJ, Weinberg CR, O'Connor JF, et al. Incidence of early pregnancy loss. N Engl J Med 1988;31:189–94.
32. Braunstein GD, Vaitukaitis JL, Carbone PP, Ross GT. Ectopic production of human chorionic gonadotropin by neoplasms. Ann Intern Med 1973;78:39–45.
33. Hoermann R, Spoettl G, Moncayo R, Mann K. Evidence for the presence of human chorionic gonadotropin (hCG) and free beta-subunit of hCG in the human pituitary. J Clin Endocrinol Metab 1990;71:179–86.

34. Odell WD, Griffin J, Bashey HM, Snyder PJ. Secretion of chorionic gonadotropin by cultured human pituitary cells. J Clin Endocrinol Metab 1990;71:
1318–21.
35. Hammond E, Griffin J, Odell WD. A chorionic gonadotropin secreting
human pituitary cell. J Clin Endocrinol Metab 1991;72:747–54.

# 22

# β-Core Fragment: Structure, Production, Metabolism, and Clinical Utility

Glenn D. Braunstein

The β-core fragment of *human chorionic gonadotropin* (hCG) is a biologically inactive, small molecular weight molecule composed of the core portion of hCGβ-subunit. The fragment appears to be a major degradation production of hCG and hCGβ metabolism and is found in large amounts in the urine of pregnant women and some patients with malignancies. Smaller quantities are found in the blood and urine of nonpregnant normal individuals and patients with benign disorders. This chapter presents a historical perspective concerning the discovery of this molecule and discusses its purification, structure, origin, and clinical applications.

## Historical Perspective

The β-core fragment was first described by Matthies and Diczfalusy in 1968 (1). These investigators carried out a series of studies to examine the differences between the hCG present in placental tissue, amniotic fluid, maternal serum, and the urine of pregnant women and women with gestational trophoblastic tumors. They chromatographed the various materials through a Sephadex G-100 column and measured hCG in the effluent tubes with both a hemagglutination-inhibition assay and a bioassay (weight increment of the accessory reproductive organs of immature male rats). They noted that all the materials contained immunoreactive hCG that was slightly retarded, with an apparent molecular size similar to serum albumin, and that this material was biologically active.

In addition, they found that the urine from pregnant women and women with chorionic tumors contained a highly retarded, immunologically active, but biologically inactive, form of hCG (Fig. 22.1). This material had a molecular weight of slightly less than 14,000 and reacted

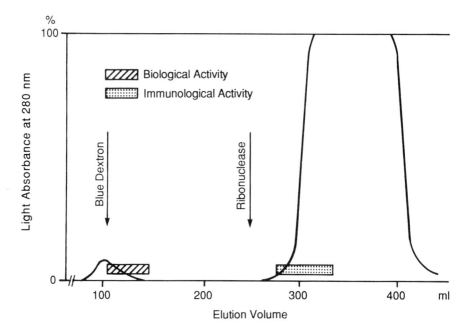

FIGURE 22.1. Sephadex G-100 chromatography pattern of pregnancy urine. Protein concentration was measured by light absorbance, hCG immunologic activity by hemagglutination inhibition assay, and hCG biological activity by rat accessory reproductive organ weight increase. Note that the high molecular weight immunologically and biologically active fraction elutes just after blue dextran and the low molecular immunologically active but biologically inactive form of hCG elutes following ribonuclease. Reprinted with permission from Matthies and Diczfalusy (2).

slightly different from the high molecular weight, biologically active hCG in their assay, suggesting that the low molecular weight substance was structurally different. They were unable to demonstrate the material in urine extracts from nonpregnant females or males, but showed that it was present in 100% of the urine samples from pregnant women and in commercial preparations of hCG. They also noted that following the injection of highly purified hCG into nonpregnant females, their urine contained the small molecular weight fraction. Finally, when they incubated the isolated low molecular hCG material with serum from nonpregnant women or men, they noted a shift of the immunoreactivity to the high molecular weight fraction of the chromatogram. These workers concluded that the small molecular weight form of immunoreactive hCG represented a renal degradation production of hCG metabolism that retained some but not all of the antigenic recognition sites of their polyclonal antiserum (1, 2).

In 1972, Franchimont et al. performed similar studies on serum and urine from pregnant women using relatively specific immunoassays for hCG, hCGα, and hCGβ (3). The biologic fluids were chromatographed on either Sephadex G-100 or G-200 and the effluent tubes examined with each of the assays and a bioassay. Pregnancy serum was shown to contain biologically active hCG and free hCGα-subunit, while pregnancy urine contained immunologically and biologically active hCG, free hCGα, and an immunologically active, but biologically inactive, form of hCGβ. The latter gave an incomplete cross-reaction in the hCGβ RIA. These investigators concluded that the small molecular weight form was a catabolic fragment of hCG (3).

In a series of studies, Vaitukaitis studied the different forms of hCG present in placental extracts and the plasma, urine, and tumor extracts of patients with gestational and nongestational hCG-secreting tumors (4–6). She noted that placental extracts contained primarily intact hCG and free α-subunit and that there were small quantities of immunoreactive hCG with a low apparent molecular size, which she felt represented incomplete forms of hCG (5). β-core was found in the urine, but not plasma or tumor tissue, of patients with hCG secreting neoplasms (4, 6). The presence of small amounts of β-core in placental extracts, large amounts in pregnancy urine, and undetectable concentrations in pregnancy serum, as measured with the hCGβ radioimmunoassay kit distributed by the National Pituitary Agency, was confirmed by Good and coworkers (7). They raised the possibility that the origin of the urinary β-core fragment was from placental secretion, rather than from peripheral metabolism of hCG or its hCGβ.

Hattori and associates examined the tumor, plasma, and urine from four patients with hCG secreting neoplasms and showed that all the urine samples contained β-core fragment that was not present in plasma or ascitic fluid (8). However, three of the four tumor specimens contained a small amount of β-core, again raising the possibility that β-core represents a secretory product.

A great deal of interest in the β-core fragment was generated by the initial publication of Papapetrou et al. in 1980 (9). These workers examined the ectopic production of hCG by neoplasms and found that 17.1% of 70 patients with malignant disease had immunoreactive hCG in their circulation, whereas 44.3% had immunoreactive hCG in their urine. From the results of column chromatography studies, they showed that the primary moiety in the serum was hCGβ, while the urine contained free hCGβ and the β-core fragment, which they designated "metabolite X". They concluded that this metabolite was a degradation production of hCGβ, probably of renal origin, a conclusion they also reached in a subsequent study (10). The high prevalence of β-core in the urine of patients with cancer was confirmed by several groups, who measured the fragment directly in urine samples through the use of various immu-

noassays, obviating the need for prior separation of the various immunoreactive moieties through column chromatography (11–23).

The characterization of the β-core fragment was primarily performed
by three major groups of investigators. Wehmann, Nisula, and Blithe
initially examined the origin of β-core as part of their studies concerning
the metabolism of hCG and its subunits (24–33). They showed the
appearance of β-core in the urine following infusion of purified hCG and
hCGβ, the rapid metabolic urinary clearance rates of the fragment, the
generation of the fragment in rat liver and kidney tissue, and described
the carbohydrate composition of purified β-core. The group led by Birken
and Canfield purified β-core and characterized its amino acid structure,
sequence, and carbohydrate composition (34, 35), while the research
team of Nishimura and associates concentrated their efforts on the carbohydrate composition of β-core and were able to clearly demonstrate the
heterogeneity of the N-linked sugar chains on the fragment (36, 37).

The first specific immunoassays for β-core were developed by Akar et
al. (38) and Krichevsky et al. in 1988 (39, 40), the former developing
a radioimmunoassay with a polyclonal antibody, and the latter group
creating radioimmunometric, two-site assays for β-core. These groups
subsequently improved on their assays through the generation of better
antibodies (33, 41). Sensitive monoclonal antibody-based immunoassays
were also developed by Kardana et al. (42), Alfthan and Stenman (43),
and deMedeiros et al. (44). These assays are now sufficiently sensitive to
detect the quantities of β-core fragment present in serum, urine, and
other biologic fluids of pregnant women, as well as in the urine of men
and nonpregnant women.

## Purification

Several purification procedures for β-core have been developed. Most use
partially purified urinary hCG obtained from commercial sources (e.g.,
Organon, Diosynth Division, Oss, and Holland) or acetone precipitates
of fresh pregnancy urine as starting material. Blithe et al. performed
Sephadex G-100 column chromatography of crude commercial hCG,
followed by *concanavalin* A (Con A)-Sepharose chromatography, with
elution of the bound β-core with 0.5 M α-methyl-D-mannoside. The
eluate was percolated through a DEAE-Sephacel column, and the
unbound material was further purified on superfine Sephadex G-75 (29).
Kato et al. also used sequential chromatographic procedures to purify the
β-core from crude, commercial urinary hCG (45). They began with
DEAE-Sephadex A-50 chromatography, followed by Con A-Sepharose
chromatography of the unbound fraction. The Con A adsorbed fraction
was eluted with 0.2 M α-methyl-D-mannoside and chromatographed on a

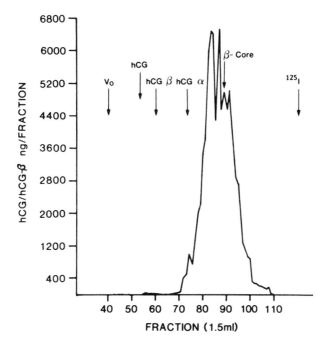

FIGURE 22.2. Sephadex G-100 chromatography of partially purified β-core fragment. Commercial urinary hCG was chromatographed on DEAE-Sephadex A-50, with the unbound fraction being subjected to concanavalin A Sephadex chromatography. The Con A-adsorbed fraction was eluted and applied to a 1.6 × 85 cm column of Sephadex G-100. The void volume (Vo) was determined by blue dextran, and $^{125}I$ was used to determine the total volume of the gel. Reprinted with permission from Kato, Kelley, and Braunstein (45).

Sephadex G-100 column. This procedure results in the production of large quantities of highly purified β-core (Fig. 22.2).

Birken et al. also began with crude urinary hCG obtained from Organon (35). The hCG was chromatographed on Sephadex G-100, followed by immunoabsorption with monoclonal antibody B201 conjugated to Sepharose. Antibody B201 binds both intact hCGβ and β-core fragment (35). β-core was eluted from the immunoabsorbant column with acetic acid, neutralized, lyophilized, and further purified by fast protein liquid chromatography gel filtration on Superose-12. This procedure results in a highly purified β-core preparation that contains both monomeric and dimeric forms (35).

Kardana et al. chromatographed Organon hCG on Sephadex G-100, followed by concentration by ultrafiltration with an Amnicon YM-5 membrane (molecular weight cutoff 5,000). Further purification was accomplished with immunoabsorption by an anti-hCGβ mouse monoclonal antibody (W14) covalently linked to cyanogen bromide-activated Sepharose CL/4B. The bound β-core fragment was eluted with 3 M ammonium thiocyanate and desalted by chromatography over Sephadex G-25 (19).

Lee et al. began with crude urinary hCG obtained from Sigma (St. Louis, MO), which was chromatographed on Sephadex G-100, followed by Sephadex G-75 (superfine) and Sephadex G-50 chromatography of the small molecular weight material retaining β-core immunoreactivity (21). Endo et al. purified the β-core from 10 L of urine obtained from a woman in her first trimester of pregnancy. The immunoreactive material was precipitated by acetone, followed by gel filtration on Sephadex G-100 (superfine), immunoabsorption of the active fractions through a column of monoclonal antibody 229, which reacts with both hCGβ and β-core fragment, conjugated to Sepharose. β-core fragment was eluted with acetic acid, and the eluate was neutralized, lyophilized, and further purified by fast protein liquid chromatography gel filtration on Superose 12 (37).

## Structure and Activity

On Sephadex or Superose gel chromatography, immunoreactive β-core migrates in the molecular weight range of 12,000–17,500 (2, 19, 24, 29, 35, 37, 46–48). Under nonreducing conditions β-core has a molecular weight of 12,000–17,000 on *sodium dodecyl sulfate-polyacrylamide gel electrophoresis* (SDS-PAGE) and under reducing conditions separates into two or three bands with molecular weights of 3000–6000 and 8000–13,000 (27, 29, 35, 49). The molecular weight determined by structural analysis is 10,479 (35).

Birken et al. sequenced β-core and showed that it contained hCGβ amino acids 6–40 disulfide bridged to amino acids 55–92 (35) (Fig. 22.3). The molecule contains N-glycosylation sites on Asn[13] and Asn[30].

Several groups have studied the carbohydrate composition of β-core fragment. Birken et al. noted the presence of 6 *mannose* (MAN), as well as 8 *N-acetylglucosamine* (GlcNAc), per β-core molecule, but did not detect *sialic acid* (NeuAc) or *fucose* (Fuc), with a low but variable content of *galactose* (Gal) (35). Blithe et al. initially studied the carbohydrate composition by both chemical analysis and lectin binding and concluded that β-core did not contain appreciable sialic acid or galactose, but did retain core fucose (29). Using direct compositional analysis, these investigators subsequently concluded that the N-linked oligosaccharides on the β-core resembled the N-linked structures of hCGβ carbohydrate, with the antennary sialic acid, galactose, and N-acetylglucosamine missing (30).

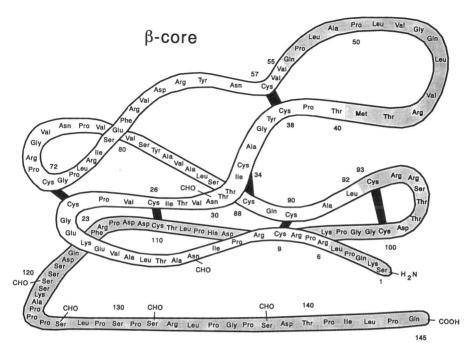

FIGURE 22.3. Amino acid sequence of β-core fragment, representing residues 6–40 disulfide bridged to residues 55–92 of hCGβ. The sequences of hCGβ that are absent in the β-core fragment are depicted in the shaded regions. The disulfide bridges are represented by black bars.

Recently Endo et al. found that the Asn-linked carbohydrates on β-core were actually a mixture of four oligosaccharides: Manα1→6(±Man α1→3)Manβ1→4GlcNAcβ1→4(±Fucα1→6)GlcNAc (Fig. 22.4) (37). Thus, during the formation of β-core, the terminal extensions of the biantennary complex containing the NeuAcα2→3Galβ1→4GlcNAcβ1→2 sequence is cleaved from the N-linked carbohydrate on hCGβ. Based on similarity in oligosaccharide structure with some lysosomal glycoproteins, Endo et al. suggested that degradation of hCG and/or hCGβ takes place within lysosomes that contain an α-mannosidase that cleaves only the Manα1→3Man linkage of the trimannosyl core portion of N-linked sugar chains (37). Since β-core does not contain the carboxyterminal portion of hCGβ, none of the O-serine-linked oligosaccharides normally located on Ser[125], Ser[127], Ser[132], and Ser[138] are present (29, 30, 35, 37).

The β-core fragment is unable to combine with the α-subunit of hCG (35), does not bind to hCG receptors (34, 46), and is devoid of biological activity in in vitro bioassays (34, 47). It exhibits a very rapid half-life in the circulation and high metabolic clearance rate in comparison to hCG, hCGβ, or hCGα (Table 22.1) (25, 26, 31, 32).

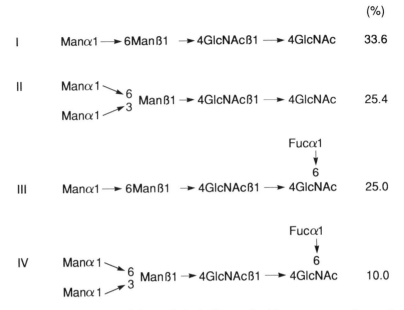

(%)

I    Manα1 ⟶ 6Manβ1 ⟶ 4GlcNAcβ1 ⟶ 4GlcNAc    33.6

II   Manα1 ⟍6
             Manβ1 ⟶ 4GlcNAcβ1 ⟶ 4GlcNAc    25.4
     Manα1 ⟋3

                                    Fucα1
                                      ↓
                                      6
III  Manα1 ⟶ 6Manβ1 ⟶ 4GlcNAcβ1 ⟶ 4GlcNAc    25.0

                                    Fucα1
                                      ↓
                                      6
IV   Manα1 ⟍6
             Manβ1 ⟶ 4GlcNAcβ1 ⟶ 4GlcNAc    10.0
     Manα1 ⟋3

FIGURE 22.4. Structures of the N-linked oligosaccharides present on β-core fragment with percent molar ratios. Reprinted with permission from Endo, Nishimura, Saito, et al. (37), © The Endocrine Society, 1992.

TABLE 22.1. Comparison of kinetic parameters and clearance rates of hCG, hCGβ, and hCGα in man.

|  | hCG | hCGβ | hCGα | β-core |
|---|---|---|---|---|
| T1/2 |  |  |  |  |
| Fast | $5.97 \pm 0.63$ h | $0.68 \pm 0.03$ h | $0.22 \pm 0.02$ h | $3.49 \pm 0.68$ min |
| Slow | $35.6 \pm 8.0$ | $3.93 \pm 0.68$ | $1.27 \pm 0.32$ | $22.4 \pm 4.2$ |
| $V_d$, ml/m$^2$ | $1665 \pm 124$ | $1958 \pm 131$ | $1729 \pm 99$ | $1950 \pm 156$ |
| MCR, ml/min/m$^2$ | $1.88 \pm 0.08$ | $19.0 \pm 0.7$ | $49.7 \pm 1.6$ | $192 \pm 8$ |
| RCR, ml/min/m$^2$ | $0.04 \pm 0.02$ | $0.13 \pm 0.02$ | $0.16 \pm 0.02$ | $13.7 \pm 1.4$ |
| UCR/MCR × 100% | $21.7 \pm 1.4$ | $0.68 \pm 0.11$ | $0.33 \pm 0.04$ | $7.05 \pm 0.55$ |

*Note:* T1/2 = serum half-life; h = hours; min = minute; $V_d$ = apparent initial volume of distribution; MCR = metabolic clearance rate; and UCR = urine clearance rate.
*Source:* Data are from Wehmann and Nisula (25) and Wehmann, Blithe, Flack, and Nisula (31).

## Origin of β-Core

An area of intense controversy concerns the origin of the β-core fragment. Some data exist that support an origin from both direct placental or tumor tissue production of β-core, while a substantial body of evidence supports the origin of β-core from the metabolic degradation of hCG and hCGβ, especially by renal tissue. This evidence is reviewed and a proposed unifying scheme presented.

## β-Core as a Tissue Secretory Product

The β-core fragment has localized to the syncytiotrophoblast layer from first-trimester placentas by immunohistochemical techniques (19). It also has been detected in placental extracts (5, 7, 49) and has been found in culture media from first-trimester placental organ cultures (49). Cole and Birken noted that hCG, hCGβ, and β-core fragment were released in a parallel fashion in placental organ culture and that in six of eight separate placentas the β-core level was greater than the hCG or the hCGβ concentrations. These investigators also noted that the highest concentrations of β-core were found in second-trimester urine, whereas the highest concentrations of hCG and hCGβ were found in first-trimester urine or serum. Therefore, they concluded that the β-core levels in the urine do not reflect the levels of hCG or hCGβ in a simple precursor-product relationship (49). Similarly, deMedeiros measured hCG and β-core in the urine and noted the same qualitative pattern of secretion, but different ratios of the two moieties, which led them to conclude that either there were changes in renal or tissue metabolism of hCG leading to β-core formation at different stages of pregnancy or there was placental production as well as peripheral conversion during pregnancy (50).

β-core fragment is found in pregnancy serum (31, 33, 43, 50), and there is a correlation between the plasma and urine β-core levels (33), suggesting that either β-core is directly secreted by placental tissue or represents leakage from liver, kidney, and possibly other tissues following formation of β-core from the degradation of hCG or hCGβ. The quantitative levels of β-core in the circulation are quite low, and this has been attributed to the very rapid half-life of the molecule (31, 34). However, recent studies by Kardana and Cole have suggested that β-core is in the circulation, masked by macromolecules that can be uncovered by treatment of the serum with 3 M ammonium thiocyanate (48). Indeed, deMedeiros et al. have recently confirmed the presence of a high molecular weight-associated form in pregnancy serum, as well as in the media from cultured first-trimester dispersed placental tissue maintained in short-term tissue culture and in amniotic fluid, granulosa cell fluid, and seminal plasma (51). These investigators were able to convert the high molecular weight form of β-core to the free β-core form through the use of 3 M ammonium thiocyanate.

As noted previously, the early studies of Matthies and Diczfalusy suggested the presence of a serum protein that bound β-core, giving it an apparently higher molecular weight (2). In contrast, Alfthan and Stenman were unable to confirm the presence of a high molecular weight β-core complex in pregnancy or trophoblastic cancer sera (22). Similarly, Wehmann et al. were unable to demonstrate substances present in pregnancy serum that masked β-core or produced a high molecular weight complex (33).

A number of studies have also suggested that nontrophoblastic tumors may secrete β-core. Kardana and associates used both a polyclonal (AK12) and monoclonal (2C2) antiserum to immunostain hydatidiform mole and choriocarcinoma tissues, as well as a variety of nontrophoblastic tumors, including squamous cell carcinoma of the cervix, clear cell carcinoma of the ovary, gastric carcinomas, and bronchial alveolar carcinomas of the lung (19). β-core immunoreactivity was also documented in parallel samples of tissue homogenates, acetic fluid, and urine from five women with ovarian carcinoma (11). Finally, several investigators have noted that urine may contain substantial quantities of β-core with undetectable serum levels of hCG or hCGβ (8, 9, 15). This finding could be the consequence of the high metabolic clearance rate of β-core, which would result in rapid clearance from the circulation of any β-core that is produced. Alternatively, if β-core were associated with other proteins to form a macromolecular complex that masks the β-core immunoreactive epitopes, β-core levels may not be detected in the serum, although the material is present.

## β-Core Origin from hCG or hCGβ Degradation

There is a substantial body of evidence indicating that β-core represents a peripheral degradation product of hCG and hCGβ metabolism. First, β-core structurally appears to be a degradation product of hCGβ. It is approximately one-third the size of hCGβ, is missing the carboxyterminal portion of hCGβ, suggesting proteolytic cleavage, and contains Asp-linked carbohydrates that are missing the terminal carbohydrate groups present in hCGβ, suggesting intralysosommal degradation (35, 37, 38).

Second, there is direct evidence in humans and animals that hCG, hCGβ, and asialo-hCG are degraded to β-core. Following the infusion of hCG, hCGβ, or asialo-hCG into nonpregnant humans, β-core appears in the urine in a precursor-product time sequence (Fig. 22.5) (24, 26, 32, 51, 52). Following a single intravenous injection of 1 mg of highly purified hCGβ into normal subjects, Wehmann et al. noted that immunoreactive hCGβ disappeared rapidly from the circulation and that the excretion of hCGβ immunoreactivity in the urine declined over the first 8 h in proportion to the serum level. However, the hCGβ immunoreactivity (mostly representing β-core) in the urine collected 12–24 h following the injection contained amounts that were out of proportion to the serum concentration, indicating the delayed appearance of the β-core fragment (24, 26). Indeed, this may explain why there is a discrepancy between serum and urine levels of hCG and β-core excretion (50), as the

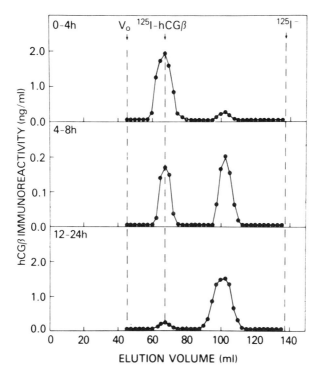

FIGURE 22.5. Elution profile of hCGβ immunoreactivity following Sephadex G-100 chromatography of urine collected over time from a normal female subject following an intravenous injection of 1.0 mg highly purified hCGβ. The predominant immunoreactive form in the 0–4 h collection represents hCGβ, while the predominant form present in urine collected between 12 and 24 h represents β-core. The urine collected between 4 and 8 h shows approximately equal concentrations of hCGβ and β-core. Reprinted with permission from Wehmann and Nisula (24), © The Endocrine Society, 1980.

metabolism of hCG and release of β-core may be delayed. This is also supported by the finding that following pregnancy resulting from artificial insemination, intact hCG appears in the urine 1–2 days before the appearance of β-core in the same samples (50). There is direct evidence that rat renal tissue degrades hCG or hCGβ to β-core (28) and that rat liver degrades hCG or asialo-hCG to β-core (27). Lefort et al. suggested that the hepatic degradation of hCG probably takes place within hepatic macrophages (27). Human granulosa cells are also capable of degrading intact hCG into β-core (51).

Finally, the measured levels of β-core in untreated serum or plasma are too low to account for the levels of β-core measured in the urine. Using the serum concentrations of β-core and the metabolic clearance rate of the fragment, Wehmann et al. concluded that approximately 1% of the urinary β-core excreted during pregnancy is derived from the clearance of β-core released into the blood stream from the placenta (31, 33). Alfthan and Stenman also concluded that the majority of β-core in pregnancy urine is formed in the urinary tract and that the relatively low serum concentrations of β-core could possibly be formed in the circulation itself through the action of proteases activated as part of coagulation and fibrinolysis, although this does not occur in vitro since there are no differences between serum and plasma β-core levels (43).

## Proposed Pathways for Generation of β-Core

Based on the above considerations, as well as recent information concerning the high prevalence of nicks in the hCGβ subunit, the pathways depicted in Figure 22.6 are proposed. The trophoblast synthesizes both α- and β-subunits from separate messenger RNAs, and following trimming of the carbohydrate moieties, the mature subunits combine noncovalently to form intact hCG (α:β in Fig. 22.6) (53). A form of α-subunit containing extra carbohydrate (Big α), which does not combine with free β-subunit, is also produced by the trophoblast and secreted into the blood-

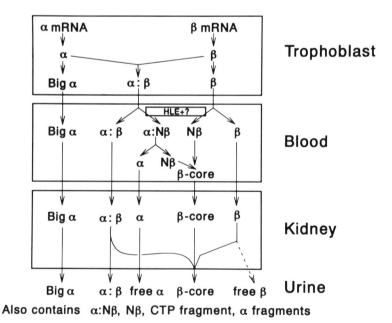

FIGURE 22.6. Proposed pathways for production of β-core and other hCG fragments found in the urine. See text for details. α:β = intact hCG; α:Nβ = hCG with nicked β-subunit; Nβ = free nicked β-subunit; and CTP fragment = carboxyterminal fragment.

stream and excreted into the urine. The intact hCG molecule is secreted and appears in the bloodstream in an intact form, some of which is cleared by the kidneys and appears unchanged in the urine (approximately 22% [25]), while much of it undergoes degradation in the kidney to β-core. Similarly, a small amount of free β-subunit is directly secreted by the trophoblast, but less than 0.7% appears in the urine unchanged (Table 22.1) (25).

A large portion of the secreted hCG, as well as free β-subunit, undergoes proteolytic cleavage in the exposed hydrophobic loop between $Cys^{38}$ and $Cys^{57}$. These nicks primarily occur at $Val^{44}$-$Leu^{45}$ and $Gly^{47}$-$Val^{48}$ (54–61). The protease *human leukocyte elastase* (hLE) is an enzyme that cleaves peptide bonds following glycine, valine, and leucine in hydrophobic regions and is present in high concentrations in leukocytes and macrophages (60). Therefore, it is likely that the production of nicks takes place within placental macrophages, as well as in the circulation, through the action of hLE. Since incubation of hCG or hCGβ with hLE initially led to nicks at $Val^{44}$-$Leu^{35}$ followed by nicks at $Ala^{51}$-$Leu^{52}$, $Leu^5$-$Arg^6$, and $Val^{48}$-$Leu^{49}$ (60), another proteolytic enzyme appears to be responsible for the cleavage at $Gly^{47}$-$Val^{48}$. Although nicked hCGβ easily combines with hCGα, the combined form (α:Nβ) disassociates rapidly in serum to form free hCGα and free nicked hCGβ (Nβ) (62). It is likely that nicked hCGβ serves as the precursor for further proteolytic and glycosidase digestion within the circulation and the kidney.

Based on the metabolic clearance rate and urinary clearance rate parameters derived from normal men and women who received a bolus injection of highly purified β-core, only 7% of the β-core in the plasma is excreted in the urine (Table 22.1) (31). This, together with the data concerning the low serum concentrations of β-core during pregnancy, indicates that most of the β-core in the urine is the result of renal metabolism of hCG and hCGβ. Following the injection of labeled hCG into male rats, the hormone is taken up through endocytosis and transported into lysosomes via large apical vesicles (63). Within the lysosomes, the hCG or hCGβ undergoes proteolytic cleavage at the amino-terminal, carboxyterminal, and hydrophobic loop regions, as well as glycosidase trimming of the biantennary extensions of the N-linked carbohydrates.

The above describes one proposed pathway for β-core production. Other pathways for the formation of β-core undoubtedly exist, as has been suggested by studies carried out with asialo-hCG, which also leads to generation of β-core following infusion in normal humans (52). Based on the kinetic parameters of desialylated hCG, it is clear that that molecule is handled differently by the kidney than fully sialylated hCG (32). Considering the complexity of various metabolic pathways for hCG, its subunits, and intermediary fragments, it is not surprising that the urine contains a myriad of different molecular forms, including intact hCG (α:β), free α, Big α, β-core fragment, hCGβ, intact hCG composed of α-subunit noncovalently linked to nicked β-subunit (α:Nβ), free nicked hCGβ-subunit (Nβ), *carboxyterminal fragments* (CPT fragments) and α fragments (Fig. 22.6).

## Origin of β-Core Present in the Nonpregnant State

In addition to pregnancy serum and urine, immunoreactive β-core has been found in the urine of normal males (19, 22, 38, 41, 64), premeno-pausal women (21–23, 38, 40, 41, 44, 64), and postmenopausal women (21–23, 38, 40, 41, 44, 51, 54, 65, 66). Three separate preparations of human menopausal gonadotropins also have been found to contain im-munoreactive β-core (65). Small quantities of β-core also are found in the circulation of healthy men and women (22, 64). It is unclear whether the source of the β-core reflects a metabolic degradation product of the hCG-like substance found in a variety of tissues, especially the pituitary (67, 68), or a metabolic degradation product of luteinizing hormone (LH-core) (65, 66). In this regard, Iles et al. performed electrophoresis and Western blotting of concentrated postmenopausal urine and demon-strated that the material migrating with the same molecular size as β-core was recognized by an anti-LH antibody, but not by a monoclonal antibody against the hCGβ-core fragment (66). Only structural studies on highly purified β-core derived from nonpregnant individuals will fully settle the issue as to whether the origin is from hCG or an hCG-like material or luteinizing hormone.

## Assays

The earliest assay used to measure β-core was the hemagglutination-inhibition assay applied to pregnancy urine that had been fractionated on Sephadex G-100 (1, 2). From 1972 until 1978 the β-core fragment was separated from intact hCG and hCGβ by column chromatography and was measured by radioimmunoassay using the National Pituitary Agency's anti-hCGβ serum, SB-6, which reacted to a hCGβ-core conformational determinant exposed on β-core, hCG, and hCGβ. In 1988, two im-munoassays were developed that were based on antibodies generated against purified β-core and were therefore relatively specific for β-core (38, 40). Subsequently, many investigators and two commercial corpor-ations described β-core assays (11, 14, 33, 39, 41–44, 64, 69). The sen-sitivities and specificities of these assays are compared in Table 22.2.

There are three caveats that are worth mentioning concerning the assay of β-core. First, β-core is highly stable in urine and is not affected by storage at 4° for 3 weeks or at −20° for 6 months, freeze-thaw cycles, pH changes between pH 5–8 and does not require the addition of preserva-tives (70). A second point is that there is a nonspecific matrix effect of

TABLE 22.2. β-core immunoassays.

| Author (ref) | Year | Type of assay | Antibody-1 | Antibody-2 | Sensitivity | hCG | hCGβ | hLH |
|---|---|---|---|---|---|---|---|---|
| | | | | | | \multicolumn{3}{c}{Cross-reaction (%)[1]} | | |

| Author (ref) | Year | Type of assay | Antibody-1 | Antibody-2 | Sensitivity | hCG | hCGβ | hLH |
|---|---|---|---|---|---|---|---|---|
| Akar et al. (38) | 1988 | RIA | Anti-β-core RW25 | — | 200 pg/ml (2.82 ng/ml)[2] | 0.2 | 0.22 | 0.009 |
| Wehmann et al. (33) | 1990 | RIA | Anti-β-core RW37 | — | 25 pg/ml (100 pg/ml)[2] | 0.2 | 0.4 | — |
| O'Connor et al. (39, 40) | 1988 | IRMA | Anti-hCG B108 | Anti-β-core B204 | 10 pg/ml (50 pg/ml)[2] | <1 | 15–25 | <1 |
| Krichevsky et al. (41) | 1991 | IRMA | Anti-β-core B210 | Anti-β-core B201 | — | — | <0.1% | — |
| Cole et al. (11, 14) | 1988, 1989 | IRMA | Anti-β-core B204 | Anti-β-HCO-514 | 21–200 pg/ml | <0.2 | 7.5 | <0.2 |
| Kardana et al. (42) | 1989 | RIA | Anti-β-core AK12 | — | 200 pg/ml | 5 | 11 | 2.6 |
| | | IRMA | Anti-β-core 2C2 | Anti-β-core AK12 | 400 pg/ml | <1 | 2 | <1 |
| Alfthan and Stenman (43, 64) | 1990 | IFMA | Anti-β-core B204 | Anti-hCG B108 | 4.4 pg/ml | 0.4 | 37 | <0.05 |
| Norman et al. (69) | 1990 | IRMA | Anti-hCGβ Ab2/6 | Anti-hCGβ 2H2 | 23 pg/ml | <0.5 | 100 | — |
| Lee et al. (21) | 1991 | RIA | Anti-β-core S504 | — | 25 pg/ml (100 pg/ml)[2] | 6.9 | 18 | <0.7 |
| deMederios (44) | 1992 | RIA | Anti-β-core DeM3 | — | 50 pg/ml (280 pg/ml)[2] | 3.8 | 10.5 | 7 |
| | | IRMA | Anti-β-core DeM3 | Anti-hCGβ 32H2 | 15 pg/ml (50 pg/ml)[2] | 2.1 | 5.3 | <0.1 |
| Triton UGP EIA kit | 1991 | EIA | Anti-β-core B210 | Anti-hCGβ polyclonal | 5 pg/ml | <0.1 | — | <0.1 |
| Wako kit | | EIA | Anti-β-core MoAb229 | Anti-hCGβ polyclonal | 50 pg/m | — | — | — |

[1] All cross-reactions are based on molar concentrations, except for Akar et al. (38), Wehmann et al. (33), and Kardana et al. (42), who used weight comparisons.

[2] Clinical sensitivity taking into account assay matrix effect.

*Note:* RIA = radioimmunoassay; IRMA = immunoradiometric assay; IFMA = time-resolved immunofluorometric assay; and EIA = enzyme immunoassay.

biologic fluids that must be taken into account in the various assays, especially radioimmunoassays (38, 44). Finally, many of the purified preparations of hCG and β-subunit, including hCG CR119, 75/537, and 75/589, as well as hCGα 75/569 and CR125 α and hCGβ 75/551, are contaminated with β-core fragment (38, 71). Therefore, the cross-reactions with hCG, hCGβ, and hCGα in some of the β-core immunoassays may have been overstated.

## Clinical Applications

### *Pregnancy*

Elevations of urine β-core fragment concentrations above those seen in the nonpregnant state are first detected 12 days after ovulation and donor insemination, while elevated urine hCG concentrations were noted to be elevated on day 10 in the same samples (50). The doubling time of β-core fragment in the urine is 24–48 h, while that for hCG is 24–36 h (50). β-core represents the major immunoreactive form of hCG in pregnancy urine (33, 38, 46, 47, 49, 50) (Fig. 22.7). The qualitative pattern of the maternal urine concentrations throughout pregnancy resembles that of hCG, although the ratio of the concentrations of β-core to hCG vary from 1.6 to 9.5, with the maximum ratio being between 10 and 20 weeks (50) (Fig. 22.8). The mean urine concentration during the first trimester is 280 ± 64 pg/ml; in the second trimester, 120 ± 81 pg/ml; and in the third trimester, 160 ± 99 pg/ml (43). During the first trimester of pregnancy, close to 2000 μg of β-core are excreted in the urine daily (33).

Early studies failed to detect β-core in the sera from pregnant women, reflecting both the insensitivity of the assays and the need to use chromatographic separation techniques in order to distinguish β-core from hCG or hCGβ. Using a sensitive radioimmunoassay and concentrated elution fractions of gel-filtered first and early second trimester pregnancy serum samples, Wehmann et al. found an average of 370 ± 80 pg of β-core/ml of serum and used this value, along with the metabolic clearance rate of β-core, to calculate the serum production rate of β-core to be 102.3 ± 22 μg/day/m$^2$ (33). In contrast, the serum production rate of hCG averages 11 mg/day/m$^2$, indicating that the serum production rate of β-core is about 100-fold less than that of hCG (33). Alfthan and Stenman used their sensitive immunofluorometric assay to measure β-core in pregnancy sera. They found an average concentration in the first trimester of 280 ± 64 pg/ml, 120 ± 81 pg/ml during the second trimester, and 160 ± 99 pg/ml during the third trimester (43). The ratio of β-core in the urine to β-core in the serum is approximately 4000:1, while that for hCG in the urine to hCG in the serum is approximately 1:2 (33).

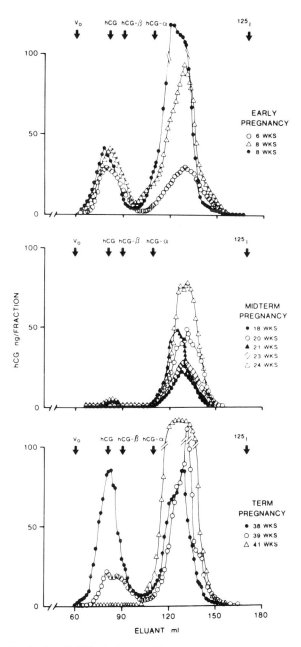

FIGURE 22.7. Sephadex G-100 elution profiles of urine concentrates from 11 early, midterm, or term pregnant women. hCG concentrations in the eluent tubes were measured with the SB-6 antiserum and expressed in terms of ng of hCG per fraction. Note that the β-core fragment represents the predominant immunoreactive moiety present in the urine during midterm pregnancy. Reprinted with permission from Kato and Braunstein (47), © The Endocrine Society, 1988.

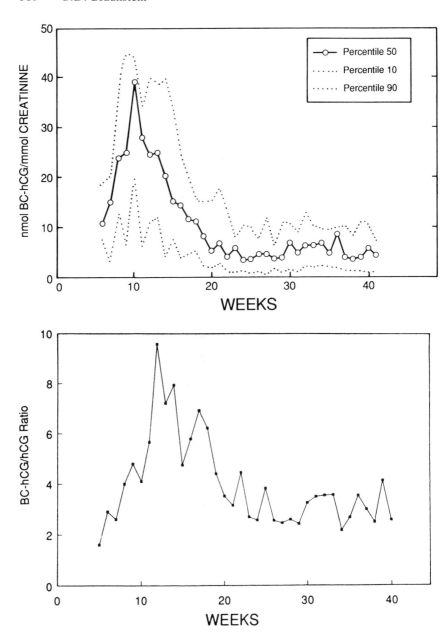

FIGURE 22.8. *Top:* Median, 10th, and 90th percentiles of weekly concentrations of β-core fragment of hCG in 741 normal pregnancies relative to the last menstrual period. *Bottom:* Relative proportion between β-core fragment and intact hCG levels during pregnancy. Reprinted with permission from The American College of Obstetricians and Gynecologists (*Obstetrics and Gynecology*, 1992, vol. 80, pp. 223–228) (50).

## Nonpregnant Controls and Patients With Benign Diseases

β-core fragment is present in the urine of nonpregnant men and women. Table 22.3 lists the percentage of males and pre- and postmenopausal females who have detectable immunoreactive β-core in their urine as measured by various assays. Two points are evident. First, the highest degree of positivity for detection of the fragment is found with the most sensitive assay (64). Second, in most studies the frequency of detection is greater in urine from postmenopausal women than that found in pre-menopausal women, reflecting an increase in the concentration of β-core fragment that is found after the menopause (21, 41, 44, 64). It is of interest that this pattern parallels the increase in hCG concentrations found in both the serum and urine of men and women after the age of 50 (64). Most studies have suggested that the upper limit of normal for urine concentration of β-core in the nonpregnant state is less than 500 pg/ml (66 pg/ml [44], 80 pg/ml [15, 18], 200 pg/ml [23, 72], and 425 pg/ml [66]). β-core fragment is only rarely measured in serum from normal individuals. Alfthan et al. detected concentrations of 22 and 15 pg/ml in 2 of 65 serum samples obtained from blood donors, but did not comment as to whether the two individuals (a 23-year-old woman and a 57-year-old man) were truly normal (64).

Patients with benign diseases tend to have a higher frequency of β-core levels in the urine that are above the upper limit of normal than do normal individuals (22, 23, 69, 72, 73). As shown in Table 22.4, which

TABLE 22.3. Detection of β-core in urine from normal control patients.

| Author (ref) | Concentration | Males | Percentage with detectable β-core in urine Premenopausal females | Postmenopausal females |
|---|---|---|---|---|
| Akar et al. (38) | ≥2.82 ng/ml | 18 | 50 | 41 |
| Cole and Nam (14) | ≥30 pg/ml | — | 3.2 | 17 |
| | ≥80 pg/ml | — | 0 | 2.9 |
| O'Conner et al. (10) | >50 pg/ml | — | 2 | 32 |
| Lee et al. (21) | >25 pg/ml | — | 54[1] | 87[2] |
| deMederios et al. (44) | ≥50 pg/ml | — | 5.5 | 19.5 |
| Kinugasa (23) | ≥50 pg/ml | — | 49.3 | 65.7 |
| | ≥200 pg/ml | — | 2.6 | 11.4 |
| Maruo et al. (72) | ≥200 pg/ml | — | 2.4 | 15.3 |
| Alfthan et al. (64) | ≥4.4 pg/ml | 83[3] | 70[4] | 80[5] |
| Norman et al. (69) | >1 ng/ml | 0 | 6.9 | — |

[1] Median concentration = 43 pg/ml.
[2] Median concentration = 260 pg/ml.
[3] Median concentration = 18 pg/ml for men <50 years and 16 pg/ml for men older than 50 years.
[4] Median concentration = 7 pg/ml.
[5] Median concentration = 16 pg/ml.

TABLE 22.4. β-core in urine from patients with gynecologic malignancies and normals.

| Author (ref) | ULN | Normals | Benign disease | Cancer | | | |
| --- | --- | --- | --- | --- | --- | --- | --- |
| | | | | Ovary | Cervix | Endometrium | Other |
| Nam et al. (15, 18) | 3 fmol/ml[1] | 17/126 (13)[2] | 7/83 (8) | 58/71 (82) | 24/52 (46) | 37/63 (59) | — |
| | 5 fmol/ml | 10/126 (8) | 4/83 (5) | 53/71 (75) | 21/52 (40) | 16/33 (48) | — |
| | 8 fmol/ml | 2/126 (2) | 1/83 (1) | 42/71 (59) | 16/52 (31) | 26/63 (41) | 9/14 (64) |
| Kinugasa et al. (23) | 200 pg/ml | 6/112 (5) | 12/111 (11) | — | 57/128 (44) | — | — |
| Maruo et al. (72) | 200 pg/ml | 15/304 (5) | 34/286 (12) | 25/39 (64) | 51/168 (30) | 27/67 (40) | 3/7 (43) |
| Negishi et al. (74) | 3.8 fmol/ml | — | 4/35 (11) | 12/18 (67) | 8/16 (50) | 6/6 (100) | — |
| Norman et al. (69) | 1 ng/ml | 2/66 (30) | 12/60 | — | 60/98 (61) | — | — |

[1] 1 fmol/ml = 10 pg β-core or 22.7 pg hCGβ.
[2] Numbers in parenthesis indicate percentage of patients above the upper limit of normal (ULN).

compares the number and percentage of normal individuals, patients with benign gynecologic diseases, and patients with gynecologic cancer who have concentrations of β-core in the urine at or above the defined upper limit of normal, patients with benign diseases have anywhere from approximately the same to a 6-fold greater degree of elevated urine β-core concentrations than do normal controls. In a study of benign and malignant biliary disease, Alfthan et al. noted that 11% of patients with benign pancreatic and hepatobiliary diseases had urinary β-core concentrations that were above the 97.5 percentile for the reference control patients (22). Similarly, Tanaka et al. reported that 19.2% of patients with benign urologic disorders had elevated urinary β-core concentrations compared to 5% of normal individuals (73). This reflects not only a greater frequency of production, but also higher concentrations of β-core than in normal individuals. Only a single report concerning serum β-core levels in patients with benign diseases has appeared in which 1 of 45 patients (2%) with benign hepatobiliary or pancreatic disorders had a level of β-core above the 97.4 percentile (22).

## Cancer

Approximately 18% of patients with nontrophoblastic neoplasms have immunoreactive hCG or hCGβ in their circulation, and 25% will have hCG or hCGβ detected by immunohistochemical techniques applied to tumor tissue (75). Although several investigators using chromatographic methods demonstrated the presence of hCG in the urine of patients with both trophoblastic and nontrophoblastic tumors (1, 4, 6, 8–10, 19, 38, 67), it was the report of Papapetrou et al. (9) that demonstrated the high prevalence of hCG in the urine of patients with cancer by showing that 31 of 70 patients (44.3%) had immunoreactive hCG in the urine, much of which was composed of β-core fragment (9). That study showed elevated levels in the urine of patients with carcinoma of the breast, ovary, cervix, lung, liver, stomach, pancreas, prostate, and rectum, as well as patients with acute leukemia and lymphoproliferative disorders.

TABLE 22.5. Urine β core in patients with gynecologic malignancies by disease stage.

| Author (ref) | ULN* | \multicolumn{5}{c}{Number/total (%) with elevated β-core} |
| | | Stage I | Stage II | Stage III | Stage IV | Recurrence |
| --- | --- | --- | --- | --- | --- | --- |
| Nam et al. (15) | 8 fmol/ml | 1/31 (3) | 3/15 (20) | 26/54 (48) | 19/24 (79) | 30/46 (65) |
| Kinugasa et al. (23) | 200 pg/ml | 23/73 (32) | 21/27 (78) | 6/10 (60) | 2/2 (100) | — |
| Mauro et al. (72) | 200 pg/m | 18/14 (57) | 2/4 (50) | 8/12 (75) | 3/3 (100) | — |
| Norman et al. (69) | 1 ng/m | 16/10 (60) | 13/25 (52) | 27/35 (77) | 14/28 (50) | — |
| Total | | 38/128 (30) | 39/71 (55) | 67/111 (60) | 38/57 (67) | 30/46 (65) |

* ULN = upper limit of normal.

With the advent of more specific immunoassays, several groups examined categories of tumors in more detail. The most intensely studied have been the gynecologic neoplasms. As shown in Table 22.4, which represents the accumulation of published studies, approximately two-thirds of patients with ovarian cancer, one-third to one-half of the patients with cervical cancer, approximately 40%–50% of patients with endometrial cancer, and half of the patients with other gynecologic neoplasms have elevated levels of β-core in their urine. There is a linear relationship between the stage of gynecologic cancer and frequency of β-core elevation in urine samples (Table 22.5). There is also a positive correlation between disease stage and concentration of β-core in patients with gynecologic cancer (11, 12, 14, 15, 17, 18), although not all investigators have found such a relationship (69). In the case of ovarian carcinoma, serial measurements of urine β-core concentrations revealed changes that were concordant with the clinical course in 20 of 21 patients (95%) who had β-core concentrations above 3 fmol/ml (14). Similarly, serial samples in patients with endometrial carcinoma or uterine mixed mullerian tumors showed a correlation between the clinical course and changes in β-core levels in 13 of 15 patients (87%) (18). In patients with carcinoma of the cervix, Kinugasa et al. noted that 23 of 26 patients with cervical cancer had a rapid decline in β-core concentrations following surgery, while three showed a transient increase followed by a later decrease (23). β-core concentrations fell in the five patients who received radiation or chemotherapy concordant with the change in their disease status. Two of four patients who had a rise in β-core concentrations also had recurrence of disease. It is of interest that Negishi et al. (74) also noted a transient increase in β-core concentrations after surgery or chemotherapy, suggesting that the β-core precursors hCG or hCGβ were released into the circulation and were subsequently metabolized to β-core. In patients with gynecologic malignancies, all tumor cell types appear to be associated with elevated excretion of β-core. The levels of β-core in the urine also increase, along with increasing grade of tumor, at least in patients with endometrial carcinoma and mixed uterine mullerian tumors (18).

Alfthan et al. examined both serum and urine β-core levels in patients with pancreatic and biliary cancer and compared the clinical performance of the β-core assays by *receiver operating characteristic* (ROC) curve analysis with that found with measurements of hCG and hCGβ (22). They noted that in the serum, 14% of patients with pancreatic cancer had elevated concentrations of hCG, 72% had elevated levels of hCGβ, and 45% had elevated concentrations of β-core. For biliary cancer, the percent elevations were 14%, 86%, and 57%, respectively. In the urine, 3% of patients with pancreatic cancer had an elevated hCG, 28% had elevated hCGβ levels, and 55% had elevated β-core concentrations. The frequency of elevated values in patients with biliary cancer were 0%, 57%, and 71%, respectively. Using the ROC curve analysis at cutoff levels giving 90% specificity, the sensitivity of serum hCGβ measurements was 72% and that of urine β-core measurements was 59% (22). These authors concluded that hCGβ is the major source of urine β-core activity, since there was a correlation between the two analyses. Finally, they noted a large amount of overlap in the amounts of β-core found in the urine of patients with benign pancreatic and hepatobiliary diseases vs. those with malignancies, emphasizing the limitation of urine β-core measurements as a screen for cancer.

Tanaka et al. noted that 21 of 52 patients with renal cell carcinoma (40%), 8 of 12 patients with cancer of other areas of the urinary, tract (67%), and 18 of 52 patients with bladder cancer (35%) had elevated concentrations of β-core in the urine, as compared to 5 of 26 patients (19.2%) with benign urinary tract diseases as assessed with the Triton assay (73). Similarly, Yamanaka et al. found elevated levels of urine β-core in 54% of patients with bladder tumors, 71% with patients with other urinary tract tumors, and in none of 17 patients with prostate cancer or 30 patients with benign diseases, including cystitis, urinary tract stones, and prostatitis (76). Levels of urinary β-core above 3 fmol/ml were found in 63% of a small number of patients with advanced colorectal cancer studies by McGill et al. (77), but these investigators did not examine a benign disease group for comparison—an issue that is pertinent since 8%–13% of noncancer patients have been found to have β-core levels at or above this concentration (15, 18). Finally, D'Agostino et al. found 20 of 83 patients (24%) with lung cancer, 4 of 11 patients (36%) with esophageal carcinoma, 3 of 11 (27%) with lymphoma, and 2 of 17 (12%) with other tumors had urinary β-core concentrations above 3 fmol/ml, while 6% of control patients with benign disease had concentrations above this level (78). In cancer patients with β-core in their urine, serial measurements showed changes in concentrations that were concordant with changes in tumor growth in 62.5% of the patients (78).

It is clear that the clinical value for measuring urinary β-core fragment in patients with nontrophoblastic neoplasms is limited in regard to potential for screening for the tumors. The number of false positives in the population in general would greatly overshadow the number of true positives; therefore, the predictive value of a positive test would be anticipated to be quite low. On the other hand, in a high-risk population, the prevalence of cancer may be great enough that the true-positive rate may be substantially greater than the false-positive rate, making the predictive value of a positive test high. At the present time urine β-core measurements appear to be most useful for monitoring the effect of therapy in patients with gynecologic neoplasms. As more information is obtained with highly specific assays and as any potential cross-reaction with LHβ-core is eliminated, then the specificity of urine hCGβ-core measurements may increase substantially.

## References

1. Matthies DL, Diczfalusy E. Differences in the physical properties of human chorionic gonadotrophin extracted from difference sources. Excerpta Medica (Amst) Int Congr ser no. 170, 1968:34.
2. Matthies DL, Diczfalusy E. Relationship between physico-chemical, immunological and biological properties of human chorionic gonadotrophin, I. Properties of human chorionic gonadotrophin as found in tissues and body fluids. Acta Endocrinol (Copenh) 1971;67:434–44.
3. Franchimont P, Gaspard U, Reuter A, Heynen G. Polymorphism of protein and polypeptide hormones. Clin Endocrinol 1972;1:315–36.
4. Vaitukaitis JL. Immunologic and physical characterization of human chorionic gonadotropin (hCG) secreted by tumors. J Clin Endocrinol Metab 1973; 37:505–14.
5. Vaitukaitis JL. Changing placental concentrations of human chorionic gonadotropin and its subunits during gestation. J Clin Endocrinol Metab 1974; 38:755–60.
6. Vaitukaitis JL, Ebersole ER. Evidence for altered synthesis of human chorionic gonadotropin in gestational trophoblastic tumors. J Clin Endocrinol Metab 1976;42:1048–55.
7. Good A, Ramos-Uribe A, Ryan RJ, Kempers RD. Molecular forms of human chorionic gonadotropin in serum, urine, and placental extracts. Fertil Steril 1977;28:846–50.
8. Hattori M, Yoshimoto Y, Matsukura S, Fujita T. Qualitative and quantitative analyses of human chorionic gonadotropin and its subunits produced by malignant tumors. Cancer 1980;46:355–61.
9. Papapetrou PD, Sakarelou NP, Braouzi H, Fessas Ph. Ectopic production of human chorionic gonadotropin (hCG) by neoplasms: the value of measurements of immunoreactive hCG in the urine as a screening procedure. Cancer 1980;45:2583–92.
10. Papapetrou PD, Nicopoulou S Ch. The origin of a human chorionic gonadotropin β-subunit-core fragment excreted in the urine of patients with cancer. Acta Endocrinol 1986;112:415–22.
11. Cole LA, Wang Y, Elliott M, et al. Urinary human chorionic gonadotropin free β-subunit and β-core fragment: a new marker of gynecologic cancer. Cancer Res 1988;48:1356–60.
12. Cole LA, Schwartz PE, Wang Y. Urinary gonadotropin fragments (UGF) in cancers of the female reproductive system, I. Sensitivity and specificity, comparison with other markers. Gynecol Oncol 1988;31:82–90.
13. Wang Y, Schwartz PE, Chambers JT, Cole LA. Urinary gonadotropin fragments (UGF) in cancers of the female reproductive system, II. Initial serial studies. Gynecol Oncol 1988;31:91–100.
14. Cole LA, Nam J-H. Urinary gonadotropin fragment (UGF) measurements in the diagnosis and management of ovarian cancer. Yale J Biol Med 1989; 62:367–78.
15. Nam J-H, Cole LA, Chambers JT, Schwartz PE. Urinary gonadotropin fragment, a new tumor marker, I. Assay development and cancer specificity. Gynecol Oncol 1990;36:383–90.

16. Cole LA, Nam J-H, Chambers JT, Schwartz PE. Urinary gonadotropin fragment, a new tumor marker, II. Differentiating a benign from a malignant pelvic mass. Gynecol Oncol 1990;36:391–4.

17. Nam J-H, Chang K-C, Chambers JT, Schwartz PE, Cole LA. Urinary gonadotropin fragment, a new tumor marker, III. Use in cervical and vulvar cancers. Gynecol Oncol 1990;38:66–70.

18. Nam J-H, Chambers JT, Schwartz PE, Cole LA. Urinary gonadotropin fragment, a new tumor marker, IV. Use in endometrial cancer and uterine mixed mullerian tumors. Gynecol Oncol 1990;39:352–7.

19. Kardana A, Taylor ME, Southall PJ, Boxer GM, Rowan AJ, Bagshawe KD. Urinary gonadotrophin peptide-isolation and purification, and its immunohistochemical distribution in normal and neoplastic tissue. Br J Cancer 1988;58:281–6.

20. Iles RK, Jenkins BJ, Oliver RTD, Blandy JP, Chard T. Beta human chorionic gonadotrophin in serum and urine. A marker for metastatic urothelial cancer. Br J Urol 1989;64:241–4.

21. Lee CL, Iles RK, Shepherd JH, Hudson CN, Chard T. The purification and development of a radioimmunoassay for β-core fragment of human chorionic gonadotrophin in urine: application as a marker of gynaecological cancer in premenopausal and postmenopausal women. J Endocrinol 1991; 130:481–9.

22. Alfthan H, Haglund C, Roberts P, Stenman U-H. Elevation of free β subunit of human choriogonadotropin and core β fragment of human choriogonadotropin in the serum and urine of patients with malignant pancreatic and biliary disease. Cancer Res 1992;52:4628–33.

23. Kinugasa M, Nishimura R, Hasegawa K, et al. Assessment of urinary β-core fragment of hCG as a tumor marker of cervical cancer. Acta Obstet Gynaecol Jpn 1992;44:188–94.

24. Wehmann RE, Nisula BC. Characterization of a discrete degradation production of the human chorionic gonadotropin β-subunit in humans. J Clin Endocrinol Metab 1980;51:101–4.

25. Wehmann RE, Nisula BC. Metabolic and renal clearance rates of purified human chorionic gonadotropin. J Clin Invest 1981;68:184–94.

26. Wehmann RE, Amr S, Rosa C, Nisula BC. Metabolism, distribution and excretion of purified human chorionic gonadotropin and its subunits in man. Ann Endocrinol (Paris) 1984;45:291–5.

27. Lefort GP, Stolk JM, Nisula BC. Evidence that desialylation and uptake by hepatic receptors for galactose-terminated glycoproteins are immaterial to the metabolism of human choriogonadotropin in the rat. Endocrinology 1984; 115:1551–7.

28. Lefort GP, Stolk JN, Nisula BC. Renal metabolism of the β-subunit of human choriogonadotropin in the rat. Endocrinology 1986;119:924–31.

29. Blithe DL, Akar AH, Wehmann RE, Nisula BC. Purification of β-core fragment from pregnancy urine and demonstration that its carbohydrate moieties differ from those of native human chorionic gonadotropin-β. Endocrinology 1988;122:173–80.

30. Blithe DL, Wehmann RE, Nisula BC. Carbohydrate composition of β-core. Endocrinology 1989;125:2267–72.

31. Wehmann RE, Blithe DL, Flack MR, Nisula BC. Metabolic clearance rate and urinary clearance of purified β-core. J Clin Endocrinol Metab 1989; 69:510–7.
32. Nisula BC, Blithe DL, Akar A, Lefort G, Wehmann RE. Metabolic fate of human choriogonadotropin. J Steroid Biochem 1989;33:733–7.
33. Wehmann RE, Blithe DL, Akar AH, Nisula BC. Disparity between β-core levels in pregnancy urine and serum: implications for the origin of urinary β-core. J Clin Endocrinol Metab 1990;70:371–8.
34. Masure HR, Jaffee WL, Sickel MA, Birken S, Canfield RE, Vaitukaitis JL. Characterization of a small molecular size urinary immunoreactive human chorionic gonadotropin (hCG)-like substance produced by normal placenta and by hCG-secreting neoplasms. J Clin Endocrinol Metab 1981; 53:1014–20.
35. Birken S, Armstrong EG, Kolks MAG, et al. Structure of the human chorionic gonadotropin β-core fragment from pregnancy urine. Endocrinology 1988;123:572–83.
36. Nishimura R, Endo Y, Tanabe K, Ashitaka Y, Tojo S. The biochemical properties of urinary human chorionic gonadotropin from the patients with trophoblastic diseases. J Endocrinol Invest 1981;4:349–58.
37. Endo T, Nishimura R, Saito S, et al. Carbohydrate structures of β-core fragment of human chorionic gonadotropin isolated from a pregnant individual. Endocrinology 1992;130:2052–8.
38. Akar AH, Wehmann RE, Blithe DL, Blacker C, Nisula BC. A radioimmunoassay for the core fragment of the human chorionic gonadotropin β-subunit. J Clin Endocrinol Metab 1988;66:538–45.
39. Krichevsky A, Armstrong EG, Schlatterer J, et al. Preparation and characterization of antibodies to the urinary fragment of the human chorionic gonadotropin β-subunit. Endocrinology 1988;123:584–93.
40. O'Connor JF, Schlatterer JP, Birken S, et al. Development of highly sensitive immunoassays to measure human chorionic gonadotropin, its β-subunit, and β core fragment in the urine: application to malignancies. Cancer Res 1988;48:1361–6.
41. Krichevsky A, Birken S, O'Connor J, et al. Development and characterization of a new, highly specific antibody to the human chorionic gonadotropin-β fragment. Endocrinology 1991;128:1255–64.
42. Kardana A, Taylor ME, Rowan AJ, Read DA, Bagshawe KD. Characterisation of antibodies to urinary gonadotrophin peptide. J Immunol Methods 1989;118:53–8.
43. Alfthan H, Stenman U-H. Pregnancy serum contains the β-core fragment of human choriogonadotropin. J Clin Endocrinol Metab 1990;70:783–7.
44. deMedeiros SF, Amato F, Matthews CD, Norman RJ. Comparison of specific immunoassays for detection of the β-core human chorionic gonadotropin fragment in body fluids. J Endocrinol 1992;135:161–74.
45. Kato Y, Kelley L, Braunstein GD. The beta-core fragment of human chorionic gonadotropin. In: Tomoda Y, Mizutani S, Narita O, Klopper A, eds. Placental and endometrial proteins: basic and clinical aspects. Utrecht, The Netherlands: VNU Science Press BV, 1988:87–90.
46. Schroeder HR, Halter CM. Specificity of human β-choriogonadotropin assays for the hormone and for an immunoreactive fragment present in urine during normal pregnancy. Clin Chem 1983;29:667–71.

47. Kato Y, Braunstein GD. β-core fragment is a major form of immunoreactive urinary chorionic gonadotropin in human pregnancy. J Clin Endocrinol Metab 1988;66:1197–201.

48. Kardana A, Cole LA. Serum hCG β-core fragment is masked by associated macromolecules. J Clin Endocrinol Metab 1990;71:1393–5.

49. Cole LA, Birken S. Origin and occurrence of human chorionic gonadotropin β-subunit core fragment. Mol Endocrinol 1988;2:825–30.

50. deMedeiros SF, Amato F, Matthews CD, Norman RJ. Urinary concentrations of beta core fragment of hCG throughout pregnancy. Obstet Gynecol 1992;80:223–8.

51. deMedeiros SF, Amato F, Bacich D, Wang L, Matthews CD, Norman RJ. Distributrion of the β-core human chorionic gonadotrophin fragment in human body fluids. J Endocrinol 1992;135:175–88.

52. Amr S, Rosa C, Birken S, Canfield R, Nisula B. Carboxyterminal peptide fragments of the beta subunit are urinary products of the metabolism of desialylated human choriogonadotropin. J Clin Invest 1985;76:350–6.

53. Hussa RO. Biosynthesis of human chorionic gonadotropin. Endocr Rev 1980;1:268–94.

54. Bidart J-M, Puisieux A, Troalen F, Foglietti M-J, Bohuon C, Bellet D. Characterization of a cleavage product in the human choriogonadotropin β-subunit. Biochem Biophys Res Commun 1988;154:626–32.

55. Puisieux A, Bellet D, Troalen F, et al. Occurrence of fragmentation of free and combined forms of the β-subunit of human chorionic gonadotropin. Endocrinology 1990;126:687–94.

56. Lustbader JW, Birken S, Pileggi NF, et al. Crystallization and characterization of human chorionic gonadotropin in chemically deglycosylated and enzymatically desialylated states. Biochemistry 1989;28:9239–43.

57. Sakakibara R, Miyazaki S, Ishiguro M. A nicked β-subunit of human chorionic gonadotropin purified from pregnancy urine. J Biochem 1990;107:858–62.

58. Cole LA, Kardana A, Ying FC, Birken S. The biological and clinical significance of nicks in human chorionic gonadotropin and its free β-subunit. Yale J Biol Med 1991;64:627–37.

59. Kardana A, Elliott MM, Gawinowicz M-A, Birken S, Cole LA. The heterogeneity of human chorionic gonadotropin (hCG), I. Characterization of peptide heterogeneity in 13 individual preparations of hCG. Endocrinology 1991;129:1541–50.

60. Birken S, Gawinowicz MA, Kardana A, Cole LA. The heterogeneity of human chorionic gonadotropin (hCG), II. Characteristics and origins of nicks in hCG reference standards. Endocrinology 1991;129:1551 8.

61. Cole LA, Kardana A, Andrade-Gordon P, et al. The heterogeneity of human chorionic gonadotropin (hCG), III. The occurrence and biological and immunological activities of nicked hCG. Endocrinology 1991;129:1559–67.

62. Cole LA, Kardana A, Park S-Y, Braunstein GD. The deactivation of hCG by nicking and dissociation. J Clin Endocrinol Metab 1993:76.

63. Markkanen SO, Rajaniemi HJ. Uptake and subcellular catabolism of human choriogonadotropin in proximal tubule cells of rat kidney. Mol Cell Endocrinol 1979;13:181–90.

64. Alfthan H, Haglund C, Dabek J, Stenman U-H. Concentrations of human choriogonadotropin, its β-subunit, and the core fragment of the β-subunit in serum and urine of men and nonpregnant women. Clin Chem 1992;38:1981–7.

65. Akar AH, Gervasi G, Blacker C, Wehmann RE, Blithe DL, Nisula BC. Human chorionic gonadotropin-like and β-core-like materials in postmenopausal urine. J Endocrinol 1990;125:477–84.
66. Iles RK, Lee CL, Howes I, Davies S, Edwards R, Chard T. Immunoreactive β-core-like material in normal postmenopausal urine: human chorionic gonadotrophin or LH origin? Evidence for the existence of LH core. J Endocrinol 1992;133:459–66.
67. Braunstein GD. Production of human chorionic gonadotropin by nontrophoblastic tumors and tissues. In: Tomoda Y, Mizutani S, Narita O, Klopper A, eds. Placental and endometrial proteins: basic and clinical aspects. Utrecht, The Netherlands: VNU Science Press BV, 1988:493–502.
68. Odell WD, Griffin J, Sawitzke A. Chorionic gonadotropin secretion in normal, nonpregnant humans. Trends Endocrinol Metab 1990;1:418–21.
69. Norman RJ, Buck RH, Aktar B, Mayet N, Moodley J. Detection of a small molecular species of human chorionic gonadotropin in the urine of patients with carcinoma of the cervix and cervical intraepithelial neoplasia: comparison with other assays for human chorionic gonadotropin and its fragments. Gynecol Oncol 1990;37:254–9.
70. deMedeiros SF, Amato F, Norman RJ. Stability of immunoreactive β-core fragment of hCG. Obstet Gynecol 1991;77:53–9.
71. Wehmann RE, Blithe DL, Akar AH, Nisula BC. β-core fragments are contaminants of the World Health Organization Reference Preparations of human choriogonadotrophin and its α-subunit. J Endocrinol 1988;117:147–52.
72. Maruo T, Kitajima T, Otani T, et al. Clinical significance of the measurement of urinary hCG-β-core fragment in patients with gynecologic malignant tumors. Obstet Gynecol Jpn 1991;7:1197–202.
73. Tanaka J, Takashi M, Okamura K, Shimochi T, Miyake K, Yamasaki H. Significance of hCGβ-core fragment in kidney lesions and urinary tract and bladder cancer [Abstract]. 12th Tumor Marker Assoc meet, Osaka, Japan, 1992.
74. Negishi N, Iwabuchi H, Sakunaga H, Okabe K. Clinical significance of UGP in gynecological tumors [Abstract]. 12th Tumor Marker Assoc meet, Osaka, Japan, 1992.
75. Braunstein GD. Placental proteins as tumor markers. In: Herberman RB, Mercer DW, eds. Immunodiagnosis of cancer. New York: Marcel Dekker, 1990:673–701.
76. Yamanaka N, Morisue K, Kawabata G, Hama M, Nishimura R. Significance of hCG β-core fragment as a tumor marker in bladder cancer [Abstract]. 12th Tumor Marker Assoc meet, Osaka, Japan, 1992.
77. McGill J, Cole L, Nam JH, et al. Urinary gonadotropin fragment (UGF): a potential marker of colorectal cancer. J Tumor Marker Oncol 1990;5:175–7.
78. D'Agostino RS, Cole LA, Ponn RB, Stern H, Schwartz PE. Urinary gonadotropin fragment measurement in patients with lung and esophageal disease. J Surg Oncol 1992;49:147–50.

# 23

# Human Pituitary Chorionic Gonadotropin Is Produced, Circulated, and Excreted in Multiple Isoforms

Arleen L. Sawitzke, Jeanine Griffin, and William D. Odell

Human *chorionic gonadotropin* (CG) is a glycoprotein hormone that was originally thought to be produced only by the placenta. Since 1976, when Chen et al. first demonstrated the presence of a CG-like material in human pituitary extracts (1), a convincing body of evidence has been compiled to support the hypothesis that CG is produced by the human pituitary (2, 3). The chemical nature of this pituitary-produced CG has not yet been fully elucidated; however, like other pituitary-produced hormones (*luteinizing hormone* [LH], *thyroid stimulating hormone* [TSH], and *follicle stimulating hormone* [FSH]), it appears to be a glycoprotein. Early work done by Chen et al., Hartree et al., and Matsurra et al. demonstrated that pituitary CG is similar in size, binding characteristics to concanavalin A, isoelectric pH, immunoactivity, and bioactivity to CG produced by the placenta (1, 4, 5). However, some more recent studies have demonstrated the existence of another CG-like material with a molecular weight of 40–43 kd (6–8), indicating a possible difference in carbohydrate content or peptide structure.

There is extensive documentation that LH, FSH, and placentally produced CG exist in multiple molecular forms known as isoforms (9–12). These isoforms are the result of heterogeneity in the carbohydrate moieties on the molecules. Wide demonstrated the existence of more than 20 isoforms of FSH and LH (13). The amino acid sequences for the alpha subunit of FSH, LH, TSH, and CG are all identical; however, Nilsson showed major differences in the carbohydrates of the alpha chains of these four hormones (14). Ulloa-Aguirre et al. (9) have shown that cultured cytotrophoblast cells produce CG with three isoforms. Lichtenberg et al. have isolated 12 isoforms of CG from placental extracts and 6 isoforms from urine (12). Graesslin et al. also showed 6 isoforms of urinary CG and felt that they resulted from microheterogeneity of the carbohydrates of the β-chain (15).

Several studies have examined the control of production of these different isoforms. Galle et al. showed that higher estrogen resulted in more basic pIs of hamster FSH (16), while Ulloa-Aguirre et al. showed that high androgens resulted in more acidic FSH isoforms in rats (17). Miller et al. and Baldwin et al. showed that GnRH added to cultured pituitary cells caused release of more basic hamster and rat FSH and LH isoforms (18, 19). In contrast, Blum and Gupta and Lichtenberg and Urban found that GnRH did not change the pIs, but that acidic iso-hormones were more prevalent in older rats (20, 21). The pI of CG produced by cultured human cytotrophoblasts became more acidic as the exposure time to cAMP increased (9).

The production, function, and degradation of these isoforms are not fully understood; however, the isoforms appear to differ in biological activity, receptor binding activity, and biological half-life (22). Because of the importance of the carbohydrate content to the biological functions of these molecules and the interesting occurrence of isoform patterns during their metabolic pathway, we have undertaken an investigation of the carbohydrate content of pituitary-produced CG. The CG was obtained from a variety of nonpregnant sources and analyzed by chromatographic methods before and after neuraminidase treatment. We compared the carbohydrate content at the levels of production, circulation, and clear-ance. We have compared these data to CG produced at three different times during pregnancy. This study adds CG to the list of pituitary glycoprotein hormones that are produced in several isoforms.

## Materials and Methods

### CG Determination

A sensitive and specific CG *immunoradiometric assay* (IRMA) using mouse monoclonal antibodies (23) was used to determine the presence of CG in the samples. This assay has been very well characterized in this laboratory and shows no detectable cross-reaction with LH ($<0.03\%$). In addition, this assay is specific for intact CG and does not cross-react with free alpha or beta subunits of CG (8, 23). The sensitivity of this assay for CG is 0.04 mIU (3 pg) per sample (10–100 μl). Because only 10–100 μl of sample are added to any assay tube, the pH of the final assay tubes in this study never varied outside of a pH range of 5–8, over which CG has been shown to be measured accurately by this IRMA. The buffers used in this study were also tested by measuring CG in their presence. No changes in the amount of CG measured were greater than 10%, even when 100 μl of the buffers were used (unpublished data). The intra- and interassay variations in this study were 10.5% and 4.0%, respectively. For samples that had a potential to contain large amounts of LH, the fractions were also assayed in a specific LH IRMA (24) to demonstrate different elution patterns for LH and CG, thus further assuring that the CG measured was not a cross-reaction artifact.

A CG preparation (CR125) with an immunopotency of 11,900 IU/mg (Second International Standard for hCG) was prepared by Dr. Robert Canfield and supplied by the *National Institute of Child and Human Development* (NICHD) and was used as the reference preparation throughout this study. This CG is purified from a crude preparation from the urine of pregnant women. Immunochemical-grade human LH (AFP-0642B), as well as individual pituitary glands, were supplied by the Hormone Distribution Program of the *National Institute of Diabetes, Digestive, and Kidney Diseases* (NIDDK).

## Isolation of CG Samples

### CG from Normal, Nonpregnant Humans

Table 23.1 summarizes the composition and use of the buffers in the chromatofocusing portion of these studies. The buffers are designated as A through E and are referred to as such in this chapter. CG was isolated from pituitaries, serum, and urine of nonpregnant humans. Each CG sample was dialyzed against buffer A, buffer B, and/or buffer C, as described in Table 23.1, in preparation for chromatofocusing and/or neuraminidase digestion.

Pituitary CG was obtained from two sources: (i) human pituitaries (approximately 500 mg apiece) obtained at postmortem (the sex, age, or endocrine status of the pituitary gland donors are not known) and (ii) immunochemical grade LH (AFP-0642B) from NIDDK. The pituitary

TABLE 23.1. Name, composition, and purpose of the buffers used in the chromatofocusing portion of these studies.

| Buffer name | Composition | Purpose |
|---|---|---|
| A. PBE 94 starting buffer[a] | 25 mmol/l imidazole-HCl, pH = 7.0 | Load and wash PBE 94 column |
| B. PBE 118 starting buffer[a] | 25 mmol/l triethanolamine-HCl, pH = 11.0 | Load and wash PBE 118 column |
| C. Acetate buffer | 50 mmol/l sodium acetate, 2 mmol/l calcium chloride, 0.2 mmol/l EDTA, pH = 5.0 | Neuraminidase digestion |
| D. Polybuffer 74-HCl[a] | pH = 3.0 | Elution from PBE 94 column |
| E. Pharmalyte[a] | pH = 7.0 | Elution from PBE 118 column |

[a] Obtained from Pharmacia (Uppsala, Sweden).

glands were homogenized in 0.9% saline and centrifuged at 10,000 g for 30 min. Aliquots of the supernatant were then dialyzed against the above three buffers and were applied directly to the chromatofocusing columns or were subjected to neuraminidase treatment and then applied to the chromatofocusing columns. Five individual pituitary glands were homogenized and independently analyzed to determine the reproducibility of our methods. Samples (25 µg) of pituitary LH (AFP-0642B) were dissolved in the above buffers and subjected to chromatofocusing. We have previously shown that this preparation of LH is contaminated with small amounts of CG (8). Three separate samples of LH (AFP-0642B) were analyzed.

Secreted CG was obtained from two sources: (i) serum from postmenopausal female donors and (ii) the supernatant of fetal pituitary cell cultures. The serum was subjected to affinity purification as described previously (8). This purification involves rotating the serum with Sepharose beads to which a monoclonal antibody against CG has been attached. The beads are then collected and the bound CG eluted with 3.5 mol/l MgCl2 (pH 3.5). The eluent is then passed through a buffer exchange column equilibrated with buffer A, buffer B, or buffer C. This method has been shown to remove greater than 85% of the CG (in intact, biologically active form) and less than 10% of the LH from serum samples (Sawitzke, Doctoral Dissertation, 1991). Approximately 50% of the removed CG is recoverable from the column for further analysis. Three independent serum samples from postmenopausal females were analyzed. The supernatant of fetal pituitary cell cultures was collected, concentrated, and dialyzed against buffer B. These cultures were the same as used in a previous study on LH isoforms (25), where details of the cultures are given. Only one sample of fetal pituitary culture supernatant was analyzed.

The urinary sources of CG were (i) *human menopausal gonadotropin* (hMG) (Pergonal, Serono), a urinary preparation high in LH, FSH, and also CG, and (ii) affinity-purified CG from fresh postmenopausal urine. Unlike some preparations, the hMG from Serono does not have pregnancy CG added to it (personal communication and Rodgers et al. [26]). Three separate samples of hMG were dissolved in buffer A or in buffer C and independently analyzed. Three liter samples of urine were collected from three menopausal females who were not taking supplemental estrogens. These samples were subjected to a 50% ammonium sulfate precipitation and the pellet resuspended and dialyzed against buffer A in preparation for chromatofocusing.

## CG from Pregnant Humans

For comparison purposes, CG was obtained from placentas, serum, and urine at the first, second, and third trimester of pregnancy. One or two of each pregnancy sample type were prepared and independently analyzed. The serum from pregnant females was subjected to affinity purification as described above and analyzed by chromatofocusing. To assess whether isoforms were lost during purification, a sample of pregnancy serum was analyzed both before and after affinity purification. Homogenates were made from placental tissue using the same protocol as was used for the pituitary glands. The homogenates were then dialyzed against buffers A and B and analyzed by chromatofocusing. Both CR-125 and fresh pregnancy urine were used as sources of excreted CG. The CR-125 was dissolved in buffer A or C. The pregnancy urine was dialyzed against buffer A.

## Neuraminidase Treatment

Samples in acetate buffer (buffer C) were subjected to treatment with neuraminidase (Sigma, St. Louis, MO) as described by Moyle et al. (27), with some modifications. One *international unit* (IU) of neuraminidase per 50 mg of pituitary homogenate protein or per 50 μg of hMG or CR-125 was allowed to incubate with the sample under nitrogen for 36 h at 37 °C. A second IU of neuraminidase was then added and allowed to incubate for a second 36 h. This treatment protocol removes 85%–90% of the sialic acid residues without altering the internal carbohydrate residues (27).

## Chromatofocusing

The samples of CG, before and after treatment with neuraminidase, were applied to a PBE 94 column with a pH range of 7–3, 1 cm × 30 cm, which had been equilibrated with buffer A. The column was washed with three ml of buffer A after application of the sample, and a pH gradient was then produced by adding buffer D. Three-ml fractions were collected and assayed for pH and CG content.

Some samples, dissolved in buffer B, were applied to a PBE 118 column with a pH range of 11–7. After washing the column to remove unbound protein, buffer E was added and 3-ml fractions were collected and assayed for pH and CG content.

## Lectin Affinity Chromatography

The samples added to the lectin columns included urine preparations from both pregnant and nonpregnant females before and after neuraminidase digestion and serum samples from pregnant females. Some fractions obtained from the chromatofocusing columns described above were pooled and applied to the Ricin lectin column. These samples included neuraminidase-treated hMG and neuraminidase-treated pituitary homogenates. All samples were loaded in 1–2 ml volumes, allowed to enter the column, and then incubated at room temperature for 1 h. Two-ml fractions were collected from all columns and analyzed for CG and LH content.

Columns (0.5 × 12 cm) of Ricin (RCA120)-Sepharose were prepared as directed by Pharmacia. Ricin Agglutinin 120 (lectin from *Ricinus communis*) was obtained from Sigma (St. Louis, MO). The samples were washed through with approximately 10 column volumes of lectin buffer (0.01 mol/l sodium phosphate, 0.15 mol/l sodium chloride, 0.1% BSA, 1 mmol/l $MgCl_2$ and 1 mmol/l $CaCl_2$, pH 7.4) and then eluted with the lectin buffer containing lactose (100 mg/ml).

Columns (0.5 × 5 cm) of Lentil-Sepharose (Lectin from *Lens culinaris*, Sepharose 4B, Sigma, St. Louis, MO) were prepared. The samples were washed through with 10 column volumes of lectin buffer and then eluted with the lectin buffer containing D-a-methyl-mannopyranoside (200 mmol/l).

Columns (0.5 × 3 cm) of *concanavalin A* (Con A)-Sepharose (Pharmacia, Uppsala, Sweden) were prepared. The samples were washed through with 10 column volumes of lectin buffer and then eluted with 80 ml of the lectin buffer containing D-a-methyl-glucopyranoside (200 mmol/l), followed by 80 ml of the lectin buffer containing D-a-methyl-mannopyranoside (200 mmol/l).

# Results

## Chromatofocusing

The goal of this study was to document significant changes in the isoforms of CG as the hormone is metabolized, rather than to document microheterogeneity of the isoforms. Therefore, the isoforms are referred to as basic if they eluted at a pH greater than 7.3, neutral/slightly acidic if they eluted between the pHs of 7.3 and 5, and acidic if they eluted at a pH less than 5. The results of this study are summarized in Table 23.2.

CG from Normal, Nonpregnant Humans

Figure 23.1 shows a representative chromatofocusing profile from the PBE 94 column of the CG contaminating pituitary LH (AFP-0642B). Only one isoform that eluted at a pH of 7.3–7.0 was detected. A very small proportion (3%) of the CG in this LH preparation eluted from the PBE 118 column (pH range = 11–7) with pHs of 8.9 and 7.9 (data not shown). The CG activity shown in Figure 23.1 represents greater than 95% of the CG that was added to the column.

Figure 23.2 shows representative profiles of CG found in pituitary homogenates. Five pituitary glands were studied in separate experiments.

TABLE 23.2. Percent of chorionic gonadotropin that eluted at a basic pH, a neutral/slightly acidic pH, and an acidic pH from nonpregnant and pregnant humans.

| Nonpregnant humans | | | | Pregnant humans | | | |
|---|---|---|---|---|---|---|---|
| | pH of elution[a] | | | | pH of elution[a] | | |
| Sample | >7.3 | 7.3–5 | <5 | Sample | >7.3 | 7.3–5 | <5 |
| Pituitary CG | | | | Placental CG | | | |
|   Type 1[b] | 0 | 96 | 4 |   1.[d] Placenta | Not available | | |
|   Type 2[c] | 23 | 45 | 32 |   2. Placenta | 46 | 44 | 11 |
|   LH(0642B) | 1 | 97 | 2 |   3. Placenta | 46 | 45 | 9 |
| Secreted CG | | | | Secreted CG | | | |
|   Fetal cultures | 0 | 70 | 30 |   1. Serum | 0 | 76 | 24 |
|   Serum | 0 | 32 | 68 |   2. Serum | 0 | 71 | 29 |
| | | | |   3. Serum | 1 | 51 | 49 |
| Excreted CG | | | | Excreted CG | | | |
|   hMG | 0 | 15 | 85 |   CR-125 | 0 | 6 | 94 |
|   Urine | 0 | 29 | 71 |   1. Urine | 0 | 0 | 100 |
| | | | |   2. Urine | 0 | 3 | 97 |
| | | | |   3. Urine | 0 | 11 | 89 |

[a] Results are given as the percent of the sample eluting at the indicated pHs (>7.3, 7.3–5, and <5).
[b] Type 1 = a pituitary homogenate that demonstrates only one isoform on a chromatofocusing column.
[c] Type 2 = a pituitary homogenate that demonstrates 3–6 isoforms on a chromatofocusing column.
[d] 1, 2, and 3 refer to the first, second, and third trimesters of pregnancy.

In three instances (Fig. 23.2A), one main isoform, eluting at a pH of 7.1, was obtained, while in two other instances (Fig. 23.2B), a more diffuse pattern was obtained, with basic, neutral/acidic, and acidic isoforms being detected. The data shown in Figures 23.1 and 23.2 demonstrate that several isoforms are produced in the pituitary and that only the isoform eluting at a pH of 7 is copurified with LH in the preparation of immunochemical-grade LH(AFP-0642B).

There was a significant difference in the amount of CG present in the two types of pituitary homogenates. The homogenates with only one isoform (type 1), as shown in Figure 23.2A, contained 169 ng/pituitary, while those with the more diffuse pattern (type 2), as shown in Figure 23.2B, contained 76 ng/pituitary, each isoform containing 10–25 ng of CG. Upon digestion with neuraminidase, the acidic isoforms were no longer present, while the more basic isoforms (pHs of 10.6, 9.5, and 7.5) were not affected.

The chromatofocusing pattern of secreted CG, obtained from the supernatant of fetal pituitary cell cultures, is shown in Figure 23.3. There are both neutral/slightly acidic and acidic isoforms, but there are no basic isoforms present. The majority (69.9%) of the CG added to the column

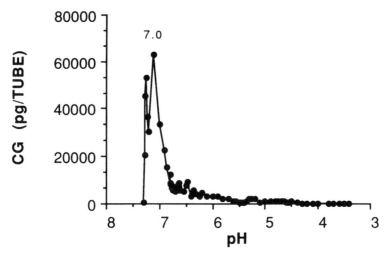

FIGURE 23.1. Fractionation of the CG found in a sample of pituitary LH (AFP-0642B) on a polybuffer chromatofocusing column (PBE 94, pH 7–3). The CG in each 3 ml fraction was measured by a CG immunoradiometric assay. The number above the curve represents the pH at which the major isoform eluted. This profile was observed in two independent experiments.

eluted between the pHs of 5 and 7.3, while 30.1% eluted at a pH less than 5. A second source of secreted CG was obtained by affinity purification of serum of postmenopausal females. The data, shown in Figure 23.4, indicated the presence of neutral/slightly acidic (32%) and acidic (68%) isoforms in the serum of postmenopausal females.

The isoforms of excreted CG are represented in Figure 23.5. This figure contains the patterns both before (A) and after (B) digestion of hMG with neuraminidase. Before neuraminidase treatment, there is one dominant acidic isoform of CG in hMG. Neuraminidase treatment shifts the elution of this CG to a pH of 7, indicating the removal of sialic acid residues. The majority of the CG from fresh menopausal urine also eluted as an acidic isoform; however, some neutral/slightly acidic isoforms (29%) were demonstrated.

CG from Pregnant Humans

For comparison with the CG from nonpregnant humans, CG from three sources of pregnant humans is shown in Figure 23.6. The isoforms from placental extracts (third trimester) are shown in Figure 23.6A. The pattern obtained with serum from the first trimester of pregnancy (5 weeks) is shown in Figure 23.6B, both before and after affinity purification of the serum. Figure 23.6C shows the pattern obtained with the CG in CR-125 and with the CG in dialyzed urine from the first trimester of pregnancy.

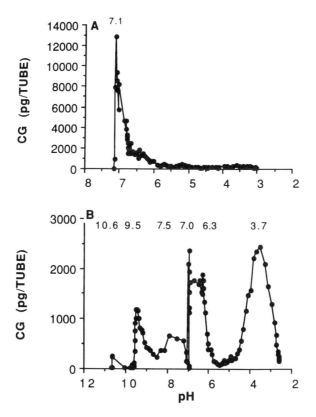

FIGURE 23.2. Fractionation of the CG in homogenized pituitary glands on chromatofocusing columns (PBE 94, pH 7–3 and PBE 118, pH 11–7) as described in Figure 23.1. *A:* A chromatofocusing profile with a single elution pH of 7.1, representative of three independent experiments. *B:* A diffuse chromatofocusing profile with elution pHs of 10.6, 9.5, 7.5, 7.0, 6.3, and 3.7, representative of two independent experiments.

The results of the chromatofocusing data are summarized in Table 23.2. In samples both from normal, nonpregnant subjects and from pregnant subjects, the presecreted CG exists in the largest number of detectable isoforms and also in the most basic isoforms. In the excreted CG samples, the most acidic isoform was predominant. Neuraminidase digestion of CG from all sources resulted in the formation of an isoform that eluted at a pH of 6.5–7.4. The more basic isoforms that elute at pHs of 10.6, 9.5, and 7.5 were not affected by the neuraminidase treatment, suggesting that their high elution pH is the result of some factor other than simply the lack of sialic acid residues.

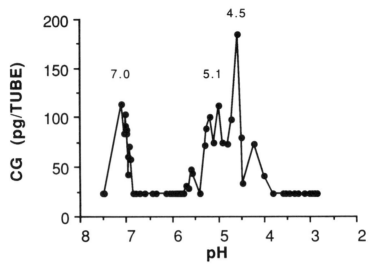

FIGURE 23.3. Fractionation of the CG in the supernatant from fetal pituitary cell cultures on a chromatofocusing column (PBE 94, pH 7–3) as described in Figure 23.1.

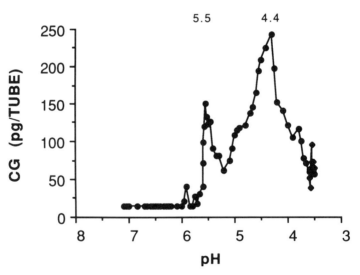

FIGURE 23.4. Fractionation of the CG in serum from postmenopausal females on a chromatofocusing column (PBE 94, pH 7–3) as described in Figure 23.1. This profile was obtained in two independent experiments.

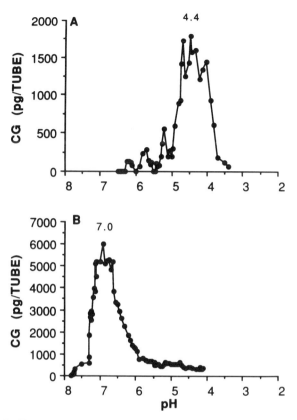

FIGURE 23.5. Fractionation of the CG in hMG (urinary human menopausal gonadotropin) on a chromatofocusing column (PBE 94, pH 7–3) as described in Figure 23.1. *A:* hMG before treatment with neuraminidase. *B:* hMG after treatment with neuraminidase. Each profile was observed in two independent experiments.

## Lectin Binding

Figure 23.7 is a diagram of CG, showing the proposed carbohydrate structures for placental CG (28, 29). Microheterogeneities of the carbohydrate structures exist; however, these basic structures are the most prevalent. Con A binds to terminal mannose units or to mannose units only substituted on carbon number two. At least two available mannose units must be present for the carbohydrate to bind to Con A, and internal carbohydrates can alter the binding behavior of molecules to Con A. Lentil lectin binds to internal fucose residues. Ricin lectin binds to terminal galactose residues. Galactose residues that are masked by neuraminic acid will not bind to the Ricin lectin.

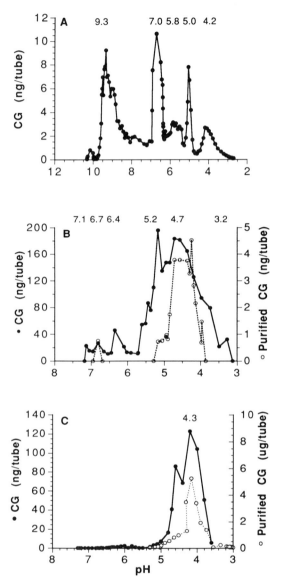

FIGURE 23.6. Fractionation of the CG in samples from pregnant females on chromatofocusing columns (PBE 94, pH 7–3 and PBE 118, pH 11–7) as described in Figure 23.1. *A:* CG obtained from a placental extract. *B:* CG obtained from the serum of a first-trimester pregnant female before (closed circles) and after (open circles) affinity purification. *C:* CG obtained from CR-125, a urinary preparation from pregnant females (closed circles), and from dialyzed urine from first-trimester pregnancy (open circles).

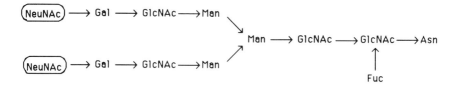

FIGURE 23.7. Accepted structures for the carbohydrate moieties attached to the asparagine (Asn) and serine (Ser) residues of placentally produced CG (28, 29). NeuNAc = N-acetyl neuraminic acid (sialic acid); Gal = galactose; GlcNAc = N-acetylglucosamine; Man = mannose; Fuc = fucose; and GalNAc = N-acetylgalactosamine.

Figure 23.8 shows the elution patterns obtained when CR-125 (Fig. 23.8A), hMG (Fig. 23.8B), and CG from pregnancy serum (Fig. 23.8C) were applied to the Con A column. All three samples were bound 100% to the Con A. When 200 mmol/l glucoside was added to the column, 74% of the CR-125, 87% of the hMG, and 79% of the pregnancy serum CG were released from the columns. The remaining amounts were released from the columns by the application of 200 mmol/l mannoside.

The three samples were not as consistent on the lentil lectin column, but they did show similar binding characteristics (Fig. 23.9). Thirty-two percent of the applied CR-125 was tightly bound, being released from the column only after the application of 200 mmol/l mannoside. Likewise, 42% of the CG in hMG and 51% of the CG in pregnancy serum were tightly bound. The percentages of unbound material from CR-125, hMG, and pregnancy serum were 12%, 14%, and 12%, respectively.

Figure 23.10 demonstrates that the CG from pregnancy serum bound tightly to the Ricin column (Fig. 23.10C), while the CG from CR-125 (Fig. 23.10A) and from hMG (Fig. 23.10B) did not bind the Ricin column. Some samples were applied to the Ricin column after digestion with neuraminidase and/or isoform separation by chromatofocusing. After removal of neuraminic acid, underlying galactose residues, if present, should be available to bind to the lectin. The binding behavior to Ricin lectin of one isoform of CG from hMG before and after neuraminidase treatment are summarized in Table 23.3. Untreated CG from hMG did not bind to the Ricin column; however, treated CG from hMG, which eluted at a pH of 7, did bind to the column, indicating that

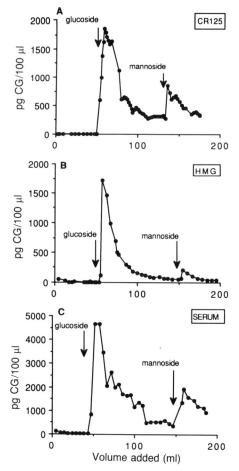

FIGURE 23.8. Elution profiles of CG from (A) CR-125, (B) hMG, and (C) the serum from a first-trimester pregnant female from a Con A-Sepharose column. Arrows indicate where glucoside and mannoside were added to the column. All three samples bound 100% to the Con A column. These profiles represent at least two independent experiments for each source.

the CG in hMG has galactose residues that can be exposed by neur-aminidase digestion.

Also summarized in Table 23.3 are the Ricin lectin binding characteristics of the isoforms of the two types of pituitary homogenates shown in Figure 23.2. Whereas 94% of the CG in a type 1 pituitary homogenate bound to the ricin column, the isoforms of the type 2 pituitary homogenate behaved quite differently. Forty percent of the most basic isoform (pH = 9.5) bound to the column, but only 1%–5% of the other two isoforms bound to the column.

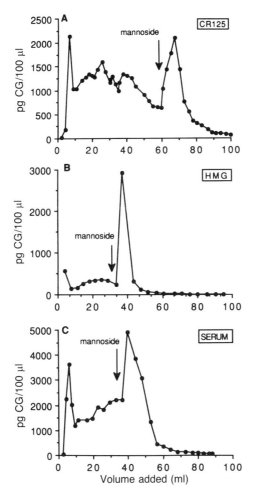

FIGURE 23.9. Elution profiles of CG from (*A*) CR-125, (*B*) hMG, and (*C*) the serum from a first-trimester pregnant female from a Lentil-Sepharose column. Arrows indicate where mannoside was added to the column. At least 86% of all three samples bound to the Lentil column. These profiles represent at least two independent experiments for each source.

## Discussion

Samples have been analyzed that include CG before it is secreted from the pituitary, CG that has been secreted from the pituitary, and CG that has been excreted in the urine. The analysis, by chromatofocusing and lectin affinity chromatography, demonstrates that there are multiple isoforms of CG produced by the pituitary gland. The chromatofocusing

336    A.L. Sawitzke et al.

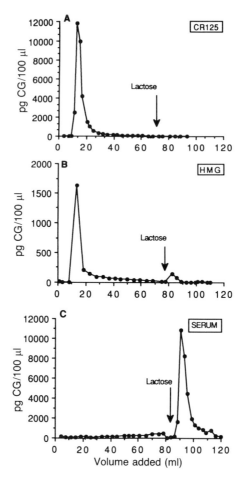

FIGURE 23.10. Elution profiles of CG from (*A*) CR-125, (*B*) hMG, and (*C*) the serum from a first-trimester pregnant female from a Ricin 120-Sepharose column. Arrows indicate where lactose was added to the column. Nearly 100% of the CG from pregnancy serum bound tightly to the Ricin column, while none of the CG from CR-125 and hMG bound to the column. These profiles represent at least two independent experiments for each source.

and lectin binding patterns of these isoforms, at all three stages of the metabolic pathway, follow very closely with those known for the other glycoproteins. These data indicate that the isoforms that elute at low pHs are due to increased sialylation of the carbohydrate moieties; however, the degree of sialylation is probably not the only difference that exists among these isoforms.

Previous reports of artifactual isoforms or the loss of genuine isoforms could result from (i) proteolysis of the CG during sample preparation,

TABLE 23.3. Percent binding to a Ricin 120-Sepharose column of samples of hMG (before and after neuraminidase-treatment) and of pituitary homogenates that had been fractionated by chromatofocusing.

| | % unbound | % bound | % tightly bound |
|---|---|---|---|
| Sample | | | |
| hMG | 80 | 15 | 5 |
| Treated hMG[a] | 15 | 77 | 9 |
| Pituitary | | | |
| Type 1[b] | | | |
| pH = 7.2 | 6 | 94 | 0 |
| Type 2[c] | | | |
| pH < 5.0 | 95 | 1 | 4 |
| pH = 5.0–7.3 | 99 | 1 | 0 |
| pH > 7.3 | 60 | 24 | 16 |

[a] Neuraminidase treated.
[b] Type 1 = a pituitary homogenate that demonstrates only one isoform on a choromatofocusing column.
[c] Type 2 = a pituitary homogenate that demonstrates 3–6 isoforms on a chromatofocusing column.

(ii) nicking and subsequent separation of small fragments (30–32), (iii) inaccurate assay of the various isoforms, or (iv) loss or preferential purification of select isoforms. We have controlled for each of these possibilities, and they would not explain the data we have obtained. It is unlikely that the isoforms shown here represent proteolysis or nicking for two reasons. First, the IRMA used here has been shown to detect only intact CG, not βCG, αCG, LH, or the carboxyl tail of CG (23). Second, the CG and the LH in some samples were analyzed by Sephadex G-100 chromatography and by polyacrylamide gel electrophoresis followed by Western blotting. Since no small molecular weight CG peaks were detected and since a nondenaturing Western blot indicated CG at a molecular weight of 50–60 kd, proteolysis, nicking, and cross-reaction with partially denatured LH were ruled out as sources of artifactual isoforms (8).

The IRMA assay used in this study was very carefully optimized for osmolality, buffer, pH effects, and potential cross-reactivity of various substances, including αCG and βCG and LH. The pregnancy samples reported herein duplicate previous work on pregnancy CG, demonstrating that the isoforms were accurately measured. Finally, isoforms were not lost or preferentially selected during purification procedures as demonstrated by the data in Figure 23.6B. The same isoform profiles were obtained before and after affinity purification of pregnancy serum. Fresh urine, hMG, and CR-125 all demonstrated acidic isoforms upon chromatofocusing; however, the fresh urine, which was affinity purified, demonstrated a significant amount of neutral/slightly acidic isoform.

The results presented here should be compared to those of Hoermann et al. (2). These workers found CG present in pituitary extracts with a pI of 4.5 (very similar to that of urinary CG from pregnancy). They also found only 75% of this pituitary CG bound to Con A. The differences in our results might be explained by the fact that they used only one of three protein peaks from their extracted pituitaries, while we used the entire pituitary homogenate. Also, they used an isoelectric focusing gel with a pH range of 3–6; thus, a fraction with a higher pI could have been overlooked. Review of their figures shows there was βCG activity at a pI of 6. Chen et al. also described pituitary CG with a pI of 4.5, but did make note of an isoform with a pI of 6–7 (33). These results compare with the pituitary homogenates reported in Figure 23.2B of the present study if the measurements were stopped at a pH of 6.

Robertson et al. studied CG from pituitaries, urine, and plasma of nonpregnant subjects in 1978 (34). They found CG with a pI of 4.5 in both the urinary and pituitary extracts. These workers saw no quantitative difference in the CG concentrations in young females, males, postmenopausal females, females receiving oral contraceptives, or subjects stimulated with GnRH. These data are in conflict with several more recent studies indicating that the detected amount of CG in nonpregnant humans depends on the technique of isolation and the physical and hormonal status of the experimental subjects (6, 35).

The reported differences in the total amounts of CG and in the isoforms detected by chromatofocusing of pituitary homogenates could be the result of (i) the endocrine status of the subjects at the time of death or (ii) assay differences, such as cross-reaction with LH and/or sensitivity. Hartree et al. demonstrated that the amount of CG in pituitary extracts could vary 50-fold (0.5–4.4 µg/gland) (4), and Chen observed 50–750 ng/gland (1), while the present work only demonstrates a twofold difference in CG content among five glands (76–169 ng/gland). The results of Ulloa-Aguirre et al. (9), which show that cultured cytotrophoblast cells produce CG with three peaks on isoelectrical focusing (7.3–7, 5.6–5.4, and 5.1–4.8), agree very well with the present data on the isoforms of secreted CG.

It is important to note that urinary excretion accounts for only 21% of the disposal of pregnancy CG (36). A molecule known as β-core (amino acids 6–40 and 55–92) actually appears to be the most abundant CG-like molecule in pregnancy urine; however, it is not detected (37–39) or is detected only in very small amounts (40) in pregnancy serum. β-core perhaps is a urinary metabolite of βCG (41) or is masked by a larger macromolecule in the serum (40). It is interesting to note that the carbohydrates of the β-core have mannose, but no glucosamine, galactose, or sialic acid (39).

The IRMA used in this study is specific for whole CG, so the isoforms that are reported here are presumably not due to the β-core metabolite. However, the behavior of type 2 pituitary homogenates on the Ricin lectin column may be consistent with a carbohydrate structure similar to that of β-core. The binding of only 40% of the isoform that eluted at a pH of 9.5 to the Ricin column indicates that 60% of these molecules contain no galactose residues. This interpretation is supported by the observation that neuraminidase digestion does not affect the isoforms that elute at pHs greater than 7; thus, the lack of binding is not due to a simple masking of the galactose residues by neuraminic acid. The behavior of the type 2 pituitary homogenate on the Ricin column might also be explained by the fact that neuraminic acid and galactose are not normally added to proteins until they reach the Golgi complex, whereas other carbohydrates are added in the endoplasmic reticulum. Perhaps the CG that did not bind to the Ricin column was still in the endoplasmic reticulum at the time the sample was obtained. Thus, the secretory activity of the pituitary at the time of sampling would alter the isoforms detected.

Further studies, including detailed carbohydrate and amino acid analysis, will help determine the true nature of the pituitary CG; however, this is the first study that has systematically examined the heterogeneity of pituitary CG along its metabolic pathway. By obtaining more information on the structure of this circulating CG, this study helps to form a base from which to study the gene regulation of the hormone glycosylation and secretion.

## References

1. Chen H-C, Hodgen GD, Matsuura S, et al. Evidence for a gonadotropin from nonpregnant subjects that has physical, immunological, and biological similarities to human chorionic gonadotropin. Proc Natl Acad Sci USA 1976;73:2885–9.
2. Hoermann R, Spoettl G, Moncayo R, Mann K. Evidence for the presence of human chorionic gonadotropin (hCG) and free β-subunit of hCG in the human pituitary. J Clin Endocrinol Metab 1990;71:179–86.
3. Odell WD, Griffin J, Sawitzke A. Chorionic gonadotropin secretion in normal, nonpregnant humans. Trends Endocrinol Metab 1990;1:418–21.
4. Hartree AS, Shownkeen RC, Stevens VC, Matsuura S, Ohashi M, Chen H-C. Studies of the human chorionic gonadotrophin-like substance of human pituitary glands and its significance. J Endocrinol 1983;96:115–26.
5. Matsuura S, Ohashi M, Chen H-C, et al. Physicochemical and immunological characterization of an hCG-like substance from human pituitary glands. Nature 1980;286:740–1.
6. Stenman UH, Alfthan H, Ranta T, Vartiainen E, Jalkanen J, Seppala M. Serum levels of human chorionic gonadotropin in nonpregnant women and men are modulated by gonadotropin-releasing hormone and sex steroids. J Clin Endocrinol Metab 1987;64:730–6.

7. Wang HY, Segal SJ, Koide SS. Classification and characterization of an incompletely glycosylated form of human chorionic gonadotropin from human placenta. Endocrinology 1988;123:795–803.

8. Sawitzke AL, Griffin J, Odell WD. Purified preparations of human luteinizing hormone are contaminated with small amounts of a chorionic gonadotropin-like material. J Clin Endocrinol Metab 1991;72:841–6.

9. Ulloa-Aguirre A, Mendez JP, Cravioto A, Grotjan E, Damian-Matsumura P, Espinoza R. Studies on the microheterogeneity of chorionic gonadotrophin secreted by the human cytotrophoblast in culture. Hum Reprod 1990;5:661–9.

10. Ulloa-Aguirre A, Espinoza R, Damian-Matsumura P, Chappel SC. Immunological and biological potencies of the different molecular species of gonadotrophins. Hum Reprod 1988;3:491–501.

11. Nwokoro N, Chen H-C, Chrambach A. Physical, biological, and immunological characterization of highly purified urinary human chorionic gonadotropin components separated by gel electrofocusing. Endocrinology 1981; 108:291–9.

12. Lichtenberg V, Holtje F, Graesslin D, Bettendorf G. Different degree of polymorphism of hCG in placenta, serum, and urine. Acta Endocrinol (Copenh) 1986;111(suppl 274):A153.

13. Wide L. Median charge and charge heterogeneity of human pituitary FSH, LH, and TSH, I. Zone electrophoresis in agarose suspension. Acta Endocrinol (Copenh) 1985;109:181–9.

14. Nilsson B, Rosen SW, Weintraub BD, Zopf DA. Differences in the carbohydrate moieties of the common α-subunits of human chorionic gonadotropin, luteinizing hormone, follicle-stimulating hormone, and thyrotropin: preliminary structural inferences from direct methylation analysis. Endocrinology 1986;119:2737–43.

15. Graesslin D, Weise HC, Braendle W. The microheterogeneity of human chorionic gonadotropin (hCG) reflected in the β-subunits. FEBS Lett 1973; 31:214–6.

16. Galle PC, Ulloa-Aguirre A, Chappel SC. Effects of oestradiol, phenobarbitone, and luteinizing hormone releasing hormone upon the isoelectric profile of pituitary follicle-stimulating hormone in ovariectomized hamsters. J Endocrinol 1983;99:31–9.

17. Ulloa-Aguirre A, Mejia JJ, Dominguez R, Guevara-Aguirre J, Diaz-Sanchez V, Larrea F. Microheterogeneity of anterior pituitary FSH in the male rat: isoelectric focusing pattern throughout sexual maturation. J Endocrinol 1986;110:539–49.

18. Miller C, Ulloa-Aguirre A, Hyland L, Chappel S. Pituitary follicle-stimulating hormone heterogeneity: assessment of biologic activities of each follicle-stimulating hormone form. Fertil Steril 1983;40:242–7.

19. Baldwin DM, Highsmith RF, Ramey JW, Krummen LA. An in vitro study of LH release, synthesis, and heterogeneity in pituitaries from proestrous and short-term ovariectomized rats. Biol Reprod 1986;34:304–15.

20. Blum W, Gupta D. Age and sex-dependent nature of the polymorphic forms of rat pituitary FSH: the role of glycosylation. Neuroendocrinol Lett 1980; 2:357–65.

21. Lichtenberg V, Urban C. Comparison of the microheterogeneity of stored and released rat pituitary lutropin (LH) after electric focusing. Acta Endocrinol (Copenh) 1983;102(suppl 253):101–2.

22. Graesslin D, Lichtenberg V, Holtje F, Schedlinski G, Rinne G. The role of sialic acid for survival of gonadotrophins and the different HCG isohormones in placenta, serum, and urine. Acta Endocrinol (Copenh) 1984;105(suppl 262):A81.

23. Griffin J, Odell WD. Ultrasensitive immunoradiometric assay for chorionic gonadotropin which does not cross-react with luteinizing hormone nor free β chain of hCG and which detects hCG in blood of nonpregnant humans. J Immunol Methods 1987;103:275–83.

24. Odell WD, Griffin J. Two-monoclonal-antibody sandwich-type assay for human luteinizing hormone which shows no cross reaction with chorionic gonadotropin. Clin Chem 1987;33:1603–7.

25. Snyder PJ, Bashey HM, Montecinos A, Odell WD, Spitalnik SL. Secretion of multiple forms of human luteinizing hormone by cultured fetal human pituitary cells. J Clin Endocrinol Metab 1989;68:1033–8.

26. Rodgers M, Mitchell R, Lambert A, Peers N, Robertson WR. Human chorionic gonadotrophin contributes to the bioactivity of Pergonal. Clin Endocrinol 1992;37:558–64.

27. Moyle WR, Bahl OP, Marz L. Role of carbohydrate of human chorionic gonadotropin in the mechanism of hormone action. J Biol Chem 1975; 250:9163–9.

28. Kessler MJ, Mise T, Ghai RD, Bahl OP. Structure and location of the O-glycosidic carbohydrate units of human chorionic gonadotropin. J Biol Chem 1979;254:7909–14.

29. Kessler MJ, Reddy MS, Shah RH, Bahl OP. Structures of N-glycosidic carbohydrate units of human chorionic gonadotropin. J Biol Chem 1979; 254:7901–8.

30. Ward DN, Reichert LE Jr, Liu W-K. Chemical studies of luteinizing hormone from human and ovine pituitaries. Recent Prog Horm Res 1973;29: 533–61.

31. Nishimura R, Ide K, Utsunomiya T, Kitajima T, Yuki Y, Mochizuki M. Fragmentation of the β-subunit of human chorionic gonadotropin produced by choriocarcinoma. Endocrinology 1988;123:420–5.

32. Puisieux A, Bellet D, Troalen F, et al. Occurrence of fragmentation of free and combined forms of the β-subunit of human chorionic gonadotropin. Endocrinology 1990;126:687–94.

33. Chen H-C, Shimohigashi Y, Dufau ML, Catt KJ. Characterization and biological properties of chemically deglycosylated human chorionic gonado-tropin. J Biol Chem 1982;257:14446–52.

34. Robertson DM, Suginami H, Montes IIII, Puri CP, Choi SK, Diczfalusy E. Studies on a human chorionic gonadotrophin-like material present in non-pregnant subjects. Acta Endocrinol (Copenh) 1978;89:492–505.

35. Kyle CV, Griffin J, Odell WD. GnRH stimulation of human chorionic gonadotropin (hCG) secretion in normal men and women [Abstract]. Clin Res 1988;36(suppl 1):181A.

36. Wehmann RE, Nisula BC. Metabolic clearance rates of the subunits of human chorionic gonadotropin in man. J Clin Endocrinol Metab 1979; 48:753–9.

37. Birken S, Armstrong EG, Gawinowicz Kolks MA, et al. Structure of the human chorionic gonadotropin β-subunit fragment from pregnancy urine. Endocrinology 1988;123:572–83.

38. Blithe DL, Akar AH, Wehmann RE, Nisula BC. Purification of β-core fragment from pregnancy urine and demonstration that its carbohydrate moieties differ from those of native human chorionic gonadotropin-β. Endocrinology 1988;122:173–80.
39. Blithe DL, Wehmann RE, Nisula BC. β-core chemical and clinical properties. Trends Endocrinol Metab 1990;1:394–8.
40. Kardana A, Cole LA. Serum HCG β-core fragment is masked by associated macromolecules. J Clin Endocrinol Metab 1990;71:1393–5.
41. Wehmann RE, Nisula BC. Characterization of a discrete degradation product of the human chorionic gonadotropin β-subunit in humans. J Clin Endocrinol Metab 1980;51:101–5.

# 24

# Thyrotropin Abnormalities in Central Hypothyroidism

PENELOPE K. MANASCO, DIANA L. BLITHE, SUSAN R. ROSE,
MARIE C. GELATO, JAMES A. MAGNER, AND BRUCE C. NISULA

The normal physiological pattern of TSH secretion is characterized by circadian variation. After its nadir in the late afternoon, the serum TSH concentration rises to a peak around midnight, remains on a plateau for several hours, and then begins its decline. The rise from afternoon nadir to nighttime peak is called the nocturnal TSH surge; it is a hallmark of the circadian variation. Not only is the thyroid gland exposed to the highest concentrations of TSH at night, the thyroid gland is most active in releasing thyroidal iodine at night (1). Thus, the thyroid gland receives the preponderance of its trophic support from TSH during the night, subsequent to the nocturnal surge of TSH.

## Abnormal Secretion of TSH in Central Hypothyroidism

Central hypothyroidism results when a hypothalamic and/or pituitary disorder leads to diminished function of the thyroid axis. Research on central hypothyroidism was expanded substantially in the 1980s, consequent to the development of ultrasensitive TSH immunoassays that permitted the accurate measurement of the full range of TSH in normal subjects (2). Such assays unequivocally demonstrated that serum TSH levels in patients with central hypothyroidism measured in blood taken in daytime clinics were not subnormal, but generally within the range of normal (2). According to classical concepts of feedback axis regulation, serum TSH levels would be expected to be subnormal in central hypothyroidism, since decreased trophic support of thyroid gland function is the cause of hypothyroidism. Research on this paradox offered the promise of not only gaining greater insight into the pathophysiology of central hypothyroidism, but also enhancing the quality of diagnostic procedures for this condition.

The search for the implied abnormality in TSH secretion in central hypothyroidism required not only using an ultra sensitive TSH immunoassay, but admitting patients to the hospital for a more extensive evaluation of the circadian pattern of TSH. Although adults were studied in the initial series (3), a larger series has involved children (4); this chapter will emphasize the latter.

The parameters of the circadian variation of TSH were established in 96 normal children by measuring serum TSH hourly around the clock. From these data, the timing of the nadir TSH and peak TSH were determined, and a method for computing the magnitude of the nocturnal TSH surge was selected (5). A series of studies of this nocturnal surge test were conducted. Both adults and children with unequivocal central hypothyroidism (based on the criterion of a subnormal free *thyroxine* [$T_4$] level) exhibited a deficient nocturnal surge of thyrotropin (3, 4). Eighteen of nineteen patients with unequivocal central hypothyroidism had a subnormal nocturnal TSH surge. The nocturnal TSH surge test showed excellent reliability in excluding thyroid axis defects. In all but one of 33 patients with hypothalamic and/or pituitary disorders whose nocturnal TSH surge was within normal limits, the free $T_4$ was also within normal limits and their mean free $T_4$ was not significantly different from that of the normal subjects. Demonstrating that hypothyroidism per se is not the cause of the deficiency of the nocturnal TSH surge in central hypothyroidism was accomplished by showing that patients with mild primary hypothyroidism exhibited an intact nocturnal TSH surge (3, 4).

As a diagnostic test for central hypothyroidism, the nocturnal TSH surge test has proven to have far greater sensitivity than the TRH test (4). Of 11 patients with unequivocal central hypothyroidism based on subnormal serum free $T_4$ values, 10 had subnormal nocturnal TSH surge tests, while only 2 had abnormal TRH tests (4). Thus, the nocturnal TSH surge test had a sensitivity of 91% in detecting patients with unequivocal central hypothyroidism, whereas the TRH test had a sensitivity of 18%. Although one group has reported that the nocturnal TSH surge test is not as sensitive as the TRH test (6), it should be clarified that they did not measure the nocturnal TSH surge using current methodology, nor did they use free $T_4$ measurements to select patients with central hypothyroidism.

The nocturnal TSH surge test may be a more sensitive diagnostic test than the free $T_4$ test. If this is the case, then the nocturnal TSH surge test should be able to select among patients with hypothalamic and/or pituitary disorders, those patients with central hypothyroidism whose free $T_4$ values are subnormal for the individual patient, albeit within the normal range of free $T_4$ values. This class of patients would be similar to those patients with primary hypothyroidism and slightly elevated TSH levels and normal thyroxine levels. As a group, such patients with hypothalamic/pituitary disorders and a subnormal TSH surge should

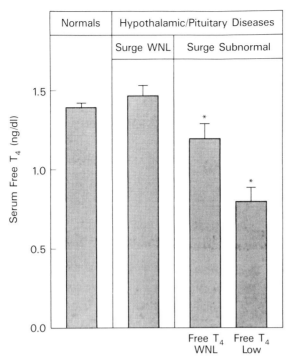

FIGURE 24.1. Mean free $T_4$ values in 56 patients with short stature due to hypothalamic and/or pituitary diseases and in 96 normal children. The 38 patients with a normal surge had a free $T_4$ similar to the normal population. Of the 13 children with an abnormal surge, 8 also had a subnormal free $T_4$ concentration. Five patients had a subnormal surge and a free $T_4$ within the normal range, but a significantly reduced mean free $T_4$.

exhibit a mean free $T_4$ value significantly lower than that of normal subjects. Indeed, among 56 children with short stature due to hypothalamic and/or pituitary disorders, 13 patients had an abnormal nocturnal TSH surge and a subnormal free $T_4$, whereas 5 patients had an abnormal nocturnal TSH surge and a free $T_4$ that was within normal limits; the mean free $T_4$ in the latter group was $1.20 \pm 0.06 \, \text{ng/dl}$, significantly less than that of normal subjects ($1.40 \pm 0.01 \, \text{ng/dl}$ [$P < 0.01$]), indicating significantly decreased thyroid function. Importantly, the free $T_4$ values of normal control subjects and patients with hypothalamic-pituitary disorders who had a normal nocturnal TSH surge are similar (Fig. 24.1). In contrast, patients with a normal free $T_4$ and an abnormal nocturnal TSH surge had significantly lower concentrations of free $T_4$ than either the normal control subjects or those patients with a normal TSH surge.

# Abnormal Structure of TSH in Central Hypothyroidism

The existence of a subset of patients with central hypothyroidism with supranormal immunoreactive serum TSH levels has prompted consideration of abnormalities in TSH structure. Studies of *adenylate cyclase stimulating activity* (B) and *receptor binding activity* (R) of immunopurified TSH from patients with central hypothyroidism revealed significantly decreased R/I and B/I ratios of the TSH in their sera (7). A single bolus of TRH was successful in increasing the R/I ratio in two of three patients, but the B/I ratio improved in only one. After 20–30 days of TRH treatment, all six patients had improvement of the B/I ratio.

A variety of biochemical and animal studies point to the carbohydrate of TSH as a likely site for structural abnormalities to occur (8–11). Assessment of alterations in carbohydrate structure in serum TSH forms has been aided by the development of lectin affinity chromatography. Lectins like *concanavalin A* (Con A) are proteins that bind specific carbohydrate moieties. Methodological issues should be emphasized in the study of thyrotropin measurements in lectin binding studies. There is a considerable matrix effect from the sugars used to elute the thyrotropin when using immunoassays to measure TSH. To minimize these effects, we found it critical to precipitate the TSH and reconstitute the samples in a common buffer.

We studied the apparent size and Con A binding properties of TSH obtained from sera from three patients with central hypothyroidism who had elevated TSH levels and from seven patients with primary hypothyroidism and similar levels of TSH. We found no significant difference between the apparent molecular weight of the serum TSH in central and primary hypothyroid patients, as measured by molecular sieve chromatography on a Sephadex G-75 column. However, Con A binding characteristics were different between the two groups , as shown in Figure 24.2. In the patients with central hypothyroidism, we found that a significantly greater percent of the intact TSH was unable to bind to Con A and that a smaller percentage was tightly bound when compared to the patients with primary hypothyroidism. The percentage of TSH that was tightly bound to Con A was similar to that found by Taylor et al. (10) and much higher than that found by Miura (11).

These disparities in the pattern of binding may be due to species variation, intrapituitary versus secreted TSH, specificity in ·processing due to different sites and types of lesions, or to technique. Immunopurification may remove some of the glycoforms of TSH; alternatively, immunoassay may not recognize all isoforms. Using intact sera in these binding assays is more consistent with the in vivo state; however, results must be interpreted as the overall binding of glycoproteins with three side chains. Selection of an appropriate control group is also critical. To

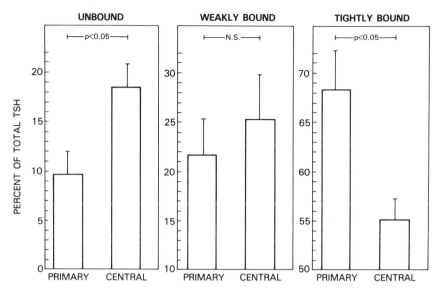

FIGURE 24.2. Con A binding of TSH in the serum of patients with primary and central hypothyroidism. One ml of serum from patients with primary hypothyroidism ($n = 7$) or central hypothyroidism ($n = 3$) was loaded on a 1 ml column of concanavalin A. The unbound fraction was eluted with PBS-BSA. The weakly bound fraction was eluted with 10 mM α-methyl-D-glucoside, and the tightly bound fraction was eluted with 0.5 M α-methyl-D-mannoside. The column was then stripped of the most tightly bound material with 1 M ammonium hydroxide and immediately neutralized to prevent destruction of the TSH. The protein was then precipitated with acetone and the TSH assayed in 10% blank TSH serum using the Serono MAIA IRMA. The tightly bound fraction includes the material eluted with α-methyl-mannose, plus the material eluted with the ammonium hydroxide.

control for the effect of hypothyroidism per se on the structure of TSH, a group of patients with primary hypothyroidism of a similar degree of severity probably provides the most appropriate control group. Methodological issues aside, it is apparent, however, that all groups agree that there are alterations in carbohydrate structure of TSH in some patients with central hypothyroidism. The role that these differences may play in bioactivity must be clarified by further structure-function studies.

In summary, differences in both the secretion and bioactivity of thyrotropin appear to be instrumental in the development of central hypothyroidism. The potential role of TRH in both of these processes is intriguing. The participation of the releasing hormones in posttranslational processing may be critical for the bioactivity and pattern of release of thyrotropin.

# *References*

1. Nicoloff JT, Fisher DA, Appelman MD. The role of glucocorticoids in the regulation of thyroid function in man. J Clin Invest 1970;49:1922–9.
2. Wehmann RE, Rubenstein HA, Pugeat MM, Nisula BC. Extended clinical utility of a sensitive and reliable radioimmunoassay of thyroid-stimulating hormone. South Med J 1983;76:969.
3. Caron PJ, Nieman LK, Rose SR, Nisula BC. Deficient nocturnal surge of thyrotropin in central hypothyroidism. J Clin Endocrinol Metab 1986;62:960–4.
4. Rose SR, Manasco PK, Pearce S, Nisula BC. Hypothyroidism and deficiency of the nocturnal thyrotropin surge in children with hypothalamic-pituitary disorders. J Clin Endocrinol Metab 1990;70:1750–5.
5. Rose SR, Nisula BC. Circadian variation of thyrotropin in childhood. J Clin Endocrinol Metab 1989;68:1086–90.
6. Adriaanse R, Romijn JA, Endert E, Wiersinga WM. The nocturnal thyroid-stimulating hormone surge is absent in overt, present in mild primary and equivocal in central hypothyroidism. Acta Endocrinol (Copenh) 1992;126:206–12.
7. Beck-Peccoz P, Amr S, Menezes-Ferreira MM, Faglia G, Weintraub B. Decreased receptor binding of biologically inactive thyrotropin in central hypothyroidism. N Engl J Med 1985;312:1085–90.
8. Amr S, Menenez-Ferrieira M, Shimohigashi Y, et al. Activities of deglycosylated thyrotropin at the thyroid membrane receptor adenylate cyclase system. J Endocrinol Invest 1985;8:537–41.
9. Taylor T, Gesundheit N, Gyves PW, Jacobowitz DM, Weintraub B. Hypothalamic hypothyroidism caused by lesions in rat paraventricular nuclei alters the carbohydrate structure of secreted thyrotropin. Endocrinology 1988;122:283–90.
10. Taylor T, Weintraub BD. Altered thyrotropin (tsh) carbohydrate structures in hypothalamic hypothyroidism created by paraventricular nuclear lesions are corrected by in vivo tsh-releasing hormone administration. Endocrinology 1989;125:2198–203.
11. Miura Y, Perkel VS, Papenberg KA, Johnson MJ, Magner JA. Concanavalin-A, lentil, and ricin lectin affinity binding characteristics of human thyrotropin: differences in the sialylation of thyrotropin in sera of euthyroid, primary, and central hypothyroid patients. J Clin Endocrinol Metab 1989;69:985–95.

# Appendix: Poster Presentations

This appendix contains 36 abstracts given as poster presentations at the Serono Symposia, USA Symposium on Glycoprotein Hormones: Structure, Function, and Clinical Implications, held March 11 to 14, 1993, in Santa Barbara, California.

DISULFIDE BOND MUTATIONS AFFECT THE FOLDING OF hCG-$\beta$ IN TRANSFECTED CHO CELLS    E Bedows, JR Huth, N Suganuma, I Boime, RW Ruddon.  The Eppley Cancer Institute, Univ of Nebraska Medical Center, Omaha, NE and Departments of Pharmacology and Obstetrics and Gynecology, Washington Univ School of Medicine, St Louis, MO

Using CHO cells transfected with mutated hCG-$\beta$ genes in which the Cys residues required for the formation of each of the six disulfide (S-S) bonds were replaced by Ala, we determined the effect that elimination of each of the native S-S bonds had on hCG-$\beta$ folding.  When the Cys required for any of the first three hCG-$\beta$ S-S linkages to form (bonds 34-88, 38-57 or 9-90) were substituted, folding did not proceed beyond the earliest detectable folding intermediate (p$\beta$1-early).  In the absence of the next S-S bond to form (bond 23-72), p$\beta$1-early was converted into a second folding intermediate (p$\beta$1-late), but conversion to the following intermediate (p$\beta$2-free) was inhibited.  When either of the final two S-S bonds (bonds 93-100 or 26-110) were removed, conversion of p$\beta$1-late to p$\beta$2-free was detected, but conversion of p$\beta$2-free to the next folding intermediate, p$\beta$2-combined was not seen.  These data support the hypothesis that individual S-S bonds are involved in discrete steps in the hCG-$\beta$ folding pathway. Supported in part by NCI Grant CA32949 and NIH grant HD23398

HCG, hCGß, hCGß CORE FRAGMENT AND hCGα IN SERUM, URINE AND HYDROCELE FLUID OF TESTICULAR CANCER (TC) PATIENTS
P. BERGER, R. GERTH, S. DIRNHOFER, C. KRATZIK*, M. MEHRABI*, M. MARBERGER*, S.MADERSBACHER* AND G.WICK
Institute for Biomedical Aging Research, Austrian Academy of Sciences, A-6020 Innsbruck; * Dept. of Urology, University of Vienna, Austria

The secretion pattern of TC and the influence of peripheral metabolism on levels of hCG derivatives in body fluids was investigated in cubital (C) or testicular vein (TV) serum, urine and hydrocele fluid (HF). Samples from 9 patients (5 seminomatous, 4 non-seminomatous) were obtained immediately before or during surgery. Analyses of hCG, hCGß, hCGß core fragment (hCGßcf) and hCGα, respectively, were performed by highly sensitive (2 pg/ml) and specific (e.g., <0.01% cross-reactivity of hCG and hCGß in the hCGßcf assay) time-resolved 2-site fluoroimmunoassays using our own panel of monoclonal antibodies. Ratios of HCG/hCGß were dramatically decreased between CV serum and HF and corresponding CV and TV sera contained similar amounts of hCG but TV hCGß levels exceeded those of CV sera. These findings point to significant influences of peripheral hormone metabolism on the composition of hCG-derived molecules in body fluids. Urinary hCGßcf was only detectable in patients with hCG-positive CV serum or HF. Surprisingly, hCGßcf was also occasionally present in HF suggesting that it might be a TC secretion product itself and is not exclusively generated by the kidney. Our observation that HF of all TC patients with negative CV serum marker levels contained particularly hCGß (290-3000pg/ml) or hCG irrespective of histology indicates that the differentiation between hCG positive and negative TC should be reconsidered.

The Structure of Chorionic Gonadotropin and its βCore Fragment from the Urine of Pregnant Chimpanzees

S. Birken, M.A. Gawinowicz, G.M. Agosto, H.C. Chen and B.C. Nisula

Columbia University College of P & S, New York, N.Y. 10032 and National Institutes of Health, Bethesda, MD. 20892

The primary amino acid sequence of des-β-COOH-terminal chimpanzee chorionic gonadotropin was determined by microsequence analysis of the protein purified from the urine of pregnant chimpanzees. It was found that most of the CG had lost its β COOH-terminal peptide (βCTP) presumably by proteolysis prior to purification by immunoaffinity. The primary structures of human and chimp CG are nearly identical. The alpha subunit has two changes, a serine/threonine heterogeneous amino terminus in the chimp as compared to homogeneous alanine in the human. A valine for glutamine substitution at position 50 was also found. The des-COOH-terminal β subunit likewise exhibited a serine/threonine amino terminus and a substitution at residue 117 of alanine for aspartic acid which exists at that position in the human sequence. In order to attempt to obtain chimp CG with its βCTP intact, urine from pregnant chimpanzees was collected by catheter. It was found that even collection under these conditions led to loss of some βCTP. The biological activity of chimp CG was approximately 5,000 IU/mg referenced to hCG. Carbohydrate analysis showed that the chimp CG was devoid of sialic acid. Collection of first morning void pregnant chimpanzee urine allowed isolation and the structural analysis of the chimp β core fragment. It was found that the relative concentration of fragment to hormone was similar to the human pattern and that the amino terminal sequence structure was identical to that of the hCG β core fragment. The production of this fragment is apparently common among primates.

STRUCTURE OF THE PITUITARY LH BETA CORE FRAGMENT

S. Birken, M.A. Gawinowicz, G. Agosto, Y. Chen, A.S. Hartree

Columbia University College of Physicians & Surgeons, Dept. Medicine, N.Y., NY 10032 and AFRC Institute of Animal Physiology and Genetics Research, Cambridge CB2 4AT, England

About a decade ago, an immunoreactive hCG beta-related fragment was reported to complicate hCG measurement in human pregnancy urine. Several groups developed specific immunoassays for this hCG beta core fragment and its structure was determined by our group to be a two chain fragment consisting of hCG beta residues 6-40 disulfide-linked to residues 55-92. Measurement of this molecule may be useful in some cancer patients but certain urine samples such as those from postmenopausal women have unexpectedly high concentrations. We now report that the amino terminal structure of an hLH beta core fragment isolated from pituitary tissue extracts is exactly analogous to the hCG beta core fragment. This LH fragment may not be the main component of beta core fragment immunoreactivity in postmenopausal women since we have isolated other proteins from postmenopausal urine by immunoaffinity using immobilized core fragment antibody. The finding of a discrete hLH beta core fragment in a tissue extract tends to support direct tissue secretion of such a fragment.

MEASUREMENT OF FREE ALPHA-SUBUNIT OF hCG IN GESTATIONAL TROPHOBLASTIC DISEASE

RH Buck, RB Haneef

Department of Chemical Pathology, University of Natal Medical School, South Africa.

The measurement of hCG and its free ß-subunit, as so-called ßhCG, for the diagnosis and monitoring of therapy in patients with gestational trophoblastic disease (GTD) is well established. However it has been demonstrated that when serum ßhCG can no longer be measured by current RIA methods, up to $10^4$ tumour cells may remain undetected. In addition, there have been isolated reports of patients with choriocarcinoma in whom ßhCG was undetectable in the serum but who appeared to be secreting only the α-subunit of hCG. This study was therefore undertaken to determine the utility of free α-subunit measurement for the diagnosis and management of patients with GTD.

Serum free α-subunit and ßhCG was measured in 61 patients with GTD, in 34 with molar disease (22 with benign molar disease [BMD] and 12 with persistent molar disease [PMD]) and 27 with choriocarcinoma. Free α-subunit was measured by an in-house immunoradiometric assay using monoclonal antibodies and ßhCG by a commercial RIA.

Increased levels of free α-subunit and ßhCG were found in the sera of all patients with GTD. There were no significant differences in the pretreatment levels of either species or in the %α/ßhCG molar ratio between the 3 subgroups tested. After evacuation of molar pregnancy and chemotherapy in PMD and choriocarcinoma, the free α-subunit levels generally decreased in parallel with ßhCG levels. However, the serum concentrations of both species were significantly higher 2-8 weeks post-evacuation in PMD compared to BMD; conversely the %α/ßhCG molar ratios were significantly higher 2-7 weeks post-evacuation in BMD compared to PMD. The rate of disappearance of ßhCG was faster than that of free α-subunit, but free α-subunit levels decreased into the reference range before ßhCG in all 3 subgroups examined because of its lower initial level.

The preliminary data from this study indicate that the measurement of free α-subunit and ßhCG may be useful for the earlier identification of patients with PMD after evacuation. Differentiation between benign and persistent molar disease as early as possible is important to allow earlier chemotherapeutic intervention and possible prevention of cerebral and liver metastases. In addition, because of its earlier disappearance, measurement of free α-subunit may be useful in the monitoring of tumour response to therapy.

# FOLLICULAR PHASE ADMINISTRATION OF hCG AND CORPUS LUTEUM RESCUE RESULTS IN A PROLONGED SUPPRESSION OF OVULATION

V. Daniel Castracane, Ph.D., G. T. Koulianos, M.D., C. A. Meeks, M.D.
Texas Tech University Health Sciences Center, Department of Obstetrics and Gynecology, Amarillo, Texas, USA

Earlier studies from our laboratory have demonstrated that follicular phase administration of hCG would rescue the corpus luteum and all the classic luteal hormones are stimulated including progesterone (P), relaxin, 17OHP, estradiol($E_2$) and inhibin. All women had a history of normal menstrual cycles and received 5 consecutive days of increasing hCG from 3000 to 5000 iu., starting at either day 1 (n=5) or day 3 (n=4) of the menstrual cycle. In all subjects, there was a clear rescue of the corpus luteum as reported earlier. In these women who are available for follow up with endocrine profiles and/or LMP, the cycles were generally prolonged with the range from 34 to 75 days (48.2 ± 4.5 days; M ± SE) before completion of the next normal luteal phase. Controls (n=5) were untreated and had a mean cycle length of 27.0 days. $E_2$ levels and P levels decline after the 5 day period of hCG treatment and remain decreased until the preovulatory estrogen surge and normal luteal phase. LH and FSH also remain at basal levels during this long hiatus. Inhibin levels reported earlier increased during luteal rescue but declined to normal early follicular phase levels and did not seem to be a mechanism to suppress FSH during this hiatus. In a last group, 3 women who were on oral contraceptives received hCG treatment in the cycle immediately following the OC cycle. All cycles were normal (28.7 ± 3.8 days) without any endocrine indication of luteal rescue, indicating that some luteal stimulation seems to be necessary to produce this prolonged follicular hiatus. This ovulatory hiatus induced by follicular hCG administration may have potential for ovulatory regulation.

## URINARY HCG EXPRESSION IN LUNG CANCER

Y. Chen, R. E. Canfield, E. Flaster and J. O'Connor

Columbia University College of P & S, New York, N.Y. 10032

The urinary content of intact hCG and hCG beta fragment was evaluated pre treatment in 43 female and 34 male patients with lung cancer. The control population consisted of 20 premenopausal women, 222 postmenopausal women and 36 normal men. The data were analyzed separately for male and females using ANOVA with a Bonferroni adjustment for the comparison of means. For males, there was a significant difference between normals and lung cancer patients for both intact hCG ($p<.03$) and beta fragment ($p<.0001$). For females both assays showed statistically significant differences ($p<.0003$) between postmenopausal women and lung cancer patients. Postmenopausal women were used in the comparison since most of the females (41/43) were over the age of 48.

The sensitivity and specificity of the assays for intact hCG and hCG beta fragment, normalized to creatinine, were as follows:

For female lung cancer patients the intact hCG assay had a sensitivity of 52.9 and a specificity of 77.8. The beta fragment assay had a sensitivity of 76.5 and a specificity of 86.1.

For males with lung cancer the intact hCG assay provided a sensitivity of 65.1 and a specificity of 57.2. The beta fragment assay had a sensitivity of 69.7 and a specificity of 68.5.

The data demonstrate that differentiation between lung cancer patients and controls is provided by both the intact hCG assay and the hCG beta fragment assay. However, the production of both intact hCG and beta fragment by normal subjects, particularly postmenopausal women, compromises the utility of these molecules as markers of malignancy.

### SELECTING AN HCG ASSAY: CONSIDERATION OF CROSSREACTING MOLECULES

*L.A. Cole, A. Kardana and G.D. Braunstein, Ob-Gyn, Yale University, New Haven CT 06510, and Dept. of Medicine, Cedars-Sinai Medical Center, U.C.L.A., Los Angeles CA 90048*

Varying pregnancy test results have been reported. The same serum or urine sample giving unlike values in different commercial hCG assays. To find the cause of this discordance, we examined the levels of the different hCG-related molecules in first trimester I.V.F. and spontaneous pregnancy fluids. hCG-related molecules include intact hCG (the bio-active hormone), nicked hCG (that missing a peptide bond at $\beta$-subunit linkage 47-48, inactive), free $\alpha$, free $\beta$ and $\beta$-core fragment. The mixtures of hCG-related molecules in serum and urine were compared with the specificities of different pregnancy test kits.

Commercial immunoassays commonly detect total hCG (intact + nicked hCG) using an anti-$\alpha$:anti-$\beta$ or similar sandwich assay combination, intact hCG (non-nicked) only using an anti hCG dimer antibody,or $\beta$hCG (total hCG + free $\beta$ and $\beta$-core fragment) using an hCG$\beta$ RIA or anti-hCG$\beta$:anti-hCG$\beta$ sandwich assay. We measured levels of intact hCG and total hCG, and calculated $\beta$hCG (as total hCG + free $\beta$ in serum, and the same + $\beta$-core fragment in urine) in 242 serum and 125 urine samples. In serum, in the two weeks following the missed period, when pregnancy tests are commonly performed, median levels of total hCG, intact hCG and $\beta$hCG were similar (<12% difference). Individual values, however, varied significantly. For intact hCG, values ranged from 41% to 145% and for $\beta$hCG from 101% to 145% of total hCG concentration. We infer, that values for indivividual serum samples can vary as much as 2½ fold between the different kinds of hCG assay. In urine individual sample variation is greater, up to 100-fold different. Individual intact hCG values varied from 1% to 148%, and $\beta$hCG values from 102 to 547% of total hCG.

A survey of 29 serum or urine hCG kits revealed that 10 were types detecting total hCG, 5 detecting intact hCG only, and 14 were $\beta$hCG assays. We conclude that use of different types of kit (specific for total, intact or $\beta$hCG) causes discordance in hCG test results. The total hCG assay has the advantage of being independent of nicking, the intact hCG assay of measuring bio-active hCG, and the $\beta$hCG assay of maximum sensitivity. All three types of assay are needed, however, values from the three types of kits should be kept separate and should not be inter-converted.

# BIOLOGICAL AND BIOCHEMICAL PROPERTIES OF RECOMBINANT HUMAN FSH ISOHORMONES

Renato de Leeuw, John Mulders, and Jan Damm

Organon International BV, Scientific Development Group, P.O. Box 20, 5340 BH Oss, The Netherlands

Gonadotropins, such as follicle stimulating hormone (FSH) and luteinising hormone (LH) exhibit considerable heterogeneity. In the present study, the respective recombinant human FSH (recFSH; Org 32489, batch 506) isohormones were characterised after separation by means of preparative iso-electric focusssing. The FSH immunoreactive profile ranged from pH 4.3 to 5.7, with a single peak at pH 5.0. The specific immunoactivity (IU/mg) of the different isohormone fractions was comparable suggesting that the different isohormones are recognised in a similar way in the applied two-site enzyme immunoassay. The receptor binding activity/immunoactivity ratio and the *in vitro* bioactivity /immunoactivity (B/I) ratio were highest for the most basic isohormones. The *in vivo* B/I ratio (Steelman Pohley bioassay) showed an opposite trend, i.e. the most acidic fractions were far more potent than the basic fractions. The *in vivo* B/I ratio increased from 0.05 at pH 5.7 to 1.7 at pH 4.3 (B/I ratio starting material 0.5). The sialic acid content correlated well with the *in vivo* bioactivity ranging from 0.9 sialic acid residue (relative to mannose = 3) at pH 5.7 to 2.8 residues at pH 4.3. It is concluded that relatively acidic recFSH isohormones have lower receptor binding and *in vitro* biological activities than the more basic isohormones. However, these acidic isoforms exhibit greater *in vivo* bioactivities because they are more heavily sialylated.

# LEVELS OF GALNAC- AND SULFOTRANSFERASE IN GONADOTROPHS ARE REGULATED TO MAINTAIN SULFATED OLIGOSACCHARIDES ON LH

Shylaja M. Dharmesh and Jacques U. Baenziger

Washington University School of Medicine, St. Louis, MO 63110

Lutropin (LH) bears Asn-linked oligosaccharides terminating with the sequence $SO_4$-4GalNAc$\beta$1,4GlcNAc$\beta$1,2Man (S4GGnM). We recently identified and characterized the GalNAc- and sulfotransferases in pituitary which account for the synthesis of these oligosaccharides. A receptor in hepatic endothelial cells specific for S4GGnM controls the circulatory half-life and *in vivo* activity of LH.

The levels of LH synthesis change dramatically at specific times during development and the ovulatory cycle. We have examined GalNAc- and sulfotransferase activities in rat pituitary following ovariectomy. Like LH, the levels of pituitary GalNAc-transferase increase to 3-fold higher than controls. The increase is reversed by treatment with estradiol. Ovariectomy and ovariectomy with estrogen replacement have similar effects on the sulfotransferase activity. In contrast, no change in the levels of galactosyltransferase is observed following ovariectomy. LH synthesized following ovariectomy has a similar content of S4GGnM-bearing structures as controls. The GalNAc- and the sulfotransferase are also expressed in the salivary gland and kidney; however, their levels do not increase in these tissues in response to ovariectomy. GalNAc- and sulfotransferase levels are tightly regulated to insure that the oligosaccharides on LH but not other glycoproteins synthesized by gonadotrophs terminate with S4GGnM.

hCG ACTIONS IN TRANSFORMED LEYDIG CELLS FOLLOWING HETEROLOGOUS
RECEPTOR DOWN-REGULATION AND DESENSITIZATION

Adviye Ergul, Tsuey-Ming Chen, and David Puett

REPSCEND Labs, Dept. of Biochem. & Mol. Biol., Univ. of Miami,
Miami, FL 33101 and Dept. of Biochem., Univ. of Georgia, Athens, GA
30602, U.S.A.

hCG of the syncytiotrophoblast stimulates maternal luteal cells and
46,XY fetal Leydig cells in early pregnancy.  The hormone is also
produced by certain neoplasms, although its effects in these
pathological conditions is not understood.  Using the transformed
murine Leydig cell line, MA-10, we have studied the interaction of hCG
binding and receptor activation, mainly via cAMP and PKA, with other
signaling pathways: EGF, which acts via a tyrosine kinase, and a
phorbol ester, which activates PKC.  hCG, EGF, and TPA stimulate, to
varying degrees, progesterone production; moreover, each increases
the mRNA levels of the protooncogenes c-fos, c-jun, and jun-B.  When
the cells were pretreated for 16 h with EGF and TPA, hCG binding was
decreased about 30% and 40%, respectfully, while the steroidogenic
response to hCG was decreased some 20% and 90%, respectively, due to
heterologous receptor down-regulation and desensitization.  Despite
the reduction in hCG receptors following EGF and TPA pretreatment,
there was no significant decrease in the level of hCG-mediated
induction of the protooncogenes jun-B and c-fos, as was found with c-
jun.  Thus, activation of three signaling pathways in Leydig tumor
cells can lead to increased steroidogenesis and protooncogene
induction, but complex interactions occur in the pathways following
heterologous receptor down regulation and desensitization.  Supported
by DK33973.

FURTHER CHARACTERIZATION OF HUMAN PITUITARY LH POLYMORPHISM

D. Graesslin, U. Hoerner, V. Lichtenberg and W. Braendle

Division of Endocrinology, Department of Obstetrics and Gynecology,
University of Hamburg, Hamburg, Germany

Starting from 600 human pituitaries 7 LH isoforms could be isolated and
highly purified using preparative isoelectric focusing as final step.
No differences with respect to number and activities of the different
hormone forms were found following investigation of pituitary extracts
from fertile and postmenopausal women as well as males of lutropin polymorphism using isoelectric focusing in sucrose density gradients contra-
ry to findings of other groups.
Although all 7 isolated species had similar intrinsic bioactivity there
was a dramatic decrease of receptor binding and in vitro biological activity (mouse Leydig cell assay) compared to immunoactivities from the alkaline to the more acidic isohormones.
Due to the low and identical NANA content of the different forms (2.2%)
short half lives between 13 and 17 min (male rats) for all isoforms were
found. Neuraminidase digestion of the isohormon complex resulted in a
shift of all forms to the more alkaline region without reduction of the
number. In addition, the complete isohormone pattern could be obtained
after digestion of each single LH form. These findings may suggest that
the charge microheterogeneity of hLH is related to the number of terminal
sulfated hexosamines of the three oligosaccharide chains.

THE IN VITRO, PROTEIN DISULFIDE ISOMERASE (PDI)-CATALYZED, AND INTRACELLULAR hCG-ß FOLDING / hCG SUBUNIT ASSEMBLY PATHWAYS ARE INDISTINGUISHABLE.
J.R. Huth, F. Perini, E. Bedows, R.W. Ruddon
Eppley Cancer Institute, Univ of Nebraska Med Ctr, Omaha, NE 68198.

It is not known whether folding enzymes in mammalian cells change the mechanism of protein folding.    Based on the order of formation of six disulfide bonds, we have found that the in vitro folding pathway of human chorionic gonadotropin ß subunit (hCG-ß) is the same as its intracellular folding pathway in JAR choriocarcinoma cells (Huth et al., (1992) J. Biol. Chem. 267, 21396) and in CHO cells transfected with the hCG-ß gene (Bedows et al., (1992) J. Biol. Chem. 267, 8880).    The same rate limiting step was found in both intracellular and in vitro folding environments; however, the $t_{1/2}$ for this step in a cell is 4 min, whereas in vitro the $t_{1/2}$ is 80 min.    Addition of PDI to the in vitro folding reactions increased the in vitro rate of this event ($t_{1/2}$ = 25 min) without changing the order of disulfide bond formation.    In intact cells, the 26-110 S-S bond of hCG-ß is not entirely formed, and thus folding is not complete until after αß assembly.    In vitro, PDI increased the rate of assembly of fully oxidized hCG-ß with hCG-α 80-fold.    During this reaction, the 93-100 and 26-110 S-S bonds of hCG-ß are reduced and then re-oxidized once assembly is completed.    These data and the intracellular data strongly suggest that incompletely folded hCG- ß is more competent for assembly than is fully folded hCG- ß.    Supported by NCI grant CA 32949.

THE ROLE OF SULFATION AND SIALYLATION ON THE BIOACTIVITY AND METABOLIC CLEARANCE RATE OF RECOMBINANT HUMAN TSH (rhTSH) AND rhTSHβ-hCGβ CARBOXY TERMINUS EXTENSION PEPTIDE   CHIMERA.
Lata Joshi, Fredric Wondisford, Yoko Murata, Mariusz Szkudlinski, Rajesh Desai, N. Rao Thotakura and Bruce D. Weintraub, Molecular and Cellular Endocrinology Branch, NIDDK, NIH, Bethesda, MD.
Previous studies from our laboratory have shown that unlike N-linked oligosaccharides of pituitary TSH which primarily terminate in N-acetylgalactosamine-sulfate, those on recombinant human TSH expressed in Chinese hamster ovary (CHO) cells terminate in galactose-sialic acid. Recently, N-acetylgalactosamine-transferase  and sulfotransferase have been demonstrated in human embryonic kidney (293) cells. We thus reasoned that rhTSH and its variants when expressed in 293 cells and possibly monkey kidney (COS-7) cells would produce TSH with N-linked oligosaccharide structures terminating in sulfate. Because previous studies with chimeric FSH produced in CHO cells showed increased half-life, we constructed a similar chimera of hTSHβ-subunit with the carboxy terminus extension peptide (CTEP) of hCGβ subunit. COS-7 and 293 cells were cotransfected with plasmids containing either WT or chimeric TSH with hCGαcDNA. Bioactivity was determined in two FRTL5 assays (cAMP production; $^3$H Thymidine uptake). Both chimeric and WT TSH expressed in 293 and COS-7 cells displayed similar bioactivity suggesting that terminal sulfation & sialylation of oligosaccharides does not alter *in vitro* bio-activity. The metabolic clearance rate (MCR) of chimeric TSH and WT TSH secreted by COS-7 and 293 cells was compared to that of WT TSH produced by CHO cells. The presence of sulfate had a dramatic effect on the MCR of WT TSH (293) which was cleared 3 times faster than WT TSH expressed in CHO cells. Interestingly, COS-7 cells had a clearance rate closer to that of CHO cells than 293 cells suggesting the presence of sialic acid. The maximum increase in circulatory half-life was demonstrated by chimeric TSH (COS-7) which showed significantly reduced MCR (3-4 fold). Despite the presence of CTEP in chimeric TSH produced by 293 cells, its MCR was identical to that of WT TSH from COS-7 cells indicating the absence of sialic acid. These results suggest that chimeric TSH produced in various cell lines can be used as a tool to delineate the role of sulfate and sialic acid in MCR and thereby the *in vivo* bioactivity.

## NICKS INCREASINGLY DEACTIVATE AND DISSOCIATE HCG AS PREGNANCY ADVANCES

A. Kardana, G.D. Braunstein, and L.A. Cole. Ob-Gyn, Yale University, New Haven CT 06510, and Dept. of Medicine, Cedars-Sinai Medical Center, U.C.L.A., Los Angeles CA 90048

A proportion of hCG molecules in pregnancy serum and urine samples have nicks or a missing peptide linkage between β-subunit residues 47-48. These nicks diminish the steroidogenic activity of hCG. We examined the source, pattern in maternal serum and stability of nicked hCG.

**Source:** Standard hCG was added to 3 samples of whole blood, and incubated 18 h at 37 C, no change was detected in extent of nicking. However, nicking was detected by gel electrophoresis (bands at Mr=17,000 and Mr=22,000) in culture fluids from placental explants and was localized in placenta tissue stained with antibodies for intact (non-nicked) and total (nicked + non-nicked) hCG. We inferred that nicking occurs in placenta tissue.

**Pattern in maternal serum:** Levels of intact and total hCG were determined in 233 serum samples from 4-40 weeks of pregnancy, and extent of nicking was estimated. A linear relationship was indicated between advancing gestation and increasing extent of nicking, with 9% in the first month rising to 21% at term.(r=0.95, P<0.00005).

**Stability:** We compared intact (standard hCG batch CR127) and nicked (C5 hCG, 100% nicked) hCG. Both were incubated in whole blood. C5 hCG (½-life 22±5.2 h) dissociated into α- and β-subunits >30 times faster than standard hCG (½-life 700 h).

Considering these values and the clearance rate of hCG the relative amount of nicked molecules produced by the trophoblast may be 3½ times higher than the 9-21% indicated in serum samples. If nicked molecules rapidly dissociate then levels of free subunits should rise in parallel. From the first to the last month of pregnancy, extent of nicking of hCG rises about 4-fold in urine (from 8% to 31%). Free α, free β-subunit and β-core fragment (degradation product of β-subunit) levels all rise by similar proportions, from 76% to 360%, 18% to 99% and from 58% to 305% of hCG levels.

**Conclusion:** We infer that, as pregnancy advances, an increasing percentage of hCG molecules is nicked; that this may be a deactivation and clearance pathway for hCG, and that nicking may contribute to diminishing hCG levels and rising proportions of free α-subunits and β-core fragment in the second and third trimesters of pregnancy.

## HCG DEGRADATION IN THE HUMAN CHORIONIC VILLOUS CORE

Harvey J. Kliman, Kyu S. Lee, Erika L. Meaddough and Laurence A. Cole. Departments of Pathology and Obstetrics and Gynecology, Yale University School of Medicine, New Haven, CT 06510

HCG is made in high concentrations during the first trimester of pregnancy. What prevents this hCG from entering the fetal circulation and deranging the developing fetal endocrine system? We reasoned that localizing sites of intact (non-nicked) biologically active hCG, inactive nicked hCG and degraded ß-core fragment (ß-core) might help us understand the fate of hCG which diffuses into the villous core. We immunohistochemically stained 8 freshly fixed human placentas between 6 and 40 weeks gestation for intact hCG (using B109 antibody), nicked hCG (using 3468 antibody which detects both intact and nicked-hCG), ß-core (using B204 antibody) and α-hCG (using Unipath 2119 antibody). Between 6 and 11 weeks of gestation we found high levels of intact hCG in chorionic villous syncytiotrophoblasts, with no staining in other cells types. During this same period we found high levels of both nicked-hCG and ß-core within the cytoplasm of villous macrophages—the so called Hofbauer cells—but not intact hCG or free ß-subunit (using 1E5 or FBT11). Decidual macrophages were never positive for any of these markers. At 12 weeks of gestation, the amount of staining for intact hCG began to decrease in the syncytiotrophoblasts. Macrophage staining for nicked and ß-core remained high, especially in villi with syncytiotrophoblasts that were stained intensely for intact hCG. Second trimester placentas showed only scattered positive syncytiotrophoblasts for intact hCG. Nicked-hCG and ß-core staining was also scattered, but always was associated with syncytiotrophoblasts that were positive for intact hCG. Term placentas showed virtually no staining for intact hCG, nicked-hCG or ß-core. α-hCG was expressed in syncytiotrophoblasts throughout pregnancy, but was highest during the first trimester. No α-hCG was ever found in villous core macrophages.

Our results suggest that villous core macrophages protect the fetus from exposure to high levels of hCG by degrading any hCG that diffuses towards the fetal circulation. Once degraded, these inactive forms may then diffuse out of the villi and into the maternal circulation. We would suggest that only nicked hCG, ß-core and small peptides and amino acids from degraded hCG diffuse into the fetal circulation, where they are filtered into the fetal urine and eventually urinated into the amniotic cavity by the fetus.

## FEEDBACK CONTROL OF TRANSCRIPTION AND mRNA LEVELS BY hCG ACCUMULATION IN THE ENDOPLASMIC RETICULUM (ER)

W.E. Merz[1], P. Licht[1,2], C.V. Rao[2]
Department of Biochemistry II[1], University of Heidelberg, Germany and Department of Obstetrics and Gynecology[2], Med. School, University of Loiusville, KY, USA

Brefeldin A is an antibiotic which blocks transiently and reversible the anterograde transport from the ER to the Golgi apparatus. Treatment of JEG and JAR cells with BFA (0.05-4 µg/ml, 1.5-6 h) resulted in a reversible dramatic decrease of hCG and free subunit secretion to ≤10% of controls (2 µg BFA/ml), even in 8-bromo-cAMP stimulated cultures. Pulse-chase labeling experiments ($^{35}$S-Met) showed that in presence of BFA *de novo* synthesized hCG or free subunits were not secreted. Total intracellular concentration of hCG-$\alpha$ increased twofold in control and 8-bromo-cAMP treated cultures (free $\beta$-subunit concentration remained unchanged or was decreased). In presence of BFA, the blockade of the secretion of hCG and free subunits was followed by a reversible and specific drop of the $\alpha$- and $\beta$mRNA concentrations to 6-10% of control values ($\beta$-actin mRNA levels were unchanged). On the transcription level, the BFA-induced intracellular arrest of hCG and free subunits caused a rapid decrease of the hCG-$\alpha$-mRNA *de novo* synthesis as measured in nuclear run on assay. The transcription of the $\beta$-subunit genes remained unchanged or were only transiently decreased. The investigations seem to indicate the presence of a mechanism which signals information about the hCG and/or subunit levels in the ER to the transcription level.

A RECEPTOR BINDING SITE IDENTIFIED IN THE REGION 81-95 OF THE ß-SUBUNIT OF HUMAN LUTEINIZING HORMONE (LH) AND CHORIONIC GONADOTROPIN (hCG).

D.E. Morbeck, P.C. Roche, H.T. Keutmann and D.J. McCormick

Depts. of Biochemistry and Molecular Biology (D.E.M., P.C.R., and D.J.M.), Mayo Clinic, Rochester, MN, 55905, and Endocrine Unit, Dept. of Medicine (H.T.K.), Massachusetts General Hospital, Boston, MA, 02114.

Two series of synthetic overlapping peptides comprising the entire sequences of the human LHß (11 peptides) and CGß (14 peptides) subunits were synthesized by solid phase methods using N$^\alpha$-Fmoc amino acid derivatives. Each series of peptides (15 residues in length each) were tested for their ability to inhibit the binding of $^{125}$I-labelled hCG or LH to rat ovarian receptor. The most potent inhibitor of LH/hCG binding was a peptide comprising the sequence 81-95 with an IC$_{50}$ of less than 20 $\mu$M for the hCGß 81-95 sequence and approximately 30 $\mu$M for LHß 81 95. To determine important receptor binding residues within this region, a third set of 12 peptides was synthesized in which each residue of the parental sequence hCGß 81-95 was successively replaced with L-alanine. Replacement of either arginine (residues 94 or 95) or leucine 86 resulted in a 2 to 2.5 fold reduction in inhibitory activity. Substitution of either cysteine 88 or cysteine 90 (but not cysteine 93) with alanine resulted in a 9 to 17 fold reduction in inhibitory activity of the peptide ß81-95. The corresponding region in FSHß has also been found to be active in binding to its receptor (Santa Coloma and Reichert, J.Biol.Chem. **265**:5037, 1990), and in both hormones the sequence includes a segment homologous to the active center of thioredoxin (Boniface and Reichert, Science **247**:61, 1990). Preliminary testing of all CGß subunit peptides for inhibition of hCG-induced progesterone stimulation in Leydig MA-10 cell bioassays demonstrated that CGß 81-95 had inhibitory activity. The results show that region 81-95 is important along with other segments in LH/hCG receptor interaction, and that at least three residues (Leu-86, Arg-94, and Arg-95) are required for maximum binding of ß81-95 to receptor.

## THE BINDING CHARACTERISTICS OF THE RECOMBINANT HUMAN FSH RECEPTOR

J.E. Mulder, P.M. Sluss, A.L. Schneyer
Reproductive Endocrine Sciences Center, National Center for Infertility Research,
Massachusetts General Hospital, Boston, Massachusetts 02114
The objective of this study was to determine the binding characteristics of the human FSH (hFSH) receptor, cloned and stably transfected in Chinese Hamster Ovary (CHO) cells (kindly provided by Ares Advanced Technology, Inc., Randolph, MA). A radioreceptor assay (RRA) was developed using a recombinant hFSH receptor. Binding of $I^{125}$ hFSH was determined as a function of time and temperature. At 20°C, steady-state binding was achieved in 4 hours. The addition of a 1200 fold excess of unlabeled hFSH rapidly displaced a portion of the specifically bound labeled hormone, indicating that the binding reaction was only partially reversible (37-60%). As has been described for native FSH receptors, the dissociation of the hFSH/hFSH-receptor complex was dependent on temperature. The steady-state association constant ($K_A$) of binding of $I^{125}$hFSH to the recombinant receptor was 1.58 +/- 0.18 x$10^9$, which is within the range reported for native bovine and rat receptors. Using this newly developed recombinant hFSH RRA, the receptor binding characteristics of recombinant hFSH were then compared to pituitary FSH standards (1st International Standard and NIH I-3) and to urinary FSH standards (Metrodin-HP and Metrodin). In conclusion, the binding characteristics of the recently cloned recombinant hFSH receptor are similar to those of well characterized native FSH receptors, and therefore this receptor is a valid representative of native human FSH receptors.    Supported by: NIH training grant (DK07028-18) and the National Center for Infertility Research at the MGH (NIH-U54-29164).

## DIRECT LH-RH INFLUENCE ON LH RECEPTION AND RESPONSES BY PORCINE GRANULOSA CELLS CULTURE: CHANGES IN LH RECEPTOR CONTENT AND IN LH-INDUCED cAMP, PROGESTERONE AND ESTRADIOL-17ß RELEASE

NITRAY, J.[1], SIROTKIN, A. V.[1], KOLENA, J.[2]

[1]Research Institute of Animal Production, 949 92 Nitra and

[2]Institute of Experimental Endocrinology, 833 06 Bratislava, Czechoslovakia

The effects LH-RH on LH/hCG receptor content, basale and LH--induced cAMP, progesterone and estradiol-17ß secretion by porcine granulosa cells were investigated under in-vitro conditions. Incubation of granulosa cells with LH-RH (0,01, 0,1, 1,0 or 10 ug/ml$^{-1}$) increased the content of LH/hCG binding sites in comparison with the cells cultured without LH-RH. LH-RH (0,1 ug/ml$^{-1}$) stimulated also the secretion of cAMP, estradiol, but not of progesterone by granulosa cells. Treatments with porcine LH (10, 100, 1 000, 10 000 or 100 000 mIU/ml$^{-1}$) increased cAMP, estradiol and progesterone secretion by the cell culture in a dose-dependent manner. LH-RH (0,1 ug/ml$^{-1}$) added together with LH (10, 100, 1000, 10 000 or 100 000 mIU/ml) further increased cAMP, estradiol and progesterone release by the cells. The present observations may sugest a direct stimulatory influence of LH-RH on porcine ovarian LH reception and responses to LH.

COORDINATED REGULATION OF GONADOTROPIN ALPHA AND BETA SUBUNIT mRNAS AND SERUM LH AND FSH CONCENTRATIONS FOLLOWING PULSATILE ADMINISTRATION OF GNRH TO NUTRITIONALLY GROWTH-RETARDED LAMBS. Vasantha Padmanabhan, Inese Z. Beitins, Thomas D. Landefeld. Reproductive Sciences Program, Depts of Pharmacology and Pediatrics, Univ of Michigan, Ann Arbor, MI 48109.

The nutritionally growth-retarded ovariectomized lamb has a slow LH pulse frequency and reduced circulating LH and FSH concentrations. Nutritional repletion increases LH (GnRH) pulse frequency, gonadotropin subunit mRNA concentrations and circulating concentrations of LH and FSH (Endocrinology, 125:351, 1989). We tested the hypothesis that an increase in GnRH release contributes to parallel increases in gonadotropin steady state mRNAs and release. Seven nutritionally growth-retarded lambs were administered GnRH (iv; 2 ng/kg) for 36h, q 2h for first 24h and q 1h for the remaining 12h. Six others served as controls. Mean pituitary and serum LH and FSH concentrations (from samples drawn q 12 min for 4h; from 32-36h for the GnRH treated group) were determined by standard radioimmunoassays. Pituitary $\alpha$, LH$\beta$ and FSH$\beta$ mRNA concentrations were measured by solution hybridization assays. Results are summarized below.

| | Serum gonadotropins (ng/ml) | | Gonadotropin mRNAs (fmol/mg total RNA) | | |
|---|---|---|---|---|---|
| | LH | FSH | Alpha | LH$\beta$ | FSH$\beta$ |
| Control (n=6) | 0.6±0.1 | 4.9±1.8 | 168.5±39.9 | 33.4±11.4 | 17.0±3.9 |
| GnRH (n=4) | 9.9±4.5* | 16.2±6.3 | 264.4±37.6 | 140.3±57.4* | 21.7±5.2 |

Pulsatile administration of GnRH for 36h resulted in a 16-fold increase (p<0.01) in mean concentrations of serum LH as compared to the control group. Mean pituitary LH, FSH and serum FSH concentrations for the GnRH-treated lambs were not different from the control lambs. GnRH treatment of nutritionally growth-retarded lambs resulted in significant increases in LH$\beta$ mRNA concentrations and circulating LH concentrations. Gonadotropin alpha, LH$\beta$ and FSH$\beta$ subunit mRNAs were positively correlated with pituitary and circulating LH and FSH concentrations. Our results demonstrate that in the absence of ovarian feedback, pulsatile GnRH treatment to growth-retarded female lambs resulted in parallel changes in steadystate mRNAs, as well as the release of LH and FSH. This contrasts with prepubertal lambs, where a similar mode of GnRH administration in the presence of ovarian feedback resulted in discordant regulation of the steady state mRNAs and circulating gonadotropin concentrations (Biol Rep 44 [suppl 1]:62, 1991). Supported by NIH HD 23812 (VP) and USDA 90-37040-5758 (TDL).

ONTOGENY OF LH SECRETION AND ITS NEGATIVE FEEDBACK REGULATION IN THE RAT, AS STUDIED WITH A NOVEL SENSITIVE IMMUNOFLUOROMETRIC ASSAY (IFMA) OF LH

P. Pakarinen, E. Proshlyakova* and I. Huhtaniemi

Department of Physiology, University of Turku, 20520 Turku, Finland, and Institute of Developmental Biology*, Russian Academy of Sciences, Moscow 117808, Russia.

A sensitive IFMA assay (Haavisto et al., Endocrinology, 1993, in press; Delfia®, Wallac OY, Turku, Finland) allows the measurement of rat LH in 25-100 µl of serum down to 0.003 µg/l of NIH-RP-2 (vs. 0.2-0.5 µg/l with the NIH RIA). LH was measurable in sera of rats from day 19.5 of fetal life (f); males 0.0046 ± 0.0013, females 0.0086 ± 0.0013 µg/l. On day f 20.5, the respective levels were 0.049 ± 0.000 and 0.24 ± 0.027 µg/l, and on day f 21.5, 0.12 ± 0.023 and 0.22 ± 0.050 µg/l. The sex difference was significant at each age (p<0.01-0.05). Castration of male fetuses on day f 20.5 increased in one day LH to 0.16 ± 0.025 µg/l, vs. 0.072 ± 0.005 µg/l in sham-operated controls (p<0.05). Treatments of 4-day-old males with the antiandrogen flutamide (100 mg/kg BW) or the Leydig cell toxicant EDS (50 mg/kg) increased serum LH from 0.031-0.045 µg/l 9- and 30-fold, respectively (p < 0.01). When mothers on days 17.5-20.5 of gestation were treated similarly, only EDS induced a 2-fold increase in LH of male fetuses in 2 days (p < 0.05), flutamide had no effect. In conclusion: 1) IFMA is sensitive enough to detect LH in single samples of fetal serum at levels down to 0.003 µg/l, 2) serum LH is detectable in fetuses from day f 19.5, and a clear sex difference (female > male) prevails, 3) gonadal negative feedback regulation of LH secretion is apparent in male rats in utero, 4) in comparison to their effects ex utero, flutamide and EDS were ineffective in increasing fetal LH secretion.

## DOWN-REGULATION OF GRANULOSA CELL α-INHIBIN BY STEROID HORMONES DURING THE LUTEINIZING HORMONE SURGE

T.R. Rao, I. Montoya, P. Chang, W.N. Burns, and R.S.Schenken

Department of Obstetrics and Gynecology, The University of Texas Health Science Center at San Antonio, San Antonio, Texas 78284-7836, USA

Luteinizing hormone/human chorionic gonadotropin (LH/hCG) in the presence of follicle stimulating hormone (FSH) suppresses α-inhibin gene expression *in vivo*. However, *in vitro* studies of cultured granulosa cells have shown that LH stimulates α-inhibin gene expression. We examined the transient expression of α-inhibin chloramphenicol acetyltransferase (CAT) gene in rat ovarian granulosa cells to explore further the regulation of the α-inhibin gene by FSH, LH and the steroid hormones estrogen ($E_2$) and progesterone ($P_4$). Our results indicate that both FSH and LH stimulated α-inhibin expression, and the expression increased several-fold when both FSH/LH were present. CAT gene activity was also increased in the presence of LH and forskolin indicating a possible synergistic effect of LH or interaction of other transcription factors which, in turn, act on the α-inhibin promoter. Gonadotropin-enhanced CAT activity was decreased in the presence of $E_2$ and $P_4$, suggesting that LH/hCG suppression of α-inhibin gene expression *in vivo* may be due to steroid responsive transcription factors that either modulate LH-induced cAMP production or interact directly with the α-inhibin gene promoter resulting in its down-regulation during the LH surge.

## MONOCLONAL ANTIBODIES TO LUTROPIN RECEPTOR (LH-R)

P. Rathnam, M. Bhatia and B. Saxena
Cornell Univ. Med. College, New York, N.Y. 10021

Monoclonal antibodies to bovine LH-R were produced. Immunoglobulins were isolated on affinity ImmunoPure-MBP columns to obtain IgM isotypes of 78K. Antibodies were localized in the rat thecal and luteal as well as in Leydig cell membranes. The antibodies caused 40% inhibition of $^{125}$I-hCG binding to the receptor and upto 75% inhibition of testosterone production by hCG-stimulated Leydig cells. LH-R antibodies caused infertility in rats and inhibited implantation in mice. In male rats, the antibodies caused upto 50% reduction in testosterone production. Monoclonal antibodies identified eight clones of bacteriophage Lambdagt11 containing bovine ovarian cDNA inserts. The DNA from the clones was extracted. The cDNA inserts were cleaved, subcloned into the plasmid pUC18 and transformed into JM83 E.coli. Recombinant clones were selected on LB-AMP- IPTG-XGAL plates. The bacteriophages, and in some cases, the E.coli were immunoscreened using five monoclonal antibodies. The DNA was also extracted and sequenced. cDNA sequences of the inserts, expressing proteins which bind to monoclonal antibodies, may represent functional epitopes of LH-R. (Support: NIH grant# HD2054641)

## IDENTIFICATION OF SPECIFIC RESIDUES IN THE NH$_2$-TERMINUS OF THE LH RECEPTOR THAT ARE ESSENTIAL FOR HORMONE BINDING

Patrick C. Roche and Daniel J. McCormick

Departments of Laboratory Medicine/Pathology (PCR)and Biochemistry/Molecular Biology (DJM), Mayo Clinic, Rochester, Minnesota 55905.

Luteinizing hormone (LH) acts on multiple cells of the ovary to regulate steroidogenesis, trigger ovulation, and promote formation of the corpus luteum. Crucial to our understanding of the molecular mechanisms involved in LH action is knowledge of the structural basis for binding to its plasma membrane receptor. We have previously used of series of overlapping and nested peptides to identify four discontinous regions of the LH receptor that interact with hormone; $Arg_{21}$-$Pro_{38}$, $Arg_{102}$-$Thr_{115}$, $Tyr_{253}$-$Phe_{272}$, and $Lys_{573}$-$Lys_{583}$. To further delineate the specific amino acids of the most active region $Arg_{21}$-$Pro_{38}$, we have synthesized two groups of peptides that represent truncations of the NH$_2$-terminal end and the COOH-terminal end of this region. Evaluation of peptides in competitive radioreceptor assays indicates that $Arg_{21}$ and $Arg_{26}$ in the NH$_2$-terminus, and $Leu_{34}$-$Leu_{37}$ in the COOH-terminus are critical for the activity of this site. Circular dichroic spectra of $Arg_{21}$-$Pro_{38}$ show the potential for high (43%) helical content. We postulate that positively charged $Arg_{21}$ and $Arg_{26}$ are contact residues for interaction with hormone and that the COOH-terminal residues are responsible for maintaining an ordered stucture for proper orientation. Fluorescence anisotropy measurements of amino-methyl coumarin-labelled peptide $Ala_{15}$-$Pro_{38}$ provide further evidence for a critical role of this region in hormone binding. Fluorescent peptide associates specifically with hCG and hLH, to a much lesser extent with hFSH, and not all with ovalbumin. In concordance with studies of the thyrotropin (TSH) and FSH receptor, we believe that the NH$_2$-terminal region of the LH receptor, specifically $Arg_{21}$-$Pro_{38}$, is a critical determinant site for hormone binding.

## FLANKING AMINO ACIDS OF THE HUMAN FOLLITROPIN BETA SUBUNIT 33-53 REGION ARE CRITICAL FOR ASSEMBLY OF FOLLITROPIN HETERODIMER

Karen E. Roth[2], Cheng Liu[2], Barbara A. Shepard[2], Jacquelin B. Shaffer[1,2] and James A. Dias[1,2]

[1]Wadsworth Center for Laboratories and Research, New York State Department of Health, Albany, New York, [2]The School of Public Health, State University of New York at Albany, Albany, New York

We have previously analyzed human follitropin (hFSH) with monoclonal antibodies (MAb) and antipeptide antibodies. Those studies have led us to a current operating hypothesis that some amino acids within the hFSHß 33-53 region are surface oriented and others participate in subunit contact. Protein structural analysis predicts $\beta$-turns within this region and our previous studies indicate that the ends may be involved in subunit contact. In this study hF3IIβ was mutagenized to change [34]TRDL[37] to [34]AAAA[37] or [48]QKTCT[52] to [48]AAACA[52] allowing us to study the ends of the hFSHß 33-53 sequence contiguous with the hFSHß sequence. Wild-type and mutant cDNAs were co-expressed with α-subunit cDNA in CHO[Flo-5] cells. Wild-type hFSH was secreted from cells co-transfected with wild-type hFSHα and hFSHβ cDNAs as expected. However, heterodimeric hFSH was minimally detected in the media from cells transfected with the [34]TRDL[37] mutant and was not detected in the case of the [48]QKTCT[52] mutant. Analysis of cell lysates (intracellular FSH) with a conformation specific MAb 3G3 revealed that similar levels of properly folded $\beta$-subunit were produced in cells expressing wild-type or either mutated $\beta$-subunit. These data indicate that the flanking amino acids of the hFSHβ 33-53 region, in particular [48]QKTCT[52], are critical for assembly of hFSH heterodimer. Supported by NIH HD-18407.

**Expression of Bovine FSHα and FSHβ in Saccharomyces *cerevisiae***

Mitali Samaddar, James F. Catterall,* and R.R. Dighe

Center for Reproductive Biology & Molecular Endocrinology, Indian Institute of Science, Bangalore 560 012, India
*The Population Council, Center for Biomedical Research, 1230 York Avenue, New York, NY 10021

The pituitary glycoprotein hormone, Follicle Stimulating Hormone (FSH), plays a central role in mammalian reproduction. It is a heterodimer comprising an α subunit and a hormone specific β subunit. In the present study, a novel yeast expression vector was used to produce subunits of bovine FSH. The coding portions of bovine FSHα and FSHβ cDNAs (without their presequences) were amplified using PCR and cloned separately into a yest expression vector YEpsec1 in such a way that this construct would produce a protein consisting of a leader signal peptide from a yeast protein fused to FSHα or FSHβ when expressed in *S. cerevisiae* grown in the presence of galactose as the carbon source. It was found that in yeast cells transformed with these constructs, FSHα or β subunit was expressed and secreted into the medium. These subunits, when tested in homologous radioimmunoassays, were found to be immunologically similar, though not identical, to the pituitary derived FSH subunits. The yield of subunits was approximately 100 to 120 μg/lit medium. Preliminary studies indicate that FSH subunits produced by the yeast are capable of annealing with the countersubunit derived from the pituitary suggesting that these subunits are biologically active. Further work is being carried out to purify and characterize the FSH subunits. [Supported by Department of Biotechnology, Government of India and by the Rockefeller Foundation (RF 87005)].

## RECEPTOR BINDING CHARACTERISTICS OF RECOMBINANT HUMAN FSH

Alan Schneyer, Deborah O'Neil, Jean Mulder,  James Hutchison, and Patrick Sluss

Reproductive Endocrine Unit, National Center for Infertility Research, Massachusetts General Hospital, Boston, MA and Ares Advanced Technology Inc., Randolph, MA (J.H.).

Current protocols for gonadotropin induced fertility assistance utilize either Pergonal, a urinary human menopausal gonadotropin preparation, or Metrodin, in which the LH activity has been removed from Pergonal. An evolution to treatment with pure FSH will soon become feasible with recombinant human FSH (rhFSH) in a highly purified form devoid of LH or other pituitary hormone activity.

In this study, the receptor binding characteristics of rhFSH was compared to urinary FSH (Metrodin, Metrodin-HP; Serono), pituitary FSH (NIH-I-3; NHPP), and the 1st International Standard (83/575; WHO). Homogenized calf testes was used as the FSH receptor source and both NIH-I-3 and rhFSH were used as radioligands after iodination using the lactoperoxidase/PAGE methods.

All binding inhibition curves were statistically ($p < 0.05$) parallel using either radioligand. In addition, the potency of rhFSH as vialed for human application was nearly identical to that of Metrodin. Since both preparations were calibrated in IUs according to the same in vivo bioassay and they exhibit similar potencies in the FSH RRA, it would appear that the biochemical composition and receptor binding affinity are nearly identical, and predicts that the in vivo half-life of rhFSH and Metrodin are also quite similar.

Supported by: Ares Serono, NIH R01-25941 and the National Center for Infertility Research at the MGH (NIH-U54-29164).

CHANGES IN LH RECEPTORS IN PORCINE OVARIAN FOLLICLES DURING
FOLLICULOGENESIS AND AFTER hCG TREATMENT

SIROTKIN, A. V.[1], KOZIKOVA, L. V.[1], KOLODZIEWSKI, L.[2],
LAURINČÍK, J.[1], PIVKO, J.[1]
[1]Research Institute of Animal Production, 949 92 Nitra,
[2]University of Veterinary Medicine, 041 81 Košice, Czechoslovakia

The objective of our work was to investigate the changes in
LH/hCG receptors in porcine ovaries during follicular growth
and under the influence of exogenous hCG. LH receptor number
and localization in the preantral, small(0,2-0,9 mm in diameter), middle (1,0-2,9 mm) and great (2,9-15 mm) ovarian follicles of immature pigs before-, 12 and 32 h after hCG treatments was analysed autoradiographically using ($^{125}$I) hCG. It
was observed, that preantral follicles containes no- or very
small amounts of LH receptors localized exclusively in theca
layer. The increase of follicular size was associated with
the increase of proportion of follicles labeled and of LH receptor number in theca interna. In the only great follicles
($^{125}$I) labels appeared also in theca externa and in granulosa cells. hCG injection was followed by the increase of number of follicles of all classes analyzed, did not change the
proportion of follicles labelled, but decreased the receptor
number in the great (but not in the middle, small and preantral) follicles. The present observations confirmed, that
follicular development in porcine ovaries is associated with
the increase of LH receptor number, with the changes in their
localization and with the appearance of the negative feedback
mechanism regulating LH reception.

## DIFFERENTIAL ROLES OF THE TERMINAL MONOSACCHARIDES OF RECOMBINANT TSH IN ITS IN VITRO AND IN VIVO BIOACTIVITY.

N. Rao Thotakura, Mariusz Szkudlinski, Holger Leitolf, Lata Joshi, Yoko Murata and Bruce Weintraub, Molecular and Cellular Endocrinology Branch, NIDDK, NIH, Bethesda, MD.

We have previously shown that recombinant human thyrotropin (rhTSH), expressed in Chinese hamster ovary cells is biologically active and due to its highly sialylated oligosaccharide chains, has lower metabolic clearance and higher in vivo bioactivity compared to pituitary hTSH. The specific role of terminal sulfate groups and the subsequent sugar residues in pituitary-derived hTSH could not be assessed individually due to the unavailability of a suitable sulphohydrolase, since the oligosaccharides of pituitary hTSH terminate in NeuAcα2,3(or6)Galβ and SO$_4$-4GalNAcβ, whereas those in rhTSH terminate only in NeuAcα2,3(or6)Galβ. Due to this type of oligosaccharide structure in rhTSH, it is now possible to deglycosylate and assess the role of individual sugar residues in the hormonal activity. The monosaccharides were sequentially removed from rhTSH by exoglycosidase digestions. The removal of sialic acid, galactose or N-acetylglucosamine in that sequence resulted in more than a 10-fold increase in the in vitro bioactivity of rhTSH. As expected, all the derivatives showed higher metabolic clearance rates compared to the intact hormone, but the Gal-removed and GlcNAc-removed derivatives were cleared slower than NeuAc-removed rhTSH. In contrast, the in vivo bioactivity decreased progressively with each monosaccharide removal and the GalNAc-removed derivative showed no activity. Resialylation of incompletely sialylated or desialylated rhTSH using sialyl transferase attenuated the in vitro activity of the hormone and increased the in vivo bioactivity to the level comparable to the untreated TSH. These data demonstrate that sialic acid residues affect both the in vitro and the in vivo bioactivities of rhTSH. In contrast to previous studies on human chorionic gonadotropin, sequential removal of sugar residues does not result in a progressive decrease in the in vitro activity of rhTSH. Thus, glycosidases and glycosyltransferases can be used as powerful tools to study the role of carbohydrate and to alter the in vivo potency of glycoprotein hormones.

## EFFECT OF LHRH ON THE DEVELOPMENT OF LACTOTROPHS AS STUDIED BY DOT BLOT AND *IN SITU* HYBRIDIZATION OF PRL mRNA IN RAT PITUITARY REAGGREGATE CELL CULTURES.

A. Van Bael, R. Huygen and C. Denef

Laboratory of Cell Pharmacology, University of Leuven, School of Medicine, Campus Gasthuisberg, B-3000 Leuven, Belgium

Based on our previous observations that LHRH, a primary regulator, as well as NPY, a secundary regulator of gonadotrophs, are capable of stimulating DNA replication in lactotrophs in pituitary reaggregates from 14-day-old female rats, we have studied the effect of LHRH on PRL mRNA levels by means of in situ hybridization and northern blot. Both the total cell population and cells processing through the cell cycle were examined. As estimated by computer-image analysis, addition of LHRH at day 5 for 40 hours resulted in a 37% increase of the total cytoplasmic areas of cells containing PRL mRNA. A similar increase was found by dot blot hybridization of extracted RNA. PRL mRNA expression was not affected by NPY. In the population of cells processing through the cell cycle, visualized by the incorporation of $^3$H-thymidine into DNA during 16 hours, LHRH increased the number of cells expressing PRL mRNA with 35%. NPY did not influence the number of $^3$H-thymidine positive cells expressing PRL mRNA and blocked completely the effect of LHRH on the latter population. The present data suggest that LHRH, presumably via a paracrine action of gonadotrophs, stimulates the recruitement of new lactotrophs, an action wich is negatively modulated by NPY. Since the magnitude of this effect was the same in the total pituitary cell population as in cells processing through the cell cycle and presumably mitosis, recruitement of lactotrophs seems to be based on differentiation of progenitor cells into PRL expressing cells rather than a mitogenic action on preexisting lactotrophs.

## HOW DO THE OLIGOSACCHARIDES IN hCG AFFECT LH-RECEPTOR ACTIVATION ?

H.J. van Loenen[1], J.M. Bidart[2] and F.F.G. Rommerts[1].
1. Dep. of Endocrinology & Reproduction, Erasmus University Rotterdam, The Netherlands.
2. Unite de Biochemie Clinique, Institute Gustave-Roussy, 94805 Villejuif, France.

In MA-10 cells deglycosylated hCG (DhCG) is 50-fold less active for stimulation of cAMP production than intact hCG (see other abstract). This raises the question whether the oligosaccharides are directly involved in LH-receptor activation or whether they merely affect the 3-D structure of the hormone thereby influencing the interaction between protein domains in hormone and receptor. We have investigated whether a panel of anti-hCG monoclonal antibodies (mAb's) can induce conformational changes in DhCG and alter the bioactivity. Two different kinds of experiments were carried out, A: mAb/DhCG complexes were added to fresh MA-10 cells or B: mAb's were added to MA-10 cells that were preincubated with DhCG.
- Two mAb's raised the DhCG stimulated cAMP production from 5 to 250 fold without any effect on the activity of intact hCG (A and B). Two mAb's blocked the bioactivity of both intact hCG and DhCG (A). One mAb which bound to hCG had no effect on either intact hCG or DhCG (A and B).
- Fab fragments of stimulating mAb's also enhanced the bioactivity of DhCG although they were less efficient than intact mAb's.
We have concluded from these data that receptor activation is mediated by a perfect matching of protein domains in hormone and receptor and that the oligosaccharides only induce the active form of protein.

## FSH RECEPTOR ACTIVATION WITHOUT DETECTABLE HORMONE BINDING.

H.J. van Loenen, J.F. Flinterman and F.F.G. Rommerts.
Dep. of Endocrinology and Reproduction, Erasmus University Rotterdam, The Netherlands.

Many FSH binding studies have been carried out in hypotonic buffers containing low concentrations of sodium, probably to optimize the binding assay. We have been able to reproduce these studies by measuring FSH binding to membrane preparations from calf testis or rat Sertoli cells (Kd's: 30 pM). However hypotonic buffers cannot be used to correlate hormone binding and receptor activation in viable cells. On the other hand in normal culture medium we could not measure any specific FSH binding, whereas the non-specific binding was not affected. We have therefore searched for a buffer that allows measurement of both hormone binding and receptor activation in intact cells.

In a buffer containing sucrose instead of sodium chloride rat Sertoli cells remained viable for at least 6 hours. Under these conditions the Kd of the FSH receptor was increased to approximately 100 pM. Moreover FSH was 10 times more potent in stimulation of cAMP production than in normal culture medium. Since the hormone/receptor complex (HR') found under low sodium conditions does not dissociate in the presence of normal sodium concentrations we postulate the existence of two complexes, HR at high sodium concentration and HR' at low sodium concentration in equilibrium with hormone (H) and receptor (R), as follows:     $H + R \rightleftharpoons HR \rightleftharpoons HR'$

Thus under normal physiological conditions when the HR complex prevails, the affinity and thus the occupancy of the FSH receptor may be more than 20 fold less than in low sodium containing buffers. This may explain why specific FSH binding could not be detected under normal physiological conditions.

## THE RELATIVE IMPORTANCE OF THE OLIGOSACCHARIDES IN hCG FOR LH-RECEPTOR ACTIVATION.

H.J. van Loenen[1], J.M. Saez[2], W.E. Merz[3] and F.F.G. Rommerts[1].
1.Dep. of Endocrinology & Reproduction, Erasmus University Rotterdam, The Netherlands.
2.INSERM U 307, Hopital Debrousse, 69322 Lyon Cedex 05, France.
3.Dep. of Biochemistry II, University of Heidelberg, Germany.

Deglycosylated hCG (DhCG) which shows normal or increased affinity for the LH-receptor has often been considered as an antagonist of intact hCG. We have studied LH-receptor activation by measuring cAMP and steroid production in Leydig cells from rat (RLC), murine (MA-10) or porcine (PLC) origin, employing four hCG preparations. We used two purified hormones and two forms of hCG obtained by stable transfection of CHO cells with the cDNA's for the hCG subunits (Dr. I. Boime).

- HF-DhCG obtained by HF treatment of purified hCG and CHO-DhCG obtained by site directed mutagenesis showed no major differences in bioactivity.
- The relative bioactivity of the DhCG's when compared to intact hCG varied from 2 to 100% depending on bioassay and celltype used.
- CHO-DhCG and intact CHO-hCG stimulated the long term steroid production in PLC in a similar fashion. In this assay CHO-hCG was 3 fold more potent as purified hCG.
- In RLC DhCG was 5-10 fold less potent as intact hCG in stimulation of the steroid and cAMP production.
- In MA-10 cells the DhCG's were 2 fold less potent in the stimulation of steroidogenesis but were 50 fold less potent in the stimulation of the cAMP production.

Although it is difficult to make general statements about the importance of the oligosaccharides in hCG for LH-receptor activation we propose that the matching between protein domains in hCG and LH-receptor without a direct interaction of the oligosaccharides determines the degree of receptor activation (see other abstract).

GRAVES' AUTOIMMUNE SERUM INHIBITS STEROIDOGENESIS IN
LEYDIG CELLS AT A SITE DISTAL TO HORMONE/RECEPTOR BINDING

KP Willey, N Hunt, R Ivell, F Leidenberger

Institute for Hormone & Fertility Research at the University of Hamburg, Germany

Thyroid hyperstimulation in Graves' disease has been attributed to chronic stimulation by TSH receptor autoantibodies, measured in assays of TSH-binding inhibition (TSH-BI). TSH-BI does not correlate with disease severity, suggesting that molecules accessory to glycoprotein hormone-induced activity are also affected. We have therefore analysed Graves' sera in homologous and heterologous systems.

Human cell lines expressing the transfected hTSH receptor were established to screen sera for TSH-BI. We assessed the ubiquity of the Graves' serum effects using hCG-stimulated steroidogenesis by Leydig cells. In the Leydig cell bioassays, both basal and LH/hCG-stimulated steroidogenesis was stimulated by normal sera but specifically inhibited by Graves' sera. Binding of hCG to the LH receptor was not perturbed and neither was there evidence of a testicular TSH receptor, suggesting that TSH receptor autoantibodies were not involved. The inhibition was not related to clinical thyroid parameters or TSH-BI and appeared to be linked to the release of steroidogenic substrate from the plasma membrane. The participating molecules are being isolated, cloned and their role elucidated.

EFFECT OF PEPTIDE NICKING IN THE HUMAN CHORIONIC GONADOTROPIN
$\beta$-SUBUNIT ON STIMULATION OF RECOMBINANT HUMAN TSH RECEPTORS

M. Yoshimura, A. E. Pekary , X. P. Pang, L. A. Cole, A. Kardana, and J. M. Hershman
West Los Angeles VA Medical Center and UCLA School of Medicine, Los Angeles, CA, and Yale University, New Haven, CT, USA

Human chorionic gonadotropin (hCG) has thyroid stimulating activity. However, little is known about the effect of peptide nicking in the hCG $\beta$-subunit on its thyrotropic potency. Using CHO cells expressing functional human TSH receptors, we examined cyclic AMP (cAMP) production induced by hCG preparations extracted from urine of normal pregnancy and trophoblastic disease. Two preparations (C2, 50% nicked and M4, 100% nicked in $\beta$ 44-49 region) showed about 1.5 fold potency of standard hCG CR-127 which is also 20 % nicked in the same region. Non-nicked hCG (P8) had the weakest potency among all of the samples tested. HCG digested by human leukocyte elastase, that may cause some nicking, increased cAMP more than standard hCG. These results suggest that nicks in the hCG $\beta$-subunit may in part modulate the activation of human TSH receptors by hCG.

# Author Index

369

# Subject Index

ISBN 0-387-94165-7